CENTER FOR JAPANESE STUDIES

BIBLIOGRAPHICAL SERIES

NUMBER 6

JAPANESE GEOGRAPHY:

A GUIDE TO JAPANESE REFERENCE AND RESEARCH MATERIALS

Library of Congress Cataloging in Publication Data

Hall, Robert Burnett, 1896-
Japanese geography.

Reprint of the ed. published by University of Michigan Press, Ann Arbor, which was issued as no. 6 of the Bibliographical series of the Center for Japanese Studies.

1. Japan--Description and travel--Bibliography.
I. Nō, Toshio, 1912- joint author. II. Title.
III. Series: Michigan. University. Center for Japanese Studies. Bibliographical series ; no. 6.
[Z3306.H3 1978] [DS806] 016.9152 78-5578
ISBN 0-313-20434-9

Reprinted in 1978 by Greenwood Press, Inc.
51 Riverside Avenue, Westport, CT. 06880

Printed in the United States of America

10 9 8 7 6 5 4 3 2 1

EDITOR'S FOREWORD ON THE BIBLIOGRAPHICAL SERIES

The Bibliographical Series of the Center for Japanese Studies has for its main purpose the listing and evaluating of the major Japanese works pertaining to the humanities and social sciences, particularly as they deal with Japan and the areas immediately adjacent to Japan. It is assumed that Western materials pertaining to Japan are adequately covered in the bibliographies of Pagés, von Wenckstern, Nachod, Praesent-Haenisch, Pritchard, Gaskill, etc., and that Western specialists in the several fields will know how to get at the Western materials in their respective fields.

The bibliographies in the present series are intended to serve as an introduction to the native research materials in the several disciplines and hence as an aid to research for teachers and students. In each case an attempt has been made to describe or to evaluate each work that is listed, or at least to justify the inclusion of each item. The authors have also attempted to indicate the American libraries at which each item may be found. Scholars and librarians will perhaps find that the several bibliographies in this series will serve as useful guides to buying programs which they may wish to initiate.

The bibliographies are selective. Each item listed is believed to be of some value or interest to the scholarly user. In those cases in which it has been impossible to examine a book or article of known value, it still is included. A book or article is thus included if it is written by a competent scholar, if it is included in a bibliography which is itself competently compiled, if it appears to treat its subject matter in detail and with an approach to completeness, if it is frequently quoted, if it is well reviewed, or if it is referred to as being authoritative. Wherever possible, notes as to why an item seems to be of value have been given.

The scope of each bibliography is defined by the compiler or compilers in their introductions, but in general each of the bibliographies lists (a) important source materials, and (b) secondary sources dating from a fixed date in the recent past, as, for instance, the Meiji Restoration, 1900, 1910, etc.

Although the materials in most cases deal with the Japanese islands, each compiler has set the limits of the geographical area which his materials cover. In certain cases expansion into areas that are outside Japan appears to be justified by the fact that Japanese research has been the dominant research for these areas. Hence one or more of the bibliographies will cover Japanese materials on Formosa, Korea, Manchuria, and the Mandated Islands.

The format is uniform within each volume. In general the name of each author or compiler is given both in romanization and characters. The surnames are given first and the given names next, as the practice is in Japan. The names of corporate authors, such as government offices, are given in romanization and characters; they are then translated.

The title of each book or article is given in romanization and characters; it is then translated. The place of publication and the name of the publisher are given in romanization alone, but a separate listing within each bibliography gathers together the names of the publishers, with the characters used in writing their names.

1. Long ā, ō, and ū are indicated by macrons over the vowels.
2. Only the first letters of initial words and proper nouns are capitalized.
3. In the bibliographical data, the compilers have given both the edition and the printing of the work cited. Significant textual variations sometimes occur between different printings of the same edition of a given work.
4. When dealing with an item composed of one volume, complete pagination is given for that volume, including all separately paged sections. If any title is in more than one volume, only the total number of volumes is given, without paging.
5. In the event that the item cited happens to be part of a series or collection, the compilers have given in brackets introduced by an equals sign the title, characters, and translated title of that series or collection and the number of the volume concerned.
6. The location of each item in American libraries is shown by means of the symbols used in the Union List of Serials. The symbols for the major libraries concerned are as follows:

   CCC - Claremont Colleges
   CSt - Stanford General Library
   Cst-H - Hoover War Library
   Cty - Yale
   CU - University of California
   DLC - Library of Congress
   ICU - Chicago
   IEN - Northwestern
   NH - Harvard
   MiU - Michigan
   NN - New York Public Library
   NNC - Columbia
   WaU - University of Washington

7. Works such as encyclopedias, dictionaries, yearbooks, series, and collections are cited by title; the name of the editor or compiler, in romanization and characters, is usually given after the title.
8. In the case of articles found in journals, quotation marks surround the Japanese title, characters, and translated title.
9. Abbreviations are explained in lists, if necessary.

10. If any volume of a journal is continuously paged, number and month may be omitted. If it has both continuous volume pagination and separate pagination for each issue, only the volume, year, and the continuous volume pagination may be given. If more than one volume appears in any single year, and each is separately and continuously paged, the procedure has been to give the volume, inclusive months of the issues in the volume, year, and continuous volume pagination.
11. In the case of a single article comprising a chapter or a section of a book which is a compilation of articles by a number of authors, this fact is shown by inserting the word "in" between the title of the article and the compilation in which it is found. Following the "in," a complete citation of the book in question is given.
12. All descriptions, evaluation, criticisms, and comments pertaining to a volume or article follow the citations in separate, indented paragraphs. These may include brief biographies of the authors or compilers, but only one biography per author or compiler is given in each bibliography. Cross-references from one citation to another are used to call attention to comments, biographies, etc.
13. A list of the standard professional journals is given whenever found to be convenient.

The following remarks may interest the growing number of scholars who are concerned with the problem of publishing materials in the lesser known fields, in which the number of copies to be distributed is necessarily small and in which special problems, such as the one here faced of giving names and titles in characters, must be met. An electric typewriter was first used to type the materials in romanization. The characters were then written in, in spaces left for the purpose. The over-all dimensions of the typed area was 10 by 14 and 1/2 inches. This was reduced to the present size in photographing the pages. The photographs thus taken were then made the basis for reproduction by offset process. Except in the case of the introductory material, no attempt was made to justify the right-hand margins.

Joseph K. Yamagiwa

Note on Japanese Geography: a Guide to Japanese Reference and Research Materials,

by Robert B. Hall and Toshio Noh

Since the inception of this bibliographical series, Far Eastern libraries in the United States have made substantial additions to their Japanese collections. Because of the wealth of these materials, it is no longer possible or necessary to indicate for each of the items listed in any bibliography the libraries which own it. The present series has therefore discontinued the practice of indicating these libraries.

The romanized titles of articles appearing in journals have been underlined in the present bibliography. This innovation will serve, it is hoped, to give greater emphasis to these titles.

J. K. Y.

THE AUTHORS' INTRODUCTION

The intent in compiling this bibliography has been to bring to the attention of Western geographers and other interested scholars those geographical writings of the Japanese which have appeared in the Twentieth Century and which are regarded by the compilers as significant contributions. Thus, only materials which fall within the scope of geography as a modern science have been included. An exception to this rule, which seems a valid one, is the inclusion of the more important anthologies of pre-Meiji geographical writings published since 1900.

The problem of selection has been a difficult one as a truly vast quantity of published research is involved. The Japanese are an unusually geographically-minded people. This is particularly true in terms of local and regional interest - an attitude which has pertained in most periods of Japanese history, as evidenced by the enormous production of local and regional descriptions and maps. More recently, studies of national scope have become increasingly more numerous. Conversely, there has been relatively little interest shown by Japanese geographers in areas other than Japan or in the compilation of systematic and exhaustive bibliographies of their own geographical materials. This general lack of standard bibliographies and guides to Japanese geographical literature has compelled the present compilers to place heavy reliance upon direct observation and upon the advice of Japanese experts. Direct observation would have been used, in any event where feasible, but neither compiler can claim expertness in all fields of geography. Furthermore, in this postwar period, Japanese scholarship has been rather sharply divided into different schools of interpretation and so it has not always been possible to secure impartial judgment. The standard for selection has been twofold: (1) which titles would prove most valuable to the general geographer, and (2) which would best bring to the attention of the Western scholar the extent and quality of Japanese geographical research? The authors have used their best judgment in selection but do not anticipate that this will be universally accepted as the wisest. In this regard, as in all other aspects of this bibliography, the hope is for criticism as to selection and evaluation. Eventually this bibliography, as the others in the Bibliographical Series of the University of Michigan's Center for Japanese Studies, will be combined into a coordinated whole and be reprinted. In this interim period criticism is essential to best quality of the final product.

The problems of organization and balance have also proved to be difficult ones, as there is little precedent, especially in light of the standard for selection employed. It was decided to allow the categories to develop as the materials selected took form and to make no attempt to balance categories but rather to let them assume proportion as the materials selected indicated.

Much of the material, valuable to geographers, in Japan as elsewhere, is not geography in the strictest sense of the word or at least is hidden in the writings of other social and physical scientists. For this reason, reference to bibliographies in other disciplines is essential. Particularly is this true of the Bibliographical Series of the University of Michigan's Center for Japanese Studies and especially: No. 1, *A Guide to Japanese Reference and Research Materials in the Field of Political Science*, by Robert E. Ward; No. 4, *Japanese History: a Guide to Japanese Reference and Research Materials*, by John W. Hall; and No. 5, *Japanese Economics: a Guide to Japanese Reference and Research Materials*, by Charles F. Remer and Saburō Kawai.

As has been previously mentioned, the vast store of pre-Meiji and early Meiji geographical writings has been excluded. This is in part because of the limitations of space in this single volume. Even more important have been the problems of interpretation, evaluation, availability, and specialized emphases. The earlier material has great geographical value and should be covered in another volume, but the task will require the close cooperation of scholars of early Japanese literature and history. Again, much of this earlier material is not geography in the strictest sense. Rather it is commonly made up of general treatises with emphasis upon local and regional history, local customs and legends, local productions, and the lore and treasures of local shrines and temples. Nevertheless, it does contain much highly important geographical data not otherwise available. Such are the oft-cited *fudoki*, the compilation of which began when the first court order pertaining to them was issued in 713 A. D. and ended about 925 A. D. Unfortunately, most of these priceless documents were lost but, later, remnants were recovered, as in the case of the *fudoki* of Izumo, Hitachi, Harima, Hizen and Bungo. Others are known today only as fragments and through references in other writings. Travel diaries also contributed valuable materials. Among these is the *Tosa nikki* of Ki-no-Tsurayuki. The *Engishiki*, 905 A. D., was in the tradition of the early *fudoki* and encompassed all of the area then under control of the central government. There is nothing to take its place for this critical period. In the middle of the Eleventh Century, another great travel diary was written. This is the *Sarashina nikki* by a daughter of Sugawara Takasue. Mention, too, should be made of the preservation of original documents in temple and private collection through these and the following centuries. The Tokugawa era accounts for the enormous Edo period (1603-1868). A bibliography or rather book listing of some 70,000 titles has been compiled by Mr. Yasui Yoshitarō 保井芳太郎 of Ōjimachi, Nara prefecture. There are several significant collections of this material extant in Japan but only one has even a sizable part of the whole. This is to be found at the Tenrikyō library at Tambaichi. Very little Edo period material is to be found abroad. Not to entirely neglect these valuable pre-Meiji contributions which are so important a part of the modern Japanese geographer's tradition, the authors have included certain transcribed compilations of these early materials, published since 1900, such as the *Dai-Nihon chishi taikei*.

In the period covered by this bibliography, both the quantity and quality of publications in physical geography have exceeded those in cultural or human geography. This is understandable in the light of the interests and training of the founders of modern geography in Japan and in the light of the security regulations and atmosphere which long prevailed. Until World War II, studies in cultural geography were largely morphological and/or purely descriptive. Currently, better interpretive studies in the cultural area are being produced and at an accelerated rate.

The research for this bibliography extends over a number of years. Independently the authors had come to know the writings in their special fields of interest. A joint publication was planned in 1951 and work was begun in Tōkyō. This was followed by a year of cooperative effort in the library of the University of Michigan's Center for Japanese Studies at Ann Arbor. During this period every aid was given by Mr. Edwin Beal of the Library of Congress. When American library resources were exhausted the project was again returned to Japan. The libraries and staffs of the Geographical Institute of Tōkyō University, of the Tōkyō Geographical Society, and the Geographical Survey Institute of the Ministry of Construction at Chiba proved most helpful. To Miss Yōko Aono, graduate student at Tōkyō University, the authors owe thanks for her faithful cooperation particularly in the later phases of this project. Mr. J. Douglas Eyre and Mr. Forrest R. Pitts, as graduate students in the Department of Geography and Center for Japanese Studies of the University of Michigan, contributed in an important way in the early days of the project.

On the campus at the University of Michigan, the authors wish to express their thanks for the help given by Mr. Yōtarō Okuno and Mr. Raymond Nunn of the library staff. Mr. Philip Weadock did much to standardize and improve the statements evaluating the different items. The authors of the preceeding bibliographies in this series gave both inspiration and the guidance of their experience. To Dr. Joseph K. Yamagiwa, the editor of this series, the authors render all gratitude for the guidance given and the hard work expended in getting this volume ready for final publication.

Robert B. Hall
Toshio Noh

TABLE OF CONTENTS

		Entries	Page
	Editor's Foreword		v
	Authors' Introduction		vii
I.	BIBLIOGRAPHIES	1-16	1
II.	ENCYCLOPEDIAS, DICTIONARIES AND GAZETTEERS, AND TRAVEL GUIDES		
	A. General	17-33	3
	B. Place Name Dictionaries	34-45	5
	C. Travel Guides	46-56	6
III.	YEARBOOKS, COLLECTIONS OF STATISTICS, AND CENSUSES	57-73	7
IV.	SETS AND COLLECTIONS	74-97	10
V.	PERIODICALS	98-123	14
VI.	ATLASES, MAPS, CARTOGRAPHY, AND AIR-PHOTO COVERAGE	124-153	17
VII.	HISTORY OF JAPANESE GEOGRAPHY		
	A. History of Japanese Geographical Thought	154-163	21
	B. History of Japanese Cartography	164-167	23
	C. History of Exploration by Japanese	168-169	23
VIII.	PHYSICAL GEOGRAPHY		
	A. Geomorphology and Geognostics	170-474	23
	B. Volcanism and Seismology	475-489	49
	C. Drainage	490-497	51
	D. Groundwater	498-517	51
	E. Flora	518-529	53
	F. Fauna	530-538	54
	G. Soils	539-549	55
	H. Climate and Weather	550-609	56
	I. Destruction by Physical Forces	610-616	62
	J. Oceanography	617-626	62
	K. Limnology	627-650	63
IX.	HISTORICAL AND CULTURAL GEOGRAPHY		
	A. Historical Geography	651-704	65
	B. Political Geography	705-709	70
	C. Population Geography		
	1. General	710-756	71
	2. Medical Geography	757-759	75
	D. Transportation and communications	760-787	76
	E. Settlement Geography		
	1. Rural	788-873	78
	2. Urban	874-916	86
	F. Linguistic Geography		
	1. Dialects	917-918	90
	2. Place Name Studies	919-935	91
	G. Religious Geography	936-938	92
	H. Geography of Folk Lore and Folk Songs	939-942	92
	I. Culture Traits and Other Cultural Distributions		
	1. General	943-945	93
	2. Specific	946-957	93
X.	ECONOMIC GEOGRAPHY		
	A. General	958-963	95
	B. Land Use	964-985	95
	C. Forestry	986-989	98
	D. Agriculture		
	1. Crops and Crop Systems		
	a. General	990-1034	98
	b. Specific	1035-1062	102

	Entries	Page
2. Irrigation	1063-1075	105
3. Reclamation	1076-1080	106
4. Animals	1081-1089	106
E. Fishing	1090-1104	107
F. Mining and Minerals	1105-1111	109
G. Industry	1112-1122	109
H. Recreation and Tourism	1123-1124	110
XI. REGIONAL DESCRIPTIVE GEOGRAPHY		
A. General	1125-1130	110
B. Traditional Regions: *dō* and *kuni*	1131	111
C. Political Regions: *ken*, *gun*, *machi*, *shi*, *fu*, *to*	1132-1172	112
D. The "Geographical" Regions	1173-1228	116
E. Other Regions	1229-1233	122
F. Areas of the Former Japanese Empire	1234-1254	122
APPENDIX - List of Publishers with Characters		125

JAPANESE GEOGRAPHY

A GUIDE TO JAPANESE REFERENCE
AND RESEARCH MATERIALS

Chapter I

Bibliographies

Bibliographical and reference work was long the main part of scholarly labor in pre-Meiji Japan. Although the tradition was thus very strong for the listing of previous written materials and the recitation of former works, the idea and the system of listing the materials was very different from modern bibliography in the Western sense. Few bibliographies of the pre-Meiji period could survive after the tests based on new standards, and the tradition in Japan has not been strong enough to produce many bibliographic publications since Meiji.

Geographical bibliographies are very scarce and those which have appeared at all have had very narrow circulation. This is no cause for wonder, since the circle of professional geographers who are concerned with purely geographical bibliographies has never been large, and their scope of interest was not confined within the limits of geography in its strictest sense.

1. Chishitsu Chōsajo 地質調査所 (Geological Survey Institute), Chishitsu Chōsajo shuppambutsu mokuroku 地質調査所出版物目録 (List of publications of the Geological Survey Institute), Kawasaki, Chishitsu Chōsajo, 1952, 6+124 pp.
This is a complete listing of the publications of the Geological Survey Institute (formerly the Imperial Geological Survey) from 1880 to 1952. A number of English and other foreign-language publications are included. With an index map of published geological charts.

2. Chūō Kishōdai 中央氣象臺 (Central Meteorological Observatory), Nihon kishōgaku bunken mokuroku (jōkan) 日本氣象學文獻目錄 (上巻) (Bibliography of Japanese meteorological publications, I), Tōkyō, Chūō Kishōdai, 1941, 1+1+375 pp.
A complete list of papers that appeared in some twenty meteorological and related periodicals, to date, such as Nihon kishōgaku bunken mokuroku (English title, "Journal of the Meteorological Society of Japan"), Chūō Kishōdai ōbun hōkoku (English title, "Bulletin of the Central Meteorological Observatory"), Chūō Kishōdai ōbun ihō (English title, "The geophysical magazine"), etc. Entries are listed by author and are arranged separately for each publication. There is no annotation.

3. Fujimoto Haruyoshi 藤本治義, Nihon chishitsu bunken mokuroku 日本地質文獻目錄 (Bibliography on Japan's geology), Tōkyō, Chijin Shokan, 1942, 4+4+337 pp.
A comprehensive but unannotated bibliography of approximately 6,000 books and articles on geology, geomorphology, and mining which were published prior to Feb., 1941. Entries are listed chronologically under 36 regional headings. Following the date, only the author, title, volume, number, and pagination are included.

4. Fujioka Kenjirō 藤岡謙二郎, Taikei chirigaku jiten 体系地理學辞典 (A systematic dictionary of geography), Tōkyō, Taimeidō, 1953, 2+299 pp.
Contents of this dictionary cover the main fields of geography, providing somewhat less than 1500 descriptive and interpretive definitions of geographical terms, topics, and concepts. Selected foreign place names are listed. The volume is designed as a general reference work for high school and college students. Some of the definitions of regional and cultural terminology peculiar to Japan will be helpful, however, to the more advanced Western student of Japanese geography.

5. Harada Jumpei 原田準平, Hompō kazan bunkenshū 本邦火山文獻集 (Bibliography of articles on Japanese volcanology), Tōkyō, Kokon Shoin, 1928, 74 pp.
This short bibliography was published as one of a series on geography and related fields. It lists selected articles only. The articles are arranged by author, according to region, and subdivided according to the volcano or volcanic chain to which they pertain. The regional breakdown is as follows: 1) Japan as a whole, 2) Honshū, 3) Kyūshū, 4) Hokkaidō, 5) Chishima, and 6) Taiwan.

6. Honjō Eijirō 本庄栄治郎, Yoshikawa Hidezō 吉川秀造, and Matsuyoshi Sadao 松好貞夫, Nihon keizaishi daisan bunken 日本經濟史第三文獻 (Bibliography of Japanese economic history, book 3), Tōkyō, Nihon Hyōron Shinsha, 1953, 12+600 pp.
Includes materials related to economic history published from 1941 through 1950 and consists of four parts: 1) general guide to the reference materials listed, 2) other historical sources useful to the student of economic history, 3) Japanese books, and 4) books in Western languages. Articles which appeared in important journals are listed, as well as books. Arrangement of entries is by title, and under each item is given the number of volumes, author or editor, date, publisher, pagination, and a brief description of the contents. Three supplements contain a bibliography of regional geography and history, a list of periodicals, and an index by title. This is the third publication in a long-term enterprise. Kaihan Nihon keizaishi bunken (A revised bibliography of Japanese economic history) appeared in 1933, and Nihon keizaishi shimbunken (New bibliography of Japanese economic history) was printed in 1942.

7. Ishimoto Mishio 石本巳四雄 and Kawasumi Hiroshi 河角廣, Jishingaku narabi ni chikyūbutsurigaku hompō bunken mokuroku I 地震學並地球物理學本邦文獻目錄 I (Bibliography of Japanese seismology and geophysics, I), Tōkyō, Kokon Shoin, 1938, 3+96+84+16 pp.

Lists writings by Japanese scientists on seismology, volcanology, and geophysics published in Japanese and foreign journals between 1870 and 1923. In three parts, the first lists articles; the second, notes and reports; the third, books. Entries are arranged chronologically by author. Unannotated.

8. Jimbun Chiri Gakkai 人文地理學會 (Association of Human Geography), Chirigaku bunken mokuroku 地理學文獻目錄 (Bibliography of geography), Kyōto, Yanagihara Shoten, 1953, 4+4+120+28 pp.

This is a bibliography of geographical books and articles published during the period 1945-51. It also lists publications in fields closely associated with geography. The compilation is not too accurate and there is need for an adequate revision. The first half deals with the physical and natural sciences: geomorphology, soil science, meteorology, climatology, oceanography, limnology, general physical geography, and animal and plant geography. The second half, concerning human geography and related topics, is subdivided as follows: anthropological, population, settlement, economic, resource, communications, political, social, cultural, historical and regional geography, and cartography.

9. Kikō Danwakai 氣候談話會 (Climatological Symposium), Kikōgaku kankei bunken mokuroku (hōbun no bu) 氣候學関係文獻目錄（邦文の部）(Bibliography of climatological writings in Japanese), Tōkyō, Kikō Danwakai, 1953, 2+2+37 pp.

This is a listing of climatological articles and books published in the period 1940-50. Emphasis is placed upon climatological references useful to the geographical researcher. However, writings representative of all phases of climatology are included. In two parts, the first covering monographs, the second books.

10. Kokuritsu Kokkai Toshokan Shibu Toshokambu 國立國會圖書館支部圖書館部 (National Diet Library, Branch Library, Libraries Section), Kanchō kankōbutsu sōgō mokuroku daiikkan 官廳刊行物總合目錄 第一巻 (A general bibliography of government publications, volume 1), Tōkyō, Kanchō Toshokan Kenkyūkai, 1952, 7+650 pp.

Lists publications up to 1950. The bibliography is organized according to the following categories: 1) general, 2) philosophy, 3) history, 4) social sciences, 5) natural sciences, 6) technology and engineering, 7) industry, 8) fine arts, 9) linguistics, and 10) literature. Entries are listed by author or editor's name, followed by title etc., and usually tell where the publication is on file. Volume two appeared in 1954 and covers the period 1950 through 1953.

11. Kokushō Iwao 黒正巖 and Kikuta Tarō 菊田太郎, Keizai chirigaku bunken sōran 經濟地理學文獻總覽 (Bibliography of economic geography), Tōkyō, Sōbunkaku, 1937, 2+2+33+320 pp.

A bibliography of books and papers printed prior to 1935 available in the library of Kyōto University's Keizai Chirigaku Kenkyūjo (Institute of Economics and Geography). The first half covers Japanese works, and the latter Western literature. The classification of the Japanese materials is as follows: a) bibliography, b) dictionaries, yearbooks and statistics, c) sets and collected papers, d) general geography, human geography, and physical geography, e) economic geography, f) geography of communications, g) industry, h) commercial geography and science of merchandise, i) world economy, international economy, colonization, trade, and international finance, j) cartography, maps, and tables, k) regional geography of Japan, and economic geography of Japan, l) regional geography of the world, and economic geography of the world.

12. Kōsaki Masao 耕崎正男, Saikin chirigaku bunken mokuroku, daiisshū 最近地理學文獻目錄 第一輯 (Bibliography of recent papers in geography, no. 1), Tōkyō, Kokon Shoin, 1931, 16+12+240 pp.

Some 3,200 articles and papers are listed, gathered from professional journals and other periodicals. A few books of collected papers are cited. This unannotated bibliography is divided into four sections, namely: 1) the nature of geography, 2) methodology, 3) regional geography (subdivided — home islands, overseas territories, the world), and 4) general geography in two parts, physical and human. Entries are listed by title and include author, name of publication, volume, number, date, and supplement.

13. Naimushō Chirikyoku 内務省地理局 (Geographical Section, Home Ministry), Chishi mokuroku 地誌目錄 (Bibliography of regional descriptive geography), Tōkyō, Naimushō Chirikyoku, 1885, 105 pp.

An unannotated list of some 2400 old books and manuscripts entirely or in part concerned with regional description. The majority were written during the Edo period. Classified according to the traditional kuni names, entries are listed by title, followed by author or compiler, number of volumes, and method of reproduction or printing. Now very rare, this bibliography is a valuable reference work for historian and historical geographer.

14. Takagi Toshita 高木利太, Kazō Nihon chishi mokuroku 家藏日本地誌目錄 (Bibliography of Takagi's library of Japanese regional geography), Ōsaka, privately published, 1927, 11+43+844+25 pp.

A bibliography of Takagi's private collection of books on the regional geography of Japan, but including many more titles related to history, folklore, place-name studies, etc. Many of the works listed are old Tokugawa regional descriptions. Publications are classified by traditional region and listed by title, followed by name of editor or author, date, and publisher. There is no additional annotation, but an index is included. In 1930 Takagi published a sequel (zokuhen) to this bibliography, listing the many volumes added to his library after 1927.

15. Uyama Yasuo 宇山保雄, Jimbun chiri bunken mokuroku 人文地理文献目録 (A bibliography of human geography), Yamaguchi, privately published, 1951, 1+1+3+179 pp.

A good summary of the content of the principal geographic periodicals in Japan. Edited as a reference guide for pre-college students, this is a listing of about 3,000 articles selected from the 18 major professional magazines. Entries are listed by title under 23 categories which conform to the units of instruction followed in the high school geography program, e. g.: 1) human geography, general, 2) population, 10) fisheries, 16) coal, 21) traffic and communication, etc.

16. Wada Mankichi 和田萬吉, *Kohan chishi gedai* 古版地誌解題 (Bibliography of old regional geographies), Tōkyō, Wada Ishirō, 1916, 2+2+5+332 pp.

An annotated listing of regional descriptions, in book or manuscript form, printed before the Meiji Restoration and contained in the private collection of Wada Ishirō. Printed on Japanese paper and now rare, this bibliography is, however, a valuable guide to primary source material of interest to the historical geographer.

Chapter II

Encyclopedias, Dictionaries and Gazetteers, and Travel Guides

Almost all Japanese encyclopedias, dictionaries, and gazetteers contain some data of interest to the geographer. It is impractical to attempt to list them all here. Only those with a definite geographical emphasis are included. In the case of specialized studies, reference should be made to the appropriate categories in the other bibliographies of this series, *e. g.*, those listed in the section on historical geography in John W. Hall's *Japanese history: a guide to Japanese reference and research materials*.

The Western geographer in his studies of Japan will find constant need for these types of reference material. This is particularly true as regards place names, local terminology, and the scholarly terminology in use at any one period. The Japanese, fortunately, have a strong tradition as compilers of such reference materials, and aids can be found in almost all areas of interest to the geographer.

The Japanese have also long been prolific producers of travel guides, items always of interest to the geographer. Many are but slightly modified copies of previous publications. Many were designed for limited and short-time circulation and are not especially useful. Some, however, are not only works of art but contain data not otherwise available. The attempt has been made here to select the most important ones for listing and comment.

A. General

17. Aono Hisao 青野壽郎 and Hoyanagi Yasuyoshi 保柳睦美, ed., *Jimbun chiri jiten* 人文地理事典 (A dictionary of human geography), Tōkyō, Kokan Shoin, 1952, 2+1+2+5+1+452 pp.

Over 500 major topics and terms are treated. Headings appear in romanization, arranged in the order of the Japanese syllabary, and the explanatory and descriptive material is in Japanese. Technical terminology was selected from all phases of human geography. Sixty-four specialists contributed to this dictionary. With a 44-page index.

18. Fujimoto Haruyoshi 藤本治義, Okada Yaichirō 岡田彌一郎, and Miwa Tomoo 三輪知雄, *Hakubutsu jiten* 博物辞典 (Dictionary of natural history), Tōkyō, Sanseidō, 1938, 1+3+1080 pp, 39 illus.

A dictionary of natural science terms compiled for high school teachers. More than 15,000 technical terms are defined, covering six principal fields: zoology, botany, physiology, hygiene, geology, and mineralogy. The appendix contains a variety of general tables on animal, plant, and mineral classification, geological time-scales, etc.

19. Heibonsha 平凡社, ed., *Daihyakka jiten* 大百科事典 (Great encyclopedia), Tōkyō, Heibonsha, 1931-34, 28 v. + supplements.

The largest and most comprehensive of the modern Japanese encyclopedias. Contents are partially revised from time to time with the publication of supplements, the third volume of which appeared in 1950. More than 1,000 scholars contributed signed articles. Volume 28 constitutes a complete index. Many of the topics treated are geographical or semi-geographical, including important place names, regional names, and economic activities such as rice farming, pearl culture, sericulture, etc.

20. Heibonsha 平凡社, ed., *Shakaika jiten* 社會科事典 (A dictionary of social studies), Tōkyō, Heibonsha, 1948-49, 10 v.

An encyclopedia-type work compiled to serve as a reference work for high school students in newly initiated programs of social studies; however, the presentation is at a relatively advanced level and the writing and editing standards are high. Therefore, the series will be of some use to the Western scholar, particularly as an aid in understanding Japanese social-science terminology. Written by about 450 specialists, 40 of them are geographers. Approximately 700 major topics are treated. A complete index is found in volume ten. Historical, geographical, and anthropological subjects and terms predominate.

21. Iimoto Nobuyuki 飯本信之, Ishii Itsutarō 石井逸太郎, and Matsuo Toshio 松尾俊郎, ed., *Jimbun chiri jiten* 人文地理事典 (A dictionary of human geography), Tōkyō, Seibundō Shinkōsha, 1951, 2+2+17+363 pp.

Designed as a reference book for students taking high school or introductory college geography, this dictionary presents some 4,000 items arranged under five headings, namely; 1) the natural environment, 2) backward and primitive areas, 3) industry in modernized areas, 4) human residences and settlements, and 5) world unity in terms of the factors of human geography. Written by twelve geographers, including the three above-named editors. With index.

22. Iwanami Shoten Henshūbu 岩波書店編集部, Kagaku no jiten 科學の事典 (Dictionary of science), Tōkyō, Iwanami Shoten, 1950, 15+1531 pp., 61 illus.

An elementary dictionary of science terms for the use of grade school students, arranged in the order of the Japanese syllabary in romanization. Definitions are clear and simple so that this book could be used as an aid in the beginning or intermediate stages of language study by the natural or social scientist.

23. Kensetsushō Chiri Chōsajo 建設省地理調査所 (The Geographic Survey Institute), Gun-shi-chō-sommei benran 郡市町村名便覧 (List of the names of gun, shi, machi, and mura), Tōkyō, Chiri Chōsajo, 1948, 2+10+16+197 pp.

A useful index of place names appearing on the Geographic Survey Institute's 1:2,000,000 base maps of gun and shi boundaries, 1:500,000 base maps showing shi, machi, and mura, and 1:50,000 topographic map sheets. Listed in the order of the Japanese syllabary, the names are in kanji with accompanying pronunciations in kana which are based on the Institute's 1948 place name survey. The listed pronunciations conform with current local usage. The name of the topographic sheet on which the place name is shown is also listed.

24. Kobayashi Fusatarō 小林房太郎 and Okada Eisaku 岡田贏策, Chirigaku jiten 地理學辭典 (Dictionary of geography), Tōkyō, Bungakusha, 1926, 2+48+316+54 pp.

The first of the Japanese geographical dictionaries, this is a non-illustrated, encyclopedia-like publication limited in scope to shizen chirigaku or "natural geography" which is, more accurately, general physical geography excluding the biological phases of that field. Human geography is entirely omitted. Items included correspond approximately to those treated in Watanabe's Chigaku jiten (No. 32) but the definitions and descriptions are, on the whole, expressed more clearly than in that volume. General definitions often cite specific illustrative examples of Japan in regard to natural features, geological phenomena, etc. This is a handy reference work, reprinted almost yearly since its initial appearance and widely utilized by both professional geographers and students in Japan. Both Japanese and English indexes are provided.

25. Kokon Shoin Henshūbu 古今書院編輯部 (Kokon Shoin Editorial Section), Chirigaku shōjiten 地理學小辭典 (Concise dictionary of geography), Tōkyō, Kokon Shoin, 1936, 379 pp.

A dictionary of geographical terms covering all phases of the field. Place names and personal names are excluded, except in a few instances where they are pertinent to the definition presented. Chigaku jiten, issued by the same publisher, treats proper nouns (see No. 32).

26. Minzokugaku Kenkyūjo 民俗學研究所 (Institute of Ethnology), Minzokugaku jiten 民俗學辭典 (A dictionary of folklore), Tōkyō, Tōkyōdō, 1951, 22+714 pp., 4 maps, 48 photos, and 54 illus.

Forty scholars collaborated on this encyclopedia which contains a descriptive commentary on some 897 selected items pertaining to such matters as beliefs, customs, festivals, social practices, social and family organization, etc. Although mythological background and historical development are traced wherever pertinent, the emphasis is upon present-day or living folklore in Japan. Articles contain a note on suggested readings in most cases. The work is arranged in the order of the Japanese syllabary under 24 general headings, e. g., house types, food, clothing, the fishing village, etc.

27. Ōsaka Shiritsu Daigaku Keizai Kenkyūjo 大阪市立大學經濟研究所 (Ōsaka Municipal University Economic Research Institute), Keizaigaku shōjiten 經濟學小辭典 (A small dictionary of economics), Tōkyō, Iwanami Shoten, 1952, 16+1264+3 pp.

Of possible interest to the economic geographer, this is a general encyclopedia-type coverage of the field of economics. Topics and terms are arranged in the order of the Japanese syllabary and a short list of references is included at the end of most articles. Indices are in Japanese and Western languages. This dictionary is the joint work of some 250 specialists in economics.

28. Ōta Tamesaburō 太田為三郎, Teikoku chimei jiten 帝國地名辭典 (Gazetteer of Japanese place names), Tōkyō, Sanseidō, 1912, 1742 pp.

Place names are arranged according to the Japanese syllabary so that both kanji and kana readings are given. A brief summary of the nature of each place is included. Not so detailed a gazetteer as Yoshida's Dai-Nippon chimei jisho, which is much superior as far as historical detail is concerned.

29. Tennensha Kishō Jiten Henshūbu 天然社氣象辭典編集部 (Tennensha Editorial Office for the Encyclopedia of Weather and Climate), Kishō jiten 氣象辭典 (The encyclopedia of weather and climate), Tōkyō, Tennensha, 1954, 7+350+20+58 pp.

This is a general encyclopedia compiled and written by 37 meteorologists and climatologists in government service and academic life. The effect of meteorological phenomena on human life and activity is emphasized. The majority of the contents pertains to Japan specifically. With a number of supplementary tables giving climatic data, typhoon tracks, past climatic disasters in Japan, meteorological associations and societies, etc.

30. Tōjō Misao 東條操, Zenkoku hōgen jiten 全國方言辭典 (Dictionary of Japanese dialects), Tōkyō, Tōkyōdō, 1951, 14+881 pp., 3 maps.

The editor based his work primarily upon written materials and utilized more than 500 sources in this compilation, including a number of books on regional geography. Dialect forms are arranged in the order of Japanese syllabary and their geographical distribution, both current and historical, is delimited. Information on etymology, grammar, and phonetics is supplemented by examples of usage. A map shows dialect distribution; it is classified into three broad categories, namely: East Honshū, West Honshū, and Kyūshū. A second map indicates major administrative divisions of Japan, old and new.

31. Wadachi Kiyoo 和達清夫, ed., Kishō no jiten 氣象の事典 (A meteorological dictionary), Tōkyō, Tōkyōdō, 1954, 2+2+5+572 pp., 41 photos, 400 illus.

A general encyclopedia on weather and climate, this is a joint work by 41 authors. Several dozen major subjects and over 1,000 minor topics are treated. The work is divided into seven parts, six of which pertain partially to Japanese meteorology. Section five deals specifically with the climate of Japan, as do section six on meteorological disasters, and section seven on scholars in meteorology and climatology. With supplementary tables of climatic statistics for Japan and the world, weather codes, a listing of principal observatories and broadcasting networks, etc.

32. Watanabe Kan 渡邊貫, Chigaku jiten 地學辭典(Dictionary of earth sciences), Tōkyō, Kokon Shoin, 1935, 14+1534+133+59+300 pp., index in separate v.

The most comprehensive one-volume encyclopedia of its kind. Topics and terms covering all phases of the earth sciences are listed in the order of the Japanese syllabary in romanization (Nihon-siki). The third section contains 44 maps and a variety of statistical tables pertaining specifically to Japan, e.g., a map showing gravity anomalies, a colored geological map, a map of the distribution of volcanic ranges and earthquake zones, etc. A list of the world's principal research institutes, universities, and organizations concerned with the earth sciences is included. The index gives equivalent English, German, and French terms and Japanese proper nouns in alphabetical order.

33. Yanagida Kunio 柳田國男, Bunrui nōson goi 分類農村語彙 (Classified farm village vocabulary), Tōkyō, Tōyōdō, 1947-48, 2 v.

A survey of rural terminology concerning farming and forestry. Words and phrases are listed according to general categories such as sowing, weeding, harvesting of rice, agricultural tools, sericulture, etc. Of interest to the geographer as a possible language aid in preparing for field work.

B. Place Name Dictionaries

34. Dai-Nihon chimei jiten narabi ni kōtsū chizu taikan 大日本地名辭典並交通地圖大鑑 (Gazetteer of Japanese place names and atlas of traffic maps), Ōsaka, Kokusai Gakujutsu Hyōronsha, 1930, 53 maps.

A prefectural atlas of Japan. One ken is shown on each sheet, in various scales. On the back of the maps place names are shown for shi, machi, mura and ōaza. The maps are well printed but the traffic data is out of date and the traffic maps are of little value.

35. Fujisawa Morihiko 藤澤衛彥, Shichōson meikan 市町村名鑑(Dictionary of shi, machi, and mura names), Tōkyō, Chihō Meikan Hakkōjo, 1918, 4+14+605 pp.

The names of shi, machi, and mura are arranged by prefecture and kana pronunciations are shown. Under mura, the names of ōaza are included but kana readings are given when the name is unusual and liable to be misread. Under shi, the names and addresses of city and other government offices are listed. An appendix gives a list of schools and Shintō shrines at that time subsidized by the state. Also included is an organizational chart of the central government.

36. Heibonsha 平凡社, Sekai chimei jiten 世界地名辭典(Place name dictionary of the world), Tōkyō, Heibonsha, 1950-52, 6 v.

The most recently published Japanese gazetteer of the world. It includes all principal Japanese place names. Volume six contains a general atlas and two indices, one in kana, the second in romanization.

37. Mozume Kōryō 物集高量, Nihon meisho jii 日本名所事彙 (Gazetteer of famous Japanese places), Tōkyō, Ikuseikai, 1910, 141+979 pp., 41 illus.

Lists the location, reasons why the place is noted, poems and songs pertaining to the place, and other miscellaneous popular information. Places are classified by kuni or traditional region. With an index giving pronunciations in kana.

38. Nihon Jichi Gyōsei Kenkyūkai 日本自治行政研究會 (Japanese Association for the Study of Self-governing Administrative Units), Zenkoku shichōson'azamei sōran 全國市町村字名總攬(Complete gazetteer of shi, machi, mura and aza names in Japan), Ōsaka, Nihon Kōtsu Tsūshin Tosho Kabushiki Kaisha, 1948, 2nd ed., 1949, 552 pp.

With the exception of aza names, all place names are given in kana, arranged in the order of the Japanese syllabary, by region or district such as Hokkaidō, Tōhoku, etc. At the end of the gazetteer there is a list of post offices and a list of place names which are particularly difficult to read.

39. Nitchi Shuppan Kabushiki Kaisha 日地出版株式會社 (Japan Geographical Printing Co.), Zenkoku shichōson-azamei taikan 全國市町村字名大鑑 (A dictionary of shi, machi, mura and aza names of Japan), Tōkyō, Nitchi Shuppan Kabushiki Kaisha, 1951, 489 pp.

Pronunciation of place names is indicated in kana, and names appear in the order of the Japanese syllabary and are arranged by prefecture. The addresses of shi, machi, and mura offices and certain other local government agencies are included.

40. Ogawa Takuji 小川琢治, Nihon chizuchō chimei sakuin 日本地圖帖地名索引 (Place name index for the "Atlas of Japan"), Tōkyō, Seishōdō, 1925, 2+5+3+487 pp.

Designed to accompany Seishōdō's atlas, this index can be used separately as a handy gazetteer. The names of all political units and most geographical features are listed in the order of the Japanese syllabary with kana readings. An index of difficult names is arranged by kanji stroke order.

41. Ogawa Takuji 小川琢治, Shichōson'ōaza yomikata meii 市町村大字讀方名彙 (Classified list of readings of the names of cities, towns, villages and ōaza), Tōkyō, Seishōdō, 1925, 403+5+43+2 pp.

Ogawa's Japanese atlas, his place name index for it, and this extensive list of place name readings constitute a three-volume set published by the same firm. This detailed compilation gives the proper kanji and kana readings for all places with postal service in the home islands, Karafuto, Taiwan, Korea, Kwantung, the South Manchurian Railway zone, and the South Seas mandate. It also includes the subdivisions in cities, towns and villages large enough to be so subdivided. The researcher will find this place name guide among the most helpful and easy-to-use publications of its kind.

42. Sawada Hisao 澤田久雄, Nihon chimei daijiten 日本地名大辞典 (Dictionary of Japanese place names), Tōkyō, Nihon Shobō, 1937-38, 6 v.

The most detailed place-name tabulation available, this set is essentially an annotated gazetteer. The names of geographical features are included, with those of large and small political units, regions, old provinces, etc. Both modern and historical names are listed. The location and/or boundaries of each place are indicated along with brief descriptive and historical remarks. Names appear in the order of the Japanese syllabary. No kanji index is provided.

43. Takazu Shikitarō 高頭式太郎, Nihon sangakushi 日本山岳志 (Mountains in Japan), Tōkyō, Hakubunkan, 1906, 5+14+27+33+48+186+54+64+46+673+37+129 pp.

A voluminous compendium giving the location, the altitude, and a description of the topography and climbing routes of each major mountain in Japan. Included are notes and comments on local flora, fauna, folklore, and history, as well as famous mountain songs and poems. In the index, mountains are listed by kuni or traditional region in the order of the Japanese syllabary and in kanji-stroke order. Takazu was the first president of the Japanese Mountain Association, the nation's leading climbing club.

44. Tanakadate Shūzō 田中館秀三, Shichōsommei no yomikata oyobi shichōson menseki, jinkō mitsudohyō 市町村名の讀み方及び市町村面積人口密度表 (Table showing the pronunciation of the names of cities, towns and villages, and their area, population, and population density), Tōkyō, Nihon Shobō, 1937, 191 pp.

A romanized listing of all shi, machi, and mura, indicating their population as of 1920, 1925, 1930, and 1935. The 1935 area, in square kilometers, and the population density for that year are included.

45. Yoshida Tōgo 吉田東伍, Dai-Nihon chimei jisho 大日本地名辞書 (Place name dictionary of Japan), Tōkyō, Fuzambō, 1900-09, 5 v.

The first thorough gazetteer published in Japan. Volumes 1 through 3 cover Japan proper, except Hokkaidō; vol. 4 is an index, listing names in the order of the Japanese syllabary; and vol. 5 is for Hokkaidō, Okinawa, and Taiwan. Place names appear in geographical sequence, classified by administrative districts. Annotation is in somewhat classical style, making reading difficult for most Western researchers, and descriptions are largely historical. Later place name dictionaries are of more practical use to the geographer.

C. Travel Guide

46. Mainichi Shimbunsha 毎日新聞社 (Mainichi Newspaper Company) and Nihon Kankōchi Sentei Kaigi 日本観光地選定會議 (Japanese Council for the Designation of Tourist Areas), Shin-Nihon kankōchi annai 新日本観光地案内 (A new guide to sight-seeing in Japan), Tōkyō, Shinkōsha, 1952, 2+4+624+16 pp.

In 1950, 138 of the most popular sight-seeing and scenic places in Japan were selected on the basis of a public poll. This is a descriptive guide to these places. It is designed for the tourist but will be of general interest to the geographer, since it contains a variety of useful and interesting travel information, numerous excellent photographs, and up-to-date route maps. Places are treated by prefecture, from north to south.

47. Matsukawa Jirō 松川二郎, Meishō onsen annai 名勝温泉案内 (Guide to sight-seeing and hot springs places), Tōkyō, Seibundō, 1927, 16+672+38 pp., 7 photos.

Although of little scientific value, this guidebook is of use to the field researcher or general traveler as a handy reference work on cultural and geographical points of interest and and accommodation available in their vicinity. It is divided into two sections, the first providing a brief description of famous sight-seeing places, and the second being a directory to hotels and inns. Out-dated, but the majority of the information remains valid.

48. Nihon Kōtsu Kōsha 日本交通公社 (Japan Travel Bureau), Ryotei to hiyō 旅程と費用 (Travel routes and expenses), Tōkyō, Nippon Kōtsu Kōsha, ed. 6, 1953, 10+940+12+29 pp.

Organized on a regional basis, this is a general guide to travel and sightseeing accommodations, transportation facilities, routes, model tours, etc., for all Japan, with notes on expenses and charges. It appeared before the war, was periodically revised from 1919 to 1938, and was re-issued again in 1952. It can be of considerable practical use to the traveler, whether scholar or ordinary tourist, and is of geographical interest as a handy route reference.

49. Takenouchi Makoto 竹内亮, Tsukushi fūkeishi 筑紫風景誌 (Landscape of Tsukushi), Tōkyō, Kokon Shoin, 1941, 6+441 pp., 6 photos.

Tsukushi is the old name for the northern Kyūshū provinces. This is mainly a description of vegetation and of mountain landforms in that region. Phyto-geographical descriptions are much generalized.

50. Tetsudōshō 鐵道省 (Ministry of Railways), Bōsō to Suikyō 房總と水鄉 (Bōsō and Suikyō), Tōkyō, Tetsudōshō, 1934, 4+8+142 pp., 5 maps and 66 photos.

Suikyō is the name of the lake district in eastern Kantō and includes some of the area along the lower Tonegawa.

51. Tetsudōshō 鉄道省 (Ministry of Railways), Hokuriku 北陸 (Hokuriku), Tōkyō, Tetsudōshō, 1938, 2+11+320 pp., 5 maps, 98 photos.
This volume is one of the most systematically compiled and edited of the several guidebooks published by the Ministry of Railways. It contains descriptions of Fukui, Ishikawa, Toyama, and a part of Niigata prefecture.

52. Tetsudōshō 鐵道省 (Ministry of Railways), Kishū 紀州 (Kishū), Tōkyō, Tetsudōshō, 1935, 2+4+180 pp., 3 maps and 96 photos.
Kishū is the old name of the southern portion of the Kii peninsula, in Wakayama prefecture, an area with numerous historical and scenic attractions.

53. Tetsudōshō 鐵道省 (Ministry of Railways), Nihon annaiki 日本案内記 (Travel guide to Japan), Tōkyō, Hakubunkan, 1929-36, 8 v., illus.
An eight-volume series of detailed travel guidebooks compiled by the Japanese government's Ministry of Railways and treating the traditional regions or districts of Japan proper. Each volume consists of four parts: introduction, general description, travel guide, and index. The general description contains information on political divisions, history, topography, climate, industries, etc. The travel guide provides a description of specific areas and places located along the principal railway lines in each district. The books contain numerous maps and photographs. Most of the maps, both large and small scale, are in color. Place-name pronunciations are given in kana. Following the initial compilation undertaken in 1929, volumes were published yearly until the set was completed in 1936. Revision was then begun but was interrupted by the outbreak of the Sino-Japanese war, and only two revised volumes appeared, in 1937 and 1938. Although the set contains a great deal of non-geographical material, it does provide a handy descriptive survey of the historically significant and important geographical areas of Japan. If not of much professional value to the geographer, it nonetheless will be of interest to him as a useful travel aid. The publication date and district treated by the individual volumes is as follows: Tōhoku, 1929 (revised, 1937); Kantō, 1930 (revised, 1938); Chūbu, 1931; Kinki (part 1), 1932; Kinki (part 2), 1933; Chūgoku and Shikoku, 1934; Kyūshū, 1935; and Hokkaidō, 1936.

54. Tetsudōshō 鐵道省 (Ministry of Railways), Shikoku 四國 (Shikoku), Tōkyō, Tetsudōshō, 1937, 2+13+306 pp., 5 maps, 92 photos., 2 colored pictures.
Shikoku Island, naturally lending itself to compact treatment, is described regionally on the basis of the four old provinces (Sanuki, Awa, Tosa, and Iyo) whose boundaries correspond approximately to the present prefectural limits. The most detail is devoted to places and areas lying along the main rail lines. The Saigoku hachijū hakkasho, or 88 famous places of pilgrimage on the island, receive predominant attention.

55. Wakimizu Tetsugorō 脇水鐵五郎, Nihon fūkeishi 日本風景誌 (The scenic beauty of Japan), Tōkyō, Kawade Shobō, 1940, 2+7+332 pp., 136 photos., 6 illus.
A good description of the geography and geology of some twenty places or areas noted for scenic beauty. The random selection includes major national parks. With numerous landscape photographs.

56. Zen-Nihon Kankō Remmei 全日本観光連盟 (All-Japan Sightseeing League), Nenjū gyōji 年中行事 (Annual functions), Tōkyō, Nippon Kōtsu Kōsha, 1953, 129 pp., 60 photos.
A compilation of 478 yearly ceremonies, festivals, and other special occasions celebrated throughout Japan, indicating the date, type and place of each event. With index.

Chapter III

Yearbooks, Collections of Statistics, and Censuses

The geographer most frequently turns to yearbooks and to statistical and census publications for his quantitative data. The breadth of geography is such that almost all publications of this sort contain some data pertinent to the subject. The selection of items here listed was based upon availability and largest value to the geographer.

The Japanese are energetic collectors of this type of quantitative data. As in other countries, the practical problems of publication compel the grouping of local statistics into larger political wholes. This is seldom satisfactory to the geographer. Fortunately, local government units usually preserve these materials and they are available for field study. This is often true at the buraku level, and almost always so in the mura offices.

One troublesome problem is that the categories for collection change very frequently and it is difficult to make comparative studies. Another is that the statistical services were interrupted or stopped by pre-war security regulations, by war time curtailment, and by change and adjustment during the occupation. If possible one should find the answer to the question, "Why and for whom were these particular statistics produced?"

57. Asahi Shimbunsha 朝日新聞社 (Asahi Newspaper Co.), Asahi nenkan 朝日年鑑 (The Asahi yearbook), Tōkyō, Asahi Shimbunsha, 1925-43, 1945-.
Similar to the Mainichi nenkan. Statistical data is presented for the previous Japanese fiscal year. The yearbook is divided into four sections: 1) the world, 2) Japan, 3) statistics, and 4) references. It contains considerable general information on cultural events and education in Japan.

58. Jiji nenkan 時事年鑑 (Jiji yearbook), Tōkyō, Jiji Shimpōsha, 1918-36; Dōmei Tsūshinsha, 1937-43; Jiji Tsūshinsha, 1947-.

Among the most detailed and best organized of the several newspaper yearbooks of general scope, this work contains a variety of information and data on Japan's economy, trade, transportation, etc., of potential use to the geographer. It also gives a broad survey of social, cultural, and political conditions and news developments of the past year.

59. Kensetsushō Jūtakukyoku 建設省住宅局 (Housing Bureau, Ministry of Construction), Jūtaku nenkan 住宅年鑑 (Housing yearbook), Tōkyō, Shōkokusha, 1951, 3+9+354 pp., 24 pages of photos.

A compilation of official facts and figures on the post-war housing situation in Japan. An excellent source of data for the urban geographer and sociologist. Information on housing needs, construction, national and local governmental projects, private building, legislation, legal regulations, war damage, fire losses, etc. Statistics cover the period from 1945 through 1950.

60. Kokusei Kyōkai 國勢協會 (Society on the State of the Nation), Kokusei sōran 國勢總覽 (An overall view of our nation's status), Tōkyō, Kokusei Rengō Tsūshinsha, 1950-.

A listing of the principal governmental and private organizations in Japan, with brief notation on their administration, activities, interests or products, identification of key personnel or officers, the number of employees, capitalization, etc. Contents are classified into four major sections, being further sub-divided by topic, with names appearing in the order of the Japanese syllabary. The sections treat: 1) governmental bureaus, offices, political organizations, etc., both national and local; 2) universities, colleges, medical schools, and other cultural organizations; 3) banks and other financial organizations, and 4) commercial and industrial concerns, including a description of major manufacturing plants. Section four occupies two-thirds of the volume.

61. Mainichi Shimbunsha 毎日新聞社 (Mainichi Newspaper Co.), Mainichi nenkan 毎日年鑑 (The Mainichi yearbook), Tōkyō, Mainichi Shimbunsha, 1919-.

A summary of important yearly events and developments, both domestic and international. This statistical yearbook contains tabular or condensed textual information on legislation, public administration, politics, economics, finance and trade, manufacturing, natural resources, labor, population and social problems, cultural affairs, etc. It includes considerable data on Japan not available in similar English-language publications.

62. Naikaku Tōkeikyoku 内閣統計局 (Statistical Bureau of the Cabinet), Daiichi [- gojūhachi] kai Dai-Nihon Teikoku tōkei nenkan 第1[-58]回日本帝國統計年鑑 (Statistical yearbook of the Japanese Empire, nos. 1 [-58]), Tōkyō, Naikaku Tōkei Kyōkai, 1882-1939.

A yearly summary of official statistics derived from various governmental agencies. Data is always one or two years older than the date of publication. Statistics cover all phases of Japan's economy and social organization. The last issue in 1939 had approximately 465 pages and contained 416 separate tables under fifteen general categories. After a ten-year interval, this series was resumed again in 1950 as Nihon tōkei nenkan, compiled and published by the Naikaku Tōkeikyoku 内閣統計局 (Statistical Bureau of the Prime Minister's Office). It is a basic reference work for the social scientist.

63. Nampō Nenkan Kankōkai 南方年鑑刊行會 (Society for the Publication of the Yearbook of the Southern Regions), Nampō nenkan 南方年鑑 (Yearbook of the southern regions), Tōkyō, Tōhōsha, 1941-44.

Generally similar to the other yearbooks on the South Seas region which appeared just prior to and after the outbreak of the Pacific War. It surveys the geographical, political, cultural and economic situation, particularly of non-Japanese areas. The economic aspects are emphasized. It is not objective and now merely of geopolitical and historical interest for the researcher seeking insight into Japan's expansionist attitudes of the war period. Compiled by some 76 writers.

64. Nan'yō Dantai Rengōkai 南洋團體聯合會 (League of Associations for the South Seas), Dai Nan'yō nenkan 大南洋年鑑 (Yearbook of the greater South Seas), Tōkyō, Nan'yō Dantai Rengōkai, 1942.

This publication summarizes from the Japanese point of view the geographic, economic, and political conditions in the Pacific region and Southeast Asia just prior to the outbreak of the Pacific War. The yearbook is organized in three parts: 1) general, 2) regional, and 3) miscellaneous. The activities of Japanese companies and official organizations in the South Seas are outlined chronologically in the last section and a variety of general information about the various concerns and agencies is included.

65. Nihon Ringyō Shiryō Kankōkai 日本林業資料刊行會 (Society for the Publication of Materials on Japanese Forestry), Nihon ringyō nenkan 日本林業年鑑 (Japan's forestry yearbook), Tōkyō, Rin'ya Kyōsaikai, 1950-.

A useful compilation aimed at supplying general information and statistical data pertaining to Japanese forestry in both its governmental and commercial aspects. Editions thus far have consisted of four sections: 1) explanation or general survey of the national forestry situation; 2) enterprises, a guide to the government agencies and principal private organizations of the lumber and timber industry, outlining their respective activities and interests; 3) statistics, on area and type of standing timber, ownership, annual cut, lumber stocking-piling, supply and demand, forest disasters, labor, prices, etc.; and 4) reference materials, listing key official and private personnel, personnel changes, forestry regulations, etc.

66. Nōrinshō 農林省 (Ministry of Agriculture and Forestry), Nōrin nenkan 農林年鑑 (Yearbook for agriculture and forestry), Tōkyō, Nippon Nōson Chōsakai, 1948-.

An official report concerning the Ministry's administration, activities, enterprises, responsibilities, and regulations in the fields of agriculture and forestry. Selected categories treated are: land reform, development, and reclamation; production of staple foods, other crops, and livestock; sericulture; fisheries; forestry; agricultural and forest economy; agricultural bureaus and associations, etc. Much of the information is in statistical form.

67. Nōrinshō Nōrin Keizai Tōkei Chōsabu 農林省農林經濟統計調査部 (Ministry of Agriculture and Forestry, Section for Inquiry into the Economic Statistics of Agriculture and Forestry), Dai --- ji Nōrinshō tōkeihyō 第……次農林省統計表 (Statistics of the Ministry of Agriculture and Forestry, No.), Tōkyō, 1924-36, 38-41, 44-.
A more detailed companion volume to the Nōrin nenkan (Yearbook of the Ministry of Agriculture and Forestry). This is an indispensable source for the agricultural and economic geographer. It provides official national statistics, tabulated and presented by prefectures. Contents are classified under 15 major headings such as: farms, crops, fisheries products, agricultural imports and exports, etc. These items are further subdivided into 68 minor units to facilitate reference.

68. Shirazaki Kyōichi 白崎享一, ed., Nihon kokusei zue 日本國勢圖會 (Statistical explanation of conditions in Japan), Tōkyō, Kokuseisha, 1927-42, annual; 1942-, irregular.
A useful reference work, this is a general statistical yearbook, privately compiled but utilizing data largely derived from official sources. It differs from governmental statistical publications in providing a textual and/or graphic summary of each table or set of figures. The volume covers domestic and foreign trade, industrial production, resources and agriculture, finance, transportation and communications, education and sanitation, labor and social conditions, geographical facts, etc. Often, comparative statistics for foreign countries are included. Only two issues have been printed in recent years but it is likely to appear again annually as before the war.

69. Sōrifu Tōkeikyoku 總理府統計局 (Statistical Bureau of the Prime Minister's Office), Nihon tōkei nenkan 日本統計年鑑 (Japan statistical yearbook), Tōkyō, Nihon Tōkei Kyōkai and Mainichi Shimbunsha, 1950-; no. 1, 1950, 6+24+1058 pp.; no. 2, 1951, 14+525 pp.
A continuation of the Dai-Nihon Teikoku tōkei nenkan series suspended in 1940. With some modification, format and content remain similar to the pre-war yearbooks (entry 62). However, the statistical categories are further subdivided to facilitate use, e.g., there are 30 sections in the 1950 edition as compared with 15 in the last number of the old series. Some privately collected data as well as official statistics are also included. Most figures are for the prefectural or national levels. Where more detail is desired, the researcher will find these yearbooks useful in indicating the types of data available in prefectural publications. No. 1 covers the ten-year statistical gap during and after World War II and has supplementary lists of war damage, losses, etc. No. 2 gives a statistical summary of the Shōwa period to date, in addition to regular contents. Later numbers, no doubt, have appeared since the compilation of this bibliography.

70. Taiwan Sōtokufu 臺灣總督府 (Taiwan Government-General), Nan'yō nenkan 南洋年鑑 (South Seas yearbook), Tōkyō, Nampō Shiryōkan, 1929-42.
The third edition of this yearbook published in 1937 was reprinted in 1942. Somewhat similar to the Dai Nan'yō nenkan (entry 64) but stressing conditions in non-Japanese territories in Southeast Asia, the East Indies, and the Philippines. With bibliography.

71. Tōkyō Shisei Chōsakai 東京市政調査會 (Tōkyō Institute of Municipal Research), Nihon toshi nenkan 日本都市年鑑 (The municipal yearbook of Japan), Tōkyō, Tōkyō Shisei Chōsakai, 1931-44, 1949-.
This yearbook is largely statistical in content although there are brief topical summaries in textual form. It presents data from both governmental and private sources on the cities of Japan including a variety of information on municipal administration, planning, economy, finance, public health and hygiene, education, housing, ports and harbors, labor, social problems, etc. The Tōkyō Institute of Municipal Research is a semi-official, authoritative research organization in the urban field and its publications offer valuable source material to the social scientist.

72. Tōkyō Temmondai 東京天文臺 (Tōkyō Astronomical Observatory), Rika nempyō 理科年表 (A yearbook of science), Tōkyō, Maruzen Shuppan Kabushiki Kaisha, 1928-, annual.
The 1954 edition with over 500 pages has five sections: 1) the calendar, 2) astronomy, 3) meteorology, 4) physics and chemistry, and 5) earth sciences. Past issues covered more or less the same subjects. Leading specialists contribute to the various parts. Staff members of the Tokyo Astronomical Observatory wrote sections one and two. This is a physical science publication, intended for the general student and teacher, summarizing standard information and covering recent developments in the respective fields. The articles on earth science pertain to earth magnetism, gravity, seismology, minerology, and other phases of geology and geophysics.

73. Tōyō Keizai Shimpōsha 東洋經濟新報社 (Oriental Economist Company), Nihon keizai nempō 日本經濟年報 (Yearbook of the Japanese economy), Tōkyō, Tōyō Keizai Shimpōsha, 1930-44, 1948-, quarterly.
Three or four major topics pertaining to the recent economic situation in Japan are considered in some detail in each issue. In addition, a chronological listing of principal domestic and international news developments of economic significance is appended, along with a statistical supplement and a table summarizing new economic legislation.

Chapter IV

Sets and Collections

The decade of 1930-40 was a highly accelerated period for geographical research and publication in Japan. Many sets of books appeared in this period covering the geography of Japan as a whole. It would seem that enough local and regional studies had been made in the preceding years to allow broader generalization to be valid. There were also sets covering the world or large parts thereof. Some were by single authors, other by several authors. These sets on Japan are still the basic sources of material for the general geography of Japan. Some new sets have appeared since the war but for the most part they show no fundamental improvement over the older ones in either quality or amount of information, except that statistical and like data are more up to date.

This category includes also sets and single volumes of collections of scholarly lectures and essays. Some are entirely by professional geographers while others include the work of other scholars whose interests penetrate or at least touch upon the field of geography. They contain some excellent work. In general these publications are not too well organized and rarely has there been any attempt at integration.

74. Aono Hisao 青野壽郎, ed., Chiri chōsahō 地理調査法 (Methods of geographic survey) [= vol. 1 of Asakura Shoten's Shin-chirigaku kōza], 1953, 6+305 pp., 112 illus. and 51 statistical tables.
Twelve geographers cooperated in the production of this volume, which is in four parts: 1) cartography and map-making; 2) techniques for the survey of physical and human geography; 3) the treatment of historical documents and statistics; 4) the techniques of regional study. Explanations and abundant examples make this a useful guide for beginners in geography.

75. Ashida Ijin 蘆田伊人, Dai-Nihon chishi taikei 大日本地誌大系 (Outline of Japanese regional geography), Tōkyō, Yūzankaku, rev. ed., 1933, 40 v.
A complete reprint of fifteen regional descriptions written during the Tokugawa period, with brief supplementary annotation by the compiler. These books represent primary sources for the historian and historical geographer. The regions and areas described are grouped by volume as follows: 1-4, Edo; 5-15, Musashi; 16-17, Yamashiro; 18, Kinki and Izumi; 19, Kamakura; 20-21, Ise, Iga, and Shima; 22-26, Ōmi, Hida, and Settsu; 27, Izumo; 28, Mikawa; 30-34, Aizu; 36-40, Sagami. The first edition of this reprint was published in 1915-17 in 3 volumes.

76. Chirigaku kōza 地理學講座 (Lectures on geography), Tōkyō, Chijin Shokan, 1930-1933, 16 v.
A collection of monographs and lectures by fifty geographers. Each volume contains six to nine studies covering all phases of the field, but the majority of them are on general topics and few of them deal with the geography of Japan specifically. The following will be of some interest to the student of Japan: Nihon no rekishi chiri (Historical geography of Japan), by T. Kida in vol. 4; Kyōdo chiri kenkyūhō (Methodology of provincial geography), by M. Odauchi in vol. 12; Kazan (Volcanoes) by F. Homma in vol. 13; and Jinkō no idō (Population movements), by Y. Takemi in vol. 14.

77. Iwanami kōza chirigaku 岩波講座地理學 (Iwanami lectures on geography), Tōkyō Iwanami Shoten, 1931-1934, 7 v.
The Iwanami firm has published many sets of lectures in several fields of science. This series on geography was published at the same time as that on geology, petrology, and mineralogy. Originally the lectures appeared in pamphlet form but were combined to form the seven volumes as indicated: vol. 1, general geography; vols. 2 and 3, physical geography; vols. 4 and 5, human geography; vol. 6, special geographic problems; vol. 7, supplement. Each group of lectures was written by a Japanese authority. However, there is no regional geography included in this series.

78. Fukui Eiichirō 福井英一郎, ed., Shizen chiri II 自然地理 II (Physical geography II) [=vol. 4 of Asakura Shoten's Shin-chirigaku kōza], 1953, 8+324 pp., 98 illus., 74 tables.
The titles of the five parts and the authors who contributed to this volume are as follows: 1) Climate, by Fukui Eiichirō, Sasakura Kōzō, and Yazawa Taiji; 2) The ocean, by Uda Michitaka; 3) Hydrography, by Yamamoto Sōki, Masuzawa Jōtarō, and Yatsu Eiju; 4) Soil, by Kamoshita Yutaka; 5) Bio-geography, by Imanishi Kinshi, and Kira Tatsuo. Each of the chapters is fairly high in its scientific standard, and the descriptions are made with abundant examples collected from Japan. A useful text-book.

79. Ishida Ryūjirō 石田龍次郎 and Watanabe Akira 渡邊光, ed., Sekai chiri taikei 世界地理大系 (Outline of world geography), Tōkyō, Kawade Shobō, 1951-52, 7 v.
A descriptive general geography of the world utilizing post-war data, this is the most recent and comprehensive Japanese series of its kind. Many leading Japanese geographers contributed. Each volume contains more than 300 pages of text, approximately 100 maps, statistical tables, and an index of place names and topical content. Volumes deal with: 1) nature and society of the world, 2) Japan, 3) western Europe, 4) the new continents, 5) Africa and south Asia, 6) the sphere of the USSR, and 7) world statistics.

80. Kaizōsha 改造社, Chiri kōza Nihon-hen 地理講座日本篇 (Geographical lectures on Japan), Tōkyō, Kaizōsha, 1934-35, 7 v.

A rather poorly systematized set of volumes covering the general geography of the Japanese empire. Emphasis is upon cultural aspects. Chapters treat such topics as landforms and geology, mining, climate, agriculture, forestry, industry, fisheries, etc. Volumes of about 400 pages each are on the following regions in the order given: Hokkaidō and Karafuto; Ōu and Kantō; Chūbu and Kinki; Chūgoku, Shikoku, and Kyūshū; Taiwan and the South Seas; Chōsen and Kwantung; general.

81. Kaizōsha 改造社, Nihon chiri taikei 日本地理大系 (Kaizōsha's Outline of the geography of Japan), Tōkyō, Kaizōsha, 1929-1931, 17 v.

Kaizōsha's Nihon chiri taikei and Shinkōsha's Nihon chiri fūzoku taikei (No. 84) are two large sets on the regional descriptive geography of Japan published in pre-war times. Twenty years have passed since the Kaizōsha set was published. While the portions upon human geography are out of date, the comments and illustrations concerning physical and historical geography are still valuable. Furthermore, under Japan's present circumstances, it would be almost impossible to duplicate the magnificence of the set. It is printed entirely upon art paper; there are maps and graphs and over half of the pages are devoted to photographs. The committee which compiled the set consisted of Ishibashi Gorō, Iimoto Nobuyuki, Katabira Jirō, Takahashi Jun'ichi, Tokuda Teiichi, Tanaka Keiji, Tanaka Shūsaku, Endō Kanehide and Shimomura Hikoichi. One or two specialists took the responsibility for preparing and designing each volume, while forty to fifty additional specialists wrote the individual descriptive items for each volume. The twelve volumes and five bekkan (special volumes) are entitled: 1-2) Sōron (General view) (vol. 1 on physical geography and vol. 2 on human geography are combined into one book); 3) Dai-Tōkyō; 4) Kantō; 5) Ōu (Tōhoku); 6A-6B) Chūbu (in two books); 7) Kinki; 8) Chūgoku, Shikoku; 9) Kyūshū; 10) Hokkaidō, Karafuto (Sakhalin); 11) Taiwan; 12) Korea; bekkan 1) Sangaku (Mountains); 2) Manshū and Nan'yō; 3-4) Kaigai Hattenchi (Overseas areas of Japanese emigration); 5) Fuji-san (Mt. Fuji). The contents of each volume on regional geography (vols. 4-12, bekkan 2) are arranged in the following approximate order: 1) several pages of superior photographs (multicolored, photo-gravure, or double-tone); 2) collections of photographs, with a short explanation for each, arranged by prefectures (sub-divisions for Hokkaidō, the eight shū or chō for Taiwan, and four areas for Korea); this section comprises over half of each volume; 3) general descriptive geography of an object region in which is mentioned physical geography (geomorphological structure, topographical outline of both land and water, and climate) and human geography (industry, communication, population, settlement and history); 4) colored general map of a region, its geological map and various other maps and charts.

82. Iimoto Nobuyuki 飯本信之 and Satō Hiroshi 佐藤弘, Nan'yō chiri taikei 南洋地理大系 (A series on the geography of the South Seas), Tōkyō, Daiyamondosha, 1942, 8v.

Eight large, well-illustrated volumes on the Pacific Islands, Southeast Asia, and India. Treatment varies considerably but includes a general or regional geographic description, or both, outlining the important physical and cultural characteristics of the regions covered. There are notes or comments on geopolitical aspects such as colonialism, nationalism, international relations and trade, particularly in relationship to Japan. Anthropological, historical, and cultural factors are considered briefly in some instances, in relative detail in others. Volumes include: 1) South Seas, general; 2) Hainan, Philippines, and Micronesia; 3) Thailand and French Indo-China; 4) Malay and Burma; 5) East Indies, I; 6) East Indies, II; 7) India and Ceylon; and 8) Australia, New Zealand, and the central Pacific islands including Hawaii.

83. Nihon chihō chishitsushi 日本地方地質誌 (Regional geology of Japan), Tōkyō, Asakura Shoten, 1950-, 9 v.

The work of seven specialists, this is an excellent regional survey of Japan's general geology, summarizing accumulated knowledge to date. The several broad geological provinces of each of the eight major traditional districts (Hokkaidō, Tōhoku, Kantō, Chūbu, etc.) are discussed in terms of present form and structure and past geological stage or era. There is one volume for each district and a general outline (sōron). With separate bibliographies and an index.

84. Nihon chiri fūzoku taikei 日本地理風俗大系 (Outline of Japanese geography and folklore), Tōkyō, Shinkōsha, 1929-1932, 19 v. illus.

A systematic, well-illustrated descriptive geographical series treating in detail the traditional regions of Japan proper and Korea, Taiwan, and Karafuto. Much of the material on cultural geography is far outdated but the compilation and writing are of high standard and the set remains of interest for general background. The sections on physical geography are of more current value and they provide excellent general summaries in most cases. The following regions are treated in more than one volume: the Kantō (3), Kinki (2), Kyūshū (2), and Korea (2). Some volumes cover two regions, i.e., Hokkaidō and Karafuto, Shikoku and Setonaikai, Chūbu and Hokuriku. This set differs from Kaizōsha's Nihon chiri taikei in having illustrations distributed throughout the text rather than concentrated in separate sections. In addition, all kanji are accompanied by kana readings, facilitating use by students not highly proficient in Japanese. Vol. 18 (Sōron) summarizes the whole work. An unnumbered index volume lists all contents in the order of the Japanese syllabary. A revised edition of this set appeared in 1936-38, combining and condensing the original 18 volumes into nine, but it had only a limited printing since the new index was not completed.

85. Nihon chiri shin-taikei 日本地理新大系 (A new outline of the geography of Japan), Tōkyō, Kawade Shobō, 1952-53, 5 v.

Similar in style and format to the Sekai chiri taikei put out by the same publisher. An excellent systematic description of the general geography of Japan. Compiled and written entirely after the war, utilizing relatively recent data, this is the first Japanese series to give a comprehensive geographical summary of post-war conditions. Volumes are: 1) Shizen (Physical), 2) Shakai to keizai (Society and economy), 3) Shigen sangyō, I (Natural resources and industry I), 4) Shigen sangyō, II, and 5) Chishi (Regional).

86. Nihon chiri taikei 日本地理大系 (Outline of the geography of Japan), Tōkyō, Kaizōsha, 1929-1931, 12+5 v., illus.

The second extensive, illustrated geographical series on Japan and her possessions published at this time, the other being Shinkōsha's Nihon chiri fūzoku taikei. An outline of physical and human geography constitutes volumes 1 and 2, combined. Volumes 3 through 12 are regional descriptions— vols. 3-9 treat traditional districts in the home islands; vol. 10, Hokkaidō and Karafuto; 11, Taiwan; and 12, Korea. Five special volumes (bekkan) concern the following additional topics and areas: 1) mountains, 2) Manchuria and the South Seas, 3 and 4) oversea areas of Japanese emigration, and 5) Mt. Fuji. Regional volumes begin with a selection of annotated photographs, arranged by prefecture. The photographs occupy about half of each book. They are followed by descriptive texts surveying both the natural and cultural environments. Included are general, geological, and other maps, tables, graphs, and diagrams. This beautifully-printed series, written and compiled by many prominent geographical specialists, continues to have considerable reference value. Those portions dealing with historical and physical geography are of special worth.

87. Oda Takeo 織田武雄, Rekishi chiri 歴史地理 (Historical geography) [= vol. 7 of Asakura Shoten's Shin-chirigaku kōza], 1953, 6+281 pp., 81 illus. and 12 statistical tables.

Nine geographers cooperated in the compilation of this volume. It consists of five parts. The first four parts are a general outline of historical geography. They are of little use as a source of information. The last part, entitled, "The historical geography of Japan," comprises about one half of the work and is written by Yonekura Jirō, Kitamura Toshio, and Asaka Yukio, who treated the ancient, medieval, and modern periods respectively. The descriptions include abundant examples of the historical geography of Japan.

88. Ogawa hakase kanreki kinen shigaku chirigaku ronsō 小川博士還歴記念史學地理學論叢 (Collected papers for the commemoration of Dr. T. Ogawa's 61st birthday), Kyōto, Kōbundō Shobō, 1930, 18+2+1064 pp.

This volume contains 26 geographical papers written in honor of Dr. Ogawa by his colleagues and a bibliography of his own writings. The bibliography includes 19 books and 250 articles, principally on the history and historical geography of China. The monographs in this collection of particular interest to the student of Japanese geography are: "On the reason for dense population in the Sanuki plain," by Terada Teiji; "A study of rural settlement in Iga province," by Muramatsu Shigeki; and "A study of the population geography of rural Kyūshū," by Ishibashi Gorō.

89. Ōtsuka Chiri Gakkai 大塚地理學會 (Ōtsuka Geographical Society), Tanaka Keiji sensei kinen Ōtsuka Chiri Gakkai rombunshū 田中啓爾先生記念 大塚地理學會論文集 (Collected papers for the commemoration of Prof. K. Tanaka's retirement, by the Ōtsuka Geographical Society), Tōkyō, Meguro Shoten, 1950, 3+4+408 pp.

Professor Tanaka founded the Department of Geography at Tōkyō Bunrika University (now Tōkyō Kyōiku University). This volume contains 43 articles on human and physical geography written by Tanaka's colleagues and students. A brief memoir by Tanaka and a list of his publications are added.

90. Shin chirigaku kōza 新地理學講座 (New lectures on geography), Tōkyō, Asakura Shoten, 1953-, 8 v.

Intended as an elementary survey of the field for the use of public school teachers, beginning college students, etc., this series is arranged by volume as follows: 1) Chiri chōsahō (Methods of geographic survey), 2) Chirigaku honshitsuron (Theories on the nature of geography), 3) Shizen chiri, part 1 (Physical geography, I), 4) Shizen chiri, part 2, 5) Jinkō to shūraku chiri (Population and settlement geography), 6) Keizai chiri (Economic geography), 7) Rekishi chiri (Historical geography), and 8) Saikin no chirigaku (Recent geography). Only volumes 1, 4, and 7 had been published by the end of 1953.

91. Shōkōshō, Nōrinshō 商工省 農林省 (Ministry of Commerce and Industry and Ministry of Agriculture and Forestry), Nampō keizaishigen sōran 南方經濟資源總攬 (General view of economic resources in the South Seas and adjacent areas), Tōkyō, Tōa Seikeisha, 1942-44, 12 v., illus.

During and prior to World War II, numerous books and articles appeared concerning the general geography and economic resources of the central and south Pacific, Southeast Asia, and adjacent regions of interest to Japan. Generally, they were highly nationalistic in slant and of a relatively low academic standard. This extensive official series, the best executed and most reliable Japanese work of its kind, was attempted in order to supply the government and military planners with accurate information on the southern regions. The individual volumes, written by regional specialists, were based on both domestic and foreign data and were compiled in a fairly objective manner. Emphasis in each is upon resources and resource development, although most of the volumes contain an outline of the physical and cultural environment. Some of the books in this series are still of potential reference value as capable descriptive surveys, although the statistical content is, of course, outdated. With maps, photographs, and tables.

92. Tanaka Keiji 田中啓爾, ed., Kyōdo shinsho 鄕土新書 (New prefectural geography), Tōkyō, Nihon Shoin, 1949-, 47 v.

Volume	Author	Published
1. Hokkaidō Shinshi (126)	Mori Sumie	1951, June
2. Aomori-ken "	Takahashi Masao	
3. Iwate-ken " (127)	Yokota Kōhachi	1950, November
4. Miyage-ken "	Kumatani Kaneo	
5. Akita-ken "	Birukawa Shōhei	
6. Yamagata-ken " (128)	Nagai Masatarō	1950, March
7. Fukushima-ken " (129)	Yasuda Hatsuo	1950, June
8. Ibaraki-ken " (130)	Yamaguchi Takayoshi	1950, September
9. Tochigi-ken "	Okuda Hisashi	1951, October
10. Gumma-ken "	Yajima Nikichi	1951, September
Saitama-ken "	Muramoto Tatsuo	1950, December
Chiba-ken "	Ozaki Torashirō	

Volume	Author	Published
13. Tōkyō-to Shinshi	Tanaka Keiji	1949, October
14. Kanagawa-ken "	Tanaka Keiji	
15. Niigata-ken "	Isobe Toshisada	
16. Toyama-ken "	Ishii Itsutarō	1949, December
17. Ishikawa-ken "	Wada Ryōji	1950, July
18. Fukui-ken "	Yoshida Mori	
19. Yamanashi-ken "	Tanaka Keiji	
20. Nagano-ken "	Itō Gōhei	1950, June
21. Gifu-ken "	Andō Masuo	1950, September
22. Shizuoka-ken "	Sasaki Seiji	
23. Aichi-ken "	Okada Shizuta	
24. Mie-ken "	Tsujii Kōtarō	1951, September
25. Shiga-ken "	Kimura Kenji	
26. Kyōto-fu "	Tsujita Usao	
27. Ōsaka-fu "	Inogi Hisaichi	
28. Hyōgo-ken "	Kōzuki Junji	1950, March
29. Nara-ken "	Horii Jin'ichirō	
30. Wakayama-ken "	Yabuuchi Yoshihiko	1951, April
31. Tottori-ken "	Iwanaga Minoru	
32. Shimane-ken "	Shōji Hisataka	
33. Okayama-ken "	Inami Etsuji	1951, October
34. Hiroshima-ken "	Funakoshi Kensaku	
35. Yamaguchi-ken "	Hamada Seikichi	1950, February
36. Tokushima-ken "	Kishimoto Minoru	1951, February
37. Kagawa-ken "	Gōda Eisaku	1950, July
38. Ehime-ken "	Murakami Settarō	
39. Kōchi-ken "	Terao Shigeru	1950, November
40. Fukuoka-ken "	Kanao Sōhei	1949, October
41. Saga-ken "	Katabuchi Tadamichi	
42. Nagasaki-ken "	Mori Sumie and Tanaka Daiji	
43. Kumamoto-ken "	Miura Yasuhisu	
44. Ōita-ken "	Shinoda Kumanda	
45. Miyazaki-ken "	Saeki Hideo	1951, June

By shinshi is meant "new descriptive geography." This series is the latest publication on prefectural descriptive geography. It is based on valuable pre-existing materials and on the new social conditions found in post-war Japan. Each number in the series has from 180 to 200 pages, with many maps. Each prefecture has been described by a geographer especially familiar with the prefecture. Provided in each volume is the order of the Japanese syllabary, a list of the books on the descriptive geography of the particular prefecture. Publication is still being continued.

93. Tanaka Keiji 田中啓爾, Zoku chirigaku rombunshū 續地理學論文集 (Collected papers on geography, continued), Tōkyō, Kokon Shoin, 1950, 2+9+404+5 pp.

The second volume of Tanaka's geographical articles, consisting of 28 papers published in professional periodicals after 1933. The majority of the monographs are concerned with regionality, physical and cultural, in Japan proper, in China, Manchuria, and adjacent areas. Five articles are on the nature and method of geography.

94. Uchida Kan'ichi Sensei Kanreki Shukugakai 内田寛一先生還暦祝賀會 (Society for Extending Congratulations to Professor Kan'ichi Uchida on his Sixty-first Birthday), Uchida Kan'ichi sensei kanreki kinen chirigaku rombunshū 内田寛一先生還暦記念地理學論文集 (Geographical essays in commemoration of Professor Kan'ichi Uchida's sixty-first birthday), Tōkyō, Teikoku Shoin, 1952, 2 v.

A collection of 57 articles on human and physical geography written mainly by the faculty members of Kyōto and Tōkyō Bunrika (now Tōkyō Kyōiku) universities. The volumes are in honor of Professor Uchida, an historical geographer who taught for many years at Bunrika University and is a graduate of Kyōto University. Contents are listed in English in volume two and a short bibliography of selected papers by Uchida is included.

95. Yamamoto Kumatarō 山本熊太郎, Shin-Nihon chishi 新日本地誌 (New regional geography of Japan), Tōkyō, Kokon Shoin, 1937-40, 5 v., 1,347 illus.

A comprehensive elementary series on the general regional geography of Japan proper, with many illustrations and tables.

96. Yamazaki Naomasa Rombunshū Kankōkai 山崎直方論文集刊行會 (Society for the Publication of the Works of Yamazaki Naomasa), Yamazaki Naomasa rombunshū 山崎直方論文集 (Collected papers of Professor Naomasa Yamazaki), Tōkyō, Kokon Shoin, 1930, 2 v.

This set comprises a partial collection of the geographical writings of Dr. Yamazaki, all of which were previously published in various scientific periodicals and geographical journals on an individual basis. The 28 articles in volume one all deal with physical geography and earth science, principally geology and geomorphology, glaciation, and volcanology. Of the 32 papers in volume two, 11 are tectonic and seismological studies, 10 deal with human and regional geography, 7 with archaeology and anthropology, and 4 with geographical education and teaching. Also included is a monograph by Tsujimura Tarō entitled "Dr. Yamazaki and geography in Japan," a bibliographical outline, and a bibliography of Dr. Yamazaki's works, totaling over 300 papers and six books.

97. Yamasaki Naomasa 山崎直方 and Satō Denzō 佐藤傳藏, Dai-Nihon chishi 大日本地誌 (Regional descriptive geography of Japan), Tōkyō, 1903-1915, Hakubunkan, 10 v.

Volume	District	Published in
1.	Kantō	1903
2.	Ōu	1904
3.	Chūbu	1904
4.	Kinki	1905
5.	Hokuriku	1906
6.	Chūgoku	1907
7.	Shikoku	1908
8.	Kyūshū	1911
9.	Hokkaidō and Karafuto (Saghalien)	1913
10.	Ryūkyū and Taiwan (Luchu Is. & Formosa)	1915

The ten volumes listed above make up this series, which was the first monumental publication on the descriptive geography of Japan. The following categories of description are common to all the volumes: Introduction; part 1, physical geography, topography by *kuni* (the traditional administrative regions of the Shogunate), geology, climate; part 2, human geography, history, administration and religion, industries; part 3, provincial descriptive geography by prefectures. Several maps — general, topographical, geological and the like — are provided. There are also various photographs covering such things as provincial customs, famous buildings (shrines, temples, prefectural offices, etc.) and industries. Since the contents of this set were written a great many years ago, much of the description of human geography is out of date. Still this set should not be neglected as a source of material on Japan's cultural and historical geography. In the chapter called "History" in part two, and in the provincial descriptions of part three, historical facts, legends, customary observances, and other interesting descriptions of each province and other places are given. Unique local industries are also described in detail in the sections entitled "Industries" in parts two and three. Lists of chapters, sub-titles, maps, and photographs are provided at the beginning of each volume but there is no index.

Chapter V

Periodicals

Japan ranks high among the nations of the world in the number and excellence of its scientific periodicals. A very considerable part of the nation's scientific writing appears in them. Most of these journals are so highly specialized that they have the support of only a very limited number of subscribers. As a result they are often short-lived. The material which appears in them is generally not reproduced elsewhere so libraries are compelled to keep many fragmentary sets of now defunct periodicals. In the field of geography, the journals associated with the foremost academic departments of geography are the most important. *Chirigaku hyōron* (Geographical review) is now the journal of the Nihon Chiri Gakkai or Association of Japanese Geographers but it started as house organ of the Geographical Institute of Tōkyō University in 1925. Its publication continues to date and it is still closely bound to Tōkyō University. The geographers of Kyōto University launched their journal, *Chikyū* (The globe) in 1924. Unfortunately, this excellent journal went out of existence in the late 1930s. Since the war Kyōto University has acquired a new geography staff which started a new journal called *Jimbun chiri* (Human geography) in 1948. Departments of geography in some other universities also have their own journals, notably Tōkyō Kyōiku Daigaku (formerly Tōkyō Bunrika Daigaku) and Tōhoku Daigaku. An exception to the rule of university affiliation for research level journals is the *Chigaku zasshi* or *Journal of geography* of the Tokyo Geographical Society.

There were a number of other geographical journals before the war which served the needs primarily of grade school teachers. These had large subscription lists and ran over a good many years until the war brought them to an end. These journals cannot be overlooked by the research geographer, for a good many excellent studies appeared in them.

98. *Chigaku ronsō* 地學論叢 (Collected papers of earth science), Tōkyō , Dai-Nihon Tosho Kabushiki Kaisha, 1908-13, irregular.

Non-periodical organ of the Tōkyō Chigaku Kyōkai 東京地學協會 (Tōkyō Geographical Society). A total of five numbers appeared. The first three contain abstracts of lectures and addresses given at the society's meetings. No. 4 commemorates Sven Hedin's visit to Japan and includes his speech translated into Japanese. The fifth number is devoted to Dr. K. Inoue's report on the 11th International Congress of Geology and that body's excursion to Spitzbergen in 1910.

99. *Chigaku zasshi* 地學雜誌 (The journal of geography), Tōkyō, Tōkyō Chigaku Kyōkai, 1889-1944 (Vols. 1-56), monthly; 1948- (Vol. 57-), quarterly.

The accumulated numbering since this journal's first publication is applied without relation to the volume number which changes year by year. By the end of 1951 the latest was no. 682 or volume 60, no. 4. This magazine is published as the organ of the Tōkyō Chigaku Kyōkai or Tokyo Geographical Society. It appeared monthly from 1889 to 1944, was stopped from 1944 to 1948 because of the war and was resumed in September, 1948 as a quarterly journal. Despite the fact that this publication is named "The journal of geography," the term chigaku means earth science in general. It contains articles on geology, minerology, and meteorology as well as on geography in a narrow sense. The aim of this journal is to present an up-to-date systhesis of the general tendencies of the scientific world. As this is the oldest journal in Japan in this field it is a convenient source of information on the general trends in Japanese science over the last 60 years. Some of its numbers were edited with a view to getting a collection of articles on topics of special interest at that time. In this regard its "China numbers" and "South Sea numbers" are of special value in obtaining an outline of the Japanese standard of regional studies at various times.

100. Chikyū 地球 (The globe), Kyōto, Chikyū Gakudan, 1924-1937, monthly.
Chikyū Gakudan (Earth Scientists Association) started its activities in 1924 with Professor Takuji Ogawa as its leader. This was one year before the appearance of Chirigaku hyōron and it is noteworthy that these two leading journals of geography began almost at the same time. Compared with the interest of the geographers sponsoring Chirigaku hyōron the interests of the Chikyū Gakudan group were much wider and included all the fields of earth-science. Monographs on oceanography, magnetics, meteorology, geology, geography (mainly physical geography), and related fields were published with the aim of fostering more mutual understanding among scholars working in those specialized pursuits. The majority of the articles concern Japan and considerable material of interest to the regional geographer is available. Chikyū averaged 80 pages per issue.

101. Chiri 地理 (Geography), Tōkyō, Ōtsuka Chiri Gakkai, 1938-, quarterly.
The Ōtsuka Chiri Gakkai or Association of Ōtsuka Geographers which had published the Ōtsuka chirigakkai rombunshū (The collected articles of the Association of Ōtsuka Geographers) started this quarterly bulletin in 1938. The principal contents are articles reporting on original geographic research in Japan, but books reviews, general news of geographic interest, and bibliographic materials are included. Most of the contributors are on the staff of Tōkyō Bunrika University and Tōkyō Higher Normal School.

102. Chiri kenkyū 地理研究 (Geographic studies), Tōkyō, Chūkōkan, 1941.
Only three numbers of this monthly journal were issued before the activities of the sponsoring association, the Chiri Kenkyūkai, were suspended because of wartime pressures. Chiri kenkyū succeeded Chiri kyōiku (Geographical education) and was intended to be of more interest to geography teachers and to serve as a publishing outlet for newcomers in the field.

103. Chiri kyōiku 地理教育 (Geographical education), Tōkyō, Chūkōkan, 1924-1941, monthly.
This journal was started as an organ for geography teachers in the public schools, although it occasionally contained material of interest to the more advanced professional geographers. Teaching aids, course outlines, and reprints of selected lectures of a general nature made up the bulk of its contents. However, a number of rather good regional studies were included, and from time to time noteworthy original articles pertaining to thought and developments in the field of geography were contributed by some of the younger Japanese geographers. In fact, for a few years, this journal enjoyed a reputation among newcomers in the field as a desirable medium for their writings. As the war approached, however, this magazine degenerated considerably, becoming very nationalistic in slant and shifting its content largely to matters of political concern. In 1941 it was superseded by the monthly publication Chiri kenkyū (Geographical studies).

104. Chiri ronsō 地理論叢 (Collected articles in geography), Tōkyō, Kokon Shoin, 1932-43, annually.
Organ of the Chirigaku Kyōshitsu or Institute of Geography at Kyōto Imperial University for 11 years before publication ceased. Early issues of this journal contained excellent articles, mainly on human geography. In later years, however, most of the papers were devoted to geo-political topics, written in a nationalistic vein; they possess minor geographical value or present-day interest. Professor Komaki, chairman of the institute, was the principal organizer of this publication.

105. Chiri to keizai 地理と經濟 (Geography and economics), Tōkyō, 1936-37, monthly.
Journal of the pre-war Nihon Keizai Chiri Gakkai (Japanese Society of Economic Geography). Sixteen numbers were issued. Articles, contributed by economists as well as geographers, covered several phases of economic and human geography, but the chief emphasis was upon industrial geography. The numbers average 120 pages and each contains about ten papers. Titles are in English.

106. Chirigaku 地理學 (Geography), Tōkyō, Kokon Shoin, 1933-1943, monthly.
As a monthly journal, Chirigaku originally aimed at a comparatively high scientific standard, with some consideration given to aids for teachers. Contents during the earlier period included original articles on both physical and human geography, travel records, geographical news and reviews. In its later period, this scientific coloration gradually faded and, finally during the Pacific War, it nearly disappeared, giving way to descriptions of the South Sea districts and similar wartime topics of a more popular nature.

107. Chirigaku hyōron 地理學評論 (The Geographical review), Tōkyō, Nihon Chiri Gakkai, 1925-, monthly.
Originally published with the English title, The geographical review of Japan, this monthly organ of the Nihon Chiri Gakkai or Association of Japanese Geographers is the oldest and most respected of the academic geographical journals in Japan. It has occupied a leading position since its appearance in 1925. The contents reflect clearly the overall trends in the Japanese geographic profession. Geomorphological studies were stressed in the journal's earlier years, until about 1935, when articles on human geography, particularly agricultural and settlement geography, became more common. The Review prints original monographs, professional reports, book reviews, notes on association news, etc. The title of each article and its author are indicated in English on the back cover of each issue. In addition, an English résumé is provided with many of

the papers, particularly in recent years. In terms of the size of individual issues, printing quality, and to a lesser degree content, the Review has not entirely regained its pre-war standards. Issues are numbered 1-12 annually and have appeared regularly except in the period 1947-49 when a few numbers were missed.

108. Chirigaku kenkyū 地理學研究 (Geographical studies), Tōkyō, Chūkōkan, 1942-1944, monthly.
Published during the war, this monthly journal was principally concerned with geographic aspects of the so-called "Greater East Asia Co-prosperity Sphere" sponsored by Japan, and hence many of the articles were of a "geo-political" nature. Emphasis was placed on applied geography and the journal was aimed at the "educated" public at large. Regional descriptions of areas occupied by the Japanese army in Southeast Asia constituted a number of the monographs but there were no first-class comprehensive regional studies, and, in the main, the academic standard of this publication was rather low. A few general descriptions of regions of Japan proper, usually on a prefectural basis, are of better quality and of some interest. Chiri kyōiku (Geographical education), the predecessor of this journal, had been issued by the same publisher for some years before the shift to the wartime format occurred.

109. Chirigaku nempō 地理學年報 (Geographical yearbook), Tōkyō, Meguro Shoten and Shōkabō, 1933-35.
Edited by G. Imamura, only three issues of this yearbook appeared. They were devoted to research reports and critical reviews of the previous year's publications in geography and related fields, including oceanography and geomorphology. Most of the reviews were written by Imamura, E. Fukui, Y. Ōtsuka, and S. Komaki. Research reports accounted only for a small portion of each number.

110. Chishitsu yōhō 地質要報 (Bulletin of the Imperial Geological Survey of Japan), Tōkyō, Shōkōshō Chishitsu Chōsajo, 1886-1923, irregular.
This irregular bulletin succeeded Chishitsu Chōsajo Nempō (The annual report of the Geological Survey), which was suspended in 1884. Some 43 numbers were issued between 1886 and 1923. After 1907, when a publication of the new bulletin (Chishitsu Chōsajo hōkoku or Report of the Geological Survey) similar in form and content was begun by the same office, this bulletin came out only once a year. The principal content, naturally, deals with geological field surveys in Japan, ore deposits, and related phenomena, but a number of soil studies which appeared from time to time should be of particular interest to the physical geographer. In earlier issues, especially, articles were contributed by a number of foreign geologists who were visiting Japan (such as Naumann and Fesca).

111. Deruta デルタ (Delta), Tōkyō, Kokon Shoin, 1937-1939, monthly.
This journal, not entirely geographic in nature, was directed towards elementary school teachers and the more general reader. Compared with Chirigaku, put out by the same publisher, its general level of scholarship and writing standards were much lower. Articles on the regional geography of Japan and descriptions of foreign countries and areas in connection with problems of the day composed its content. Averaged 100 pages per issue.

112. Jimbun chiri 人文地理 (Human geography), Kyōto, Jimbun Chiri Gakkai, 1948-, irregular.
Organ of the Jimbun Chiri Gakkai or Association of Human Geography at the Chirigaku Kyōshitsu or Institute of Geography, Kyōto University. Early numbers appeared irregularly, but in 1950 the journal became a quarterly, and after 1951 it was issued six times a year. The first four numbers concern single topics, e. g., no. 1, man and his environment; no. 2, cities and villages; no. 3, nature and agriculture; and no. 4, human geography as a social study. Later issues contain original papers, mostly on historical geography. There are also research reports, reviews, and notes on geographical activities. Since 1950, the last number of each volume has been devoted to an annotated bibliography of articles which appeared in the previous year. Numbers average 80 pages and include English language abstracts.

113. Kazan 火山 (Volcanoes), Tōkyō, Nihon Kazan Gakkai, 1932-1939, 2 v. a year.
This publication of the Nihon Kazan Gakkai or Volcanological Society of Japan is devoted strictly to volcanology and to those aspects of other sciences concerned with or related to volcanology. Articles on the petrology and structural geology of volcanic areas will be of interest to the geomorphologist or physical geographer. Not confined to Japan in scope, although reports on studies made in the home islands predominate. Includes general news items on world-wide volcanic activity and on the society's projects and interests in Japan. Went out of publication prior to the war and had not reappeared up to 1953.

114. Kyōto Teikoku Daigaku Bungakubu chirigaku kenkyū hōkoku 京都帝國大學文學部地理學研究報告 (Geographic reports from the Institute of Geography, College of Letters, Kyōto Imperial University), Kyōto, 1937-38.
Only two numbers were published, one in 1937, the other in 1938. Each contained four papers on human geography and the history of geography.

115. Ōtsuka Chiri Gakkai rombunshū 大塚地理學會論文集 (Collected papers of the Association of Ōtsuka Geographers), Tōkyō, Ōtsuka Chiri Gakkai, 1933-1935, 6 v.
The Ōtsuka Chiri Gakkai or Association of Ōtsuka Geographers was founded in 1933 by members of the geography staffs at Tōkyō Higher Normal School and Tōkyō Bunrika University. The publication lasted only two years, during which time six numbers appeared. Articles cover nearly all fields of geography but the majority are on the geography of minor regions in Japan. The number of articles in each issue varies from eight to thirty-five.

116. Rikusuigaku zasshi 陸水學雜誌 (Journal of limnology), Tōkyō, Nihon Rikusui Gakkai, 1931-1944, 1949-, 5 times per year.
As the organ of the Nihon Rikusui Gakkai or Association of Japanese Limnologists, this scientific publication has a wide range in content, covering those aspects of the several physical and natural sciences concerned with limnology. Although some technical articles, such as those on fresh water chemistry and physics, are of little interest to the geographer, a number of monographs on ecological relationships and on the geological factors explaining the location and form of several of the major Japanese lakes may be useful. Publication dates rather irregular.

117. Shakai chiri 社會地理 (The geography for social life), Tōkyō, Nitchi Shuppan Kabushiki Kaisha, 1947-50.
Organ of the Nihon Shakai Chiri Kyōkai 日本社會地理協會 which translates its name as follows: The Japanese Association of Geography for Social Life. Shakai chiri aims at being a somewhat popular magazine bringing geographical knowledge to the common people. The latest data and information are rapidly included in its contents. The special mapping techniques of the publisher, the Nitchi Shuppan Kabushiki Kaisha (Japan Map Publishing Company), are applied in the illustration of each number (36 pages). Each number also contains a special insert of several colored charts. Purely physical geographic monographs are not included. Stress is placed on geographic studies in relation to social life. The index for Nos. 1-10 appears in No. 11 and the index for Nos. 11-20 appears in No. 21. Publication was stopped after No. 30 was published and the magazine will probably not be revived although there has yet been no official declaration of the end of its publication or the extinction of the Association.

118. Shigen Kagaku Kenkyūjo ihō 資源科學研究所彙報 (Miscellaneous reports of the Research Institute for Natural Resources), Tōkyō, Shigen Kagaku Kenkyūjo, 1941-, irregular.
Included are physical geographical studies, primarily geomorphological in character.

119. Shin-chiri 新地理 (New geography), Tōkyō, Teikoku Shoin, 1947-51, irregular.
This journal began publication in 1947 and lasted until 1951. From five to ten numbers were issued each year. In 1952 it reappeared as a bulletin of the Nihon Chiri Kyōiku Gakkai 日本地理教育學會 (Association for Geographical Education in Japan). Each number contains several original papers, short reports on geographic activities, reviews, etc. Most of the articles are on Japanese historical and regional geography.

120. Suiro yōhō 水路要報 (Bulletin of hydrography), Tōkyō, Suirobu, 1922-1943, monthly.
A monthly bulletin published by the Suirobu or Japanese Hydrographic Office. Main items included are the explanation of sea charts newly published by this office and monographs on astronomy, geography, cartography and other fields of science related to hydrography or navigation. The publication has been suspended since 1943.

121. Tōhoku chiri 東北地理 (Northeastern geography), Sendai, Tōhoku Chiri Gakkai, 1948-, quarterly.
This is the bulletin of the Tōhoku Chiri Gakkai or the Association of Tōhoku Geographers which reappeared in 1948 as a quarterly journal after several years' suspension. The publication, at first not regular, is improved. Contents vary but ordinarily consist of articles on current ideas, original research and field reports, geographical reviews and notes. Emphasis has been largely on human geography with the area of study more or less confined to Tōhoku or northeastern Japan.

122. Tōkyō Daigaku chirigaku kenkyū 東京大學地理學研究 (Geographic studies, Tokyo University), Tōkyō, 1950-, irregular.
This publication appeared in 1950 as an outlet for occasional papers contributed by geographers at Tokyo University, by professors of other universities, and by graduate students. Of 23 articles in the first two numbers, half concern physical geography and half various aspects of human geography. Abstracts in English follow each article.

123. Tōkyō Daigaku Jishin Kenkyūjo ihō 東京大學地震研究所彙報 (Bulletin of the Earthquake Research Institute of Tōkyō University), Tōkyō, Tōkyō Daigaku Jishin Kenkyūjo, 1926-.
Organ of Japan's leading seismological research center, this publication has appeared very irregularly. Contents are predominantly technical monographs on earthquake analysis and related aspects of geo-physics, etc. More articles are written in English and French than in Japanese. Before 1945 printed as Tōkyō Teikoku Daigaku Jishin Kenkyūjo ihō.

Chapter VI

Atlas, Maps, Cartography, and Air-Photo Coverage

Japan has a long and interesting history of map making. Almost countless maps were printed, largely from wood block, before the Restoration and in the early Meiji period. These included maps of the world as Japan knew it, of Japan as a whole, of the great roads and the regions through which they passed, and of cities and towns. These early printed maps are intriguing collectors' items and are fast disappearing. They are generally of limited value for research, except in the history of cartography, as they were essentially pictograms and cannot be compared with the modern, scientifically compiled map. Researchwise, however, this early period produced many detailed manuscript maps of local areas, portraying irrigation systems, political holdings, land reclamation projects, and many other features. These are to be found today in the collections of temples, of great families, and in local governmental offices and are generally available to the field worker for study or reproduction.

During the period of Japanese geography as a science, there have been two major lines of cartographic production. The first is the topographical sheets of the Imperial Survey, now being carried forward by the Chiri Chōsajo or Geographical Survey Institute of the Ministry of Construction. These sheets are in magnificient detail, have long been easily and cheaply available to the public, and cover the entire country. The major series were 1 : 50,000 and 1 : 25,000, although sheets of scales to 1 : 5,000 and even less were made for critical areas of Japan. These top sheets are the greatest single source of geographical data on Japan and have been of tremendous aid to geographers. It is safe to say that these sheets have been practically the sole source of cartographic data for non-government publishers of maps and atlases.

The other main line of cartographic production has been in the publication of atlases. Most of these are small and cheap and are revised frequently for the use of school children. These number literally in the hundreds and are all very much alike, so there is little point in attempting to list them. Atlases at a more comprehensive level are few indeed, although since the war this condition has been much improved.

Air-photographs originally were under the strict control of the Japanese Army and were not available for ordinary research purposes. SCAP also restricted them, and the current Defense Force seems unlikely to alter this established policy. Immediately after the war some of the old Japanese Army air-photo material became public and a few books of air-photos were printed. However, under conditions existing now this potentially excellent research material is not available.

124. Chihō Chōsa Kikan Zenkoku Kyōgikai 地方調査機關全國協議會 Tōhoku Kaihatsu Kenkyūkai 東北開發研究會 (National Council of Regional Research Media, Association for Research into the Opening of Northeast Japan), Nihon chiiki gensei zusetsu 日本地域現勢圖說 (Tables and diagrams showing the present status of regions in Japan), Tōkyō, Kawade Shobō, 1952, 224 pp.

The Sōrifu Keizai Shingichō 總理府經濟審議廳 (Board of Economic Research, Office of the Prime Minister), supervised the publication of this compendium which presents a great variety of information and data concerning the physical geography, economics, social problems, industries, etc. of Japan, tabulated on a regional basis. The first portion is devoted to tables and diagrams, and the second part is statistics.

125. Chiiki Keizai Kenkyūjo 地域經濟研究所 (Institute of Regional Economy), Hokkaidō shakai keizai zufu 北海道社會經濟圖譜 (Atlas of the social economy of Hokkaidō), Tōkyō, Ishizaki Shoten, 1953, 19+19 pp., 19 maps.

A general economic and physical atlas of Hokkaidō compiled at the request of the Hokkaidō prefectural government. The map section is followed by an explanatory appendix. Treated are such subjects as: topography, soils, climate, general land use, agriculture, stock raising, forestry, fisheries, mineral resources, manufacturing industries, and transportation and communications.

126. Chiri Chōsajo 地理調査所 (Geographical Survey Institute), Gojūman-bun no ichi chihōzu 五十萬分之一地方圖 (1:500,000 regional maps), Chiba, Chiri Chōsajo, 1948, 9 sheets.

Standard cultural and physical maps, similar in form to the 1 : 200,000 general land form maps, but with one more color, making four colors in all. Roads are printed in red. All Japan is covered with 9 overlapping sheets, namely: 1) eastern Hokkaidō, 2) southwestern Hokkaidō, 3) Tōhoku, 4) Kantō and Shin'etsu, 5) Tōkai and Hokuriku, 6) Kinki, 7) Chūgoku and Shikoku, 8) Kyūshū, and 9) the Inland Sea region.

127. Chiri Chōsajo 地理調査所, Goman-bun no ichi chikeizu 五十萬分之一地形圖 (1 : 50,000 topographic maps of Japan), Chiba, Chiri Chōsajo, 1945.

In 1945 the Geographical Survey Institute succeeded the Japanese Imperial Land Survey (Nihon Teikoku Rikuchi Sokuryōbu) in its function of carrying on topographical surveying and cartographic production. The latter organization had begun its activities originally in 1893. All of Japan proper had been mapped at a scale of 1: 50,000 by 1926 in a series with 1,248 separate topographical sheets (no overlap), covering individually 15 minutes of longitude by 10 minutes of latitude. The map sheets, printed in black, are now revised periodically. The contour interval is 20 meters except in areas of low relief where subsidiary 10 and 5-meter intervals are added. Drainage, vegetation, and cultural detail are generally excellent, considering the scale limitations. Reproduction is also usually of high quality.

128. Chiri Chōsajo 地理調査所, Goman-bun no ichi tochi kiyō zu 五十萬分之一土地利用圖 (1 : 50,000 land use maps), Chiba, Chiri Chōsajo, 1953-.

Compiled by the Geographical Survey Institute in cooperation with prefectural governments, this cartographic project was begun under a five-year plan in 1953. Sixty-four sheets were completed by June of 1954, and were based on both recent aerial photography and field survey. Detailed land classification, culture, and topography are shown. Both agricultural crops and natural vegetation are classified. The sheets are handsomely printed in eleven colors. In 1954 they were not yet available for public sale, being obtainable only through official sources.

129. Chiri Chōsajo 地理調査所, Nihon denryokuzu 日本電力圖 (Maps for electric power in Japan), Chiba, Chiri Chōsajo, 1949, 3 sheets.

One of the 1 : 800,000 map series published by the Geographical Survey Institute showing Japan's post-war hydro-electric and thermal generating plants and transmission net. Generating facilities are classified into three grades on the basis of capacity, i. e., over 100 kv., 50-100 kv., and below 50 kv. Also shown are control stations, the boundaries of supply areas, and meteorological bureaus and stations.

130. Chiri Chōsajo 地理調査所, Nihon kōtsūzu 日本交通圖 (Communications map of Japan), Chiba, Chiri Chōsajo, 1953, 3 sheets.
A three-sheet set in the scale of 1 : 800, 000 showing Japan's road and rail network and coastal shipping lines. Freight and passenger capacities are indicated. The frequency of bus, rail, and boat service is also shown by means of a four-category classification. The time necessary to reach the nearest railroad station from rural settlements is expressed by isochronic lines. Three-color maps.

131. Chiri Chōsajo, Nihon rōdō jinkō to toshi kinō 日本勞働人口と都市機能 (Japanese labor population and its functions in urban society), Chiba, Chiri Chōsajo, 1950, 3 sheets.
This is another of the 1 : 800, 000 map series, being a new cartographic representation of Japan's predominant occupational patterns, both in rural and urban areas.

132. Chiri Chōsajo, Nihon tochi riyōzu 日本土地利用圖 (Maps of land use in Japan), Chiba, Chiri Chōsajo, 1947, 3 sheets.
These are 1 : 800, 000 general land use maps, prepared under a GHQ, SCAP directive during the occupation period, and made available for public distribution after Feb., 1950. Pre-1950 editions are in English. Land classification breakdown is as follows: single and double-cropped paddy fields, "upland" fields, mulberry, tea, orchard, pasture, forest land, waste land, marsh, and salt fields. Printed in seven colors. Another edition of this set entitled Nihon tochi riyōzu, suiden oyobi futsūbata no bumpu (Maps of land-use in Japan: the distribution of paddies and dry fields), shows only three categories of cultivated land use, namely, single-crop paddy, double-crop paddy, and upland fields.

133. Chiri Chōsajo, Nijūman-bun no ichi chiseizu 二十萬分之一地勢圖 (1 : 200, 000 general land form maps), Chiba, Chiri Chōsajo, 1899.
Except for a few minor islands, all Japan is covered by the 112 sheets of this map series. Each sheet covers 1 degree of longitude and 40 minutes of latitude, being approximately 18 x 23 inches in size. Land forms are generalized, 100-meter contours being used. Post-war editions are in three colors; contours in brown, drainage and water bodies in blue, culture in black. Cultural data are simplified but generally similar to that shown on the 1 : 50, 000 topographic sheets. Four-color pre-war editions are generally unavailable.

134. Chiri Chōsajo, Niman gosen-bun no ichi chikeizu 二萬五千分之一地形圖 (1 : 25, 000 topographic maps of Japan), Chiba, Chiri Chōsajo, 1912.
In the early Meiji period the Japanese Army General Staff began preparing 1 : 20, 000 topographic maps of important military areas, shifting later to the 1 : 25, 000 scale and recompiling the initial maps. These sheets do not cover all Japan but many of the densely populated and strategically important areas of the country are now mapped. Similar in size and format to the 1 : 50, 000 series, the individual sheets cover an area of 7 minutes and 30 seconds of longitude by 5 minutes of latitude. Contours are at 10-meter intervals except on sheets where 5 and 2. 5 subsidiary contours are necessary. By 1954, 1, 208 sheets had been published and an extension of the coverage of the series was planned.

135. Chiri Chōsajo, Nōka genkin shūnyū kara mita nōgyō chiikiteki sai to shichōson-betsu tokushoku 農家現金收入からみた農業地域的差異と市町村別特色 (Differentiation by agricultural region, as seen from the cash income of agricultural households, and characteristics by city, town and village), Chiba, Chiri Chōsajo, 1951, 3 sheets.
At a scale of 1 : 800, 000 these maps present the regional differentiation of agricultural function based on the predominant sources of farm income, using statistics derived from the census of October, 1947. Thirteen types of cash crop are classified. In multicolor.

136. Chūgoku Chihō Sōgō Kaihatsu Iinkai 中國地方綜合開發委員會 (Joint Development Committee for the Chūgoku Region), Chūgoku chihō gensei zufu 中國地方現勢圖譜 (Atlas of the present situation in the Chūgoku area), Hiroshima, Chūgoku Chihō Sōgō Kaihatsu Iinkai, 1950, 50 maps.
Many committees of the kind which published this atlas were established in post-war Japan to study regional problems and development. They often considered America's T. V. A. as a model to follow. This publication is an atlas of the Chūgoku district, most of the maps being at a scale of 1 : 800, 000. Most of the maps are multicolor, with statistical and explanatory annotations in addition to the standard legends. Topics dealt with include: economic structure, primary manufacturing, mining, forestry, agriculture, electric-power generation and transmission, communication and transportation, population, public health, education, physical geography, and future trends and prospects.

137. Chūō Kishōdai 中央氣象臺 (Central Meteorological Observatory), Nihon oyobi rimpō kikōzu 日本及隣邦氣候圖 (Climatic atlas of Japan and her neighboring countries), Tōkyō, 1929, 95 maps.
A complete revision of the classic climatic atlas of Japan first drawn in 1897 by Dr. K. Nakamura, at that time director of the Central Meteorological Observatory in Tōkyō. During the 32 years between the compilation date of the present atlas and that of the original, a vast amount of climatic data, of superior accuracy and detail, accrued and became available. Included is a detailed explanation of the compilation methods employed. A summary of the climatic data used appears in table form at the end of the volume together with a rainfall record for selected stations for the 1901-1925 period. To indicate content, a partial listing of plates follows: isothermal charts, atmospheric pressure and prevailing wind direction, mean wind velocity, vapor tension, relative humidity, cloud cover, sunshine duration, date of first and last frosts, number of frost-free days, date of first freeze, evaporation rates, and precipitation frequency, amounts, and distribution. With 95 map plates and 27 pages of explanatory material.

138. Yoshida Tōgo 吉田東伍, Dai-Nihon dokushi chizu 大日本讀史地圖 (Historical atlas of Japan), Tōkyō, Fuzambō, 1935, 4+4+141+27 pp., 82 maps.

An excellent atlas of value to the historian and historical geographer. Maps are arranged in chronological order, the latest being for 1926, the beginning of Shōwa. Standard small-scale maps, on a regional or national base, are supplemented by several special large-scale maps of early Kyōto, Edo, etc. A brief explanation of each map is given in an appendix.

139. Fukuoka Kanku Kishōdai 福岡管區氣象臺 (Fukuoka District Meteorological Observatory), Kyūshū kikō zuchō 九州氣候圖帳 (Kyūshū climatic atlas), Tōkyō, Nihon Kishō Gakkai, 1949, 117+19 pp.

A general climatic atlas based on meteorological observations recorded at numerous Kyūshū stations during a 20-year period, 1926-45. Maps show mean monthly maximum and minimum temperatures, temperature extremes, monthly precipitation, dates of first and last frost, typhoon tracks, etc. Climatic statistics are included at the end of the atlas.

140. Fujita Motoharu 藤田元春, Shin-Nihon zuchō 新日本圖帳 (A new atlas of Japan), Tōkyō, Tōkō Shoin, 1934.

Although this atlas has long been out of print, it remains generally available and serves as a useful, comparatively detailed reference work. Six general maps at a scale of 1 : 1,800,000 treat topography, geology, climate, land use, communications, and political divisions. Some 25 sheets at 1 : 250,000 give regional coverage for Japan proper. The major urban areas are shown on separate maps at larger scales. Japanese territories and spheres of influence are included on maps at scales over 1 : 1,000,000.

141. Kitada Kōzō 北田宏藏, "Nihon chizu ni taisuru tōei zuhō no sentaku 日本地圖に對する投影圖法の選擇 (The selection of a new projection for maps of Japan)," Chirigaku hyōron, v. 6 (1930), no. 7, 407-439.

Advantages and disadvantages of the various map projections are considered in terms of selecting one most suitable for constructing a base for the Japanese archipelago. Kitada favors cylindrical and conic projections with oblique axes to avoid undue distortion and, of the equal area type, he supports the use of Albers' projection.

142. Kokubu Yasukichi 國分安吉, "Bunka keikan no rittai zushiki 文化景觀の立體圖式 (Geographical stereograms of cultural landscape)," Chirigaku hyōron, v. 10 (1934), no. 2, pp. 25-40.

Although a variety of stereograms have been drawn by cartographers for the purpose of representing topographic forms, comparatively few diagrammatic methods to show cultural landscape have been advanced. Here Kokubu presents his technique of visual representation for several types of area in Japan and explains his methods.

143. Matsumoto Ken'ichi 松本賢一, Ōshū kohan Nihon chizushū 歐州古版日本地圖集 (Collection of European maps of Japan), Tōkyō, Tōichigumi Shuppambu, 1943, 4+74+8 pp., 44 maps.

An interesting set of maps portraying the chronological growth of Western cartographical knowledge of East Asia and Japan. With a historical summary, an explanation of each map, and index.

144. Mori Hideo 森秀男, Nihon nōgyō chizu (sakutsuke buai) 日本農業地圖(作付歩合) (Maps of Japanese agriculture — rates of planted area), Tōkyō, Jitsugyō Kyōkasho Kabushiki Kaisha, 1948, 165 pp.

A study of agricultural land use in Japan based on the distribution of eleven key or indicator crops. Founded on extensive statistical analysis, and some field checking, the maps show quantitative acreage allotments for the selected crops in given political divisions, with the end result revealing rather clearly the relative importance of major crops by prefecture. The "planting rates" of the indicator crops are shown on some 80 maps covering all prefectures. Accompanied by an explanation of the compilation method and the type and source of statistical data used. As a pioneer project of this kind, the study should be of particular value to those concerned with land use problems and interpretation.

145. Nishida Yoshirō 西田與四郎 and Katabira Jirō 帷子二郎, Nihon chiri zushū 日本地理圖集 (Geographical atlas of Japan), Tōkyō and Ōsaka, Tōyō Tosho Kabushiki Kaisha, 1937, 5+208 pp., 289 maps.

A school atlas of high standard. It went through several pre-war editions but has not been reprinted since World War II. Maps treat both physical and cultural features and factors on a national and regional base. Sources are cited for each map. With 208 pages of explanatory material and selected statistics.

146. Odauchi Tsūbin 小田内通敏, Nihon — fūdo to seikatsu keitai — kōkū-shashin ni yoru jimbun chirigakuteki kenkyū 日本—風土と生活形態—航空寫眞による人文地理學的研究 (Japan: its natural environment and living form — a human geographical study by aerial photography), Tōkyō, Tettō Shoin, 1931, 14+35 pp., 53 air-photos.

An early collection of air photographs surveying representative physical and cultural landscapes in Japan. The selection includes views of characteristic land forms, land use patterns, and settlement features. The settlement features range from disseminated types to urban agglomerations. Although the photographs are old and their quality not outstanding, they constitute an interesting geographical reconnaissance. There are few similar Japanese studies. Pictures are mounted on 35 special sheets and are accompanied by 14 pages of explanatory text.

147. Ogawa Takuji 小川琢治, Nihon chizuchō 日本地圖帖 (Atlas of Japan), Ōsaka, Seishōdō, 1924, 61 maps.

The main portion of this atlas consists of 23 sheets of 1 : 500,000 scale covering Japan proper with Hokkaidō, and several 1 : 2,000,000 maps of former Japanese territories. All of them were reduced from 1 : 200,000 maps and larger scale topographic maps from Imperial Surveys. Where there was no 1 : 200,000 coverage, data was taken from 1:50,000 topographic sheets. Isodepth lines are based on charts published by the Kaigun Suirobu (Coastal Survey of the Imperial Navy). A place name index, published in a separate volume, accompanies this atlas.

148. Sasaki Kyōichi 佐々木恭一 and Kimura Bunsuke 木村文助, Dai-Nihon bunken chizu awasete chimei sōran 大日本分縣地圖併地名總覽 (Atlas of Japan by prefecture, and gazetteer of place names), Tōkyō and Ōsaka, Kokusai Chigaku Kyōkai, 1952.

This is a revision of the atlas compiled by Kusaka Ihei and first published in 1937 under the same name. Each map is accompanied by an index of gun, machi, and shi, the kanji listings accompanied by kana readings as an aid to pronunciation. A single page is devoted to each prefectural map, arranged approximately from north to south starting with Hokkaidō (scales vary but most are 1 : 200,000 or 1 : 250,000); and at the end is a general gazetteer which lists all place names arranged by ken, shi, and gun. In this general index the kana readings of machi and mura are given but only the kanji for ōaza names appear. Included is a table of governmental organizations and institutions and other establishments of importance.

149. Shashin Sokuryōjo 寫眞測量所 (Photo Survey Institute), Tōkyō kōkū shashin chizu 東京航空寫眞地圖 (Aerial photo maps of Tōkyō), Tōkyō, Sōgensha, 1953-54, 3 v., over 300 photo maps.

This three-volume set provides excellent aerial-photo coverage of metropolitan Tōkyō. Photography was accomplished in March and April, 1953. Each photo map, at a scale of 1 : 5,000, has a cellophane overlay which gives ward, area, and street names, and identifies principal buildings. Each volume has an index and brief explanatory section. Volume one covers 8 wards in northeastern Tōkyō; volume two 9 northwestern wards; and volume three, the remaining 6 southern wards.

150. Tanaka Keiji 田中啓爾 and Yamamoto Kumatarō 山本熊太郎, Nihon jinkō bumpuzu 日本人口分布圖 (Distribution maps of Japanese population), Tōkyō, Kokon Shoin, 1928, 5 sheets.

An exhaustive compilation job requiring some seven years to complete. The authors correlated the 1920 census data with rural and urban settlements appearing on the latest available 1 : 50,000 topographic sheets and transferred the resulting dot pattern to a 1 : 1,000,000 map covering all Japan. Three classifications of population dot symbols were used: for 200, 5,000, and 10,000 people. Although the scale of the map is far too small to show clearly the distribution in the major urban areas, the maps give a good picture of the rural population pattern. Individual sheets are for these regions: Ōu (Tōhoku), Kantō and Chūbu, Chūbu and Kinki, Chūgoku and Shikoku, and Kyūshū.

151. Tanakadate Shūzō 田中館秀三, Tomita Yoshirō 富田芳郎, and Saitō Fumio 齋藤文雄, Tōhoku chihō shichōson-betsu jinkō-mitsudohyō oyobi dō-mitsudozu kaisetsu 東北地方市町村別人口密度表及同密度圖解說 (A table of population density by shi, machi, and mura of the Tōhoku region and an explanation of the density map) [=Saitō Hōonkai Gakujutsu Kenkyū Sōmubu 齋藤報恩會學術研究總務部 (Saitō Requital Foundation, General Affairs Section for Scientific Research), Gakujutsu kenkyū hōkoku 學術研究報告 (Scientific research report), no. 4], Sendai, 1927, 18+43 pp.

This report consists of a population density map showing number of persons per square kilometer for each shi, machi and mura in Tōhoku, as of October 1, 1920. The mapping technique employed is explained and the map is accompanied by a table indicating population density by ken, gun, and shi, and by a second table giving density figures for shi, machi, and mura.

152. Zenkoku Kyōiku Tosho Kabushiki Kaisha 全國教育圖書株式會社 (All-Japan Educational Book Company), Nihon keizai chizu 日本經濟地圖 (The economic atlas of Japan), Tōkyō, Zenkoku Kyōiku Tosho Kabushiki Kaisha, 1954, 21+61+166 pp., 61 maps.

Designed for general reference purposes, this is the first Japanese economic atlas. Maps treat the following topics: the orientation of Japan's economy in the world scene, natural conditions in Japan, structure of primary and secondary industries, agriculture and forestry, transportation and communications, trade and finance, regional planning. All maps are multicolor, beautifully reproduced, and compiled by specialists in the respective fields. In addition to complete legends and explanatory annotations accompanying each map, there is a 150-page textual summary at the back of the volume, with a 16-page statistical appendix.

153. Zenkoku Kyōiku Tosho Kabushiki Kaisha 全國教育圖書株式會社 (All-Japan Educational Book Company), Shashin Nihon chiri taikei 寫眞日本地理大系 (Pictorial geography of Japan), Tōkyō, Zenkoku Kyōiku Tosho Kabushiki Kaisha, 1953, 297+67 pp., 618 photos.

A photographic study of the cultural and physical geography of Japan, arranged on a regional basis. For each of the eight traditional districts treated, a map is provided, and photographs are grouped by prefecture. Each photograph has an English caption and a longer explanatory annotation in Japanese. At the end of the volume there is a textual summary of the physical characteristics of each district and its natural resources and industrial pattern, including generalizations on regional economic development.

Chapter VII

History of Japanese Geography

Surprisingly little has been published on the history of Japanese geography. The probable reasons for this are (1) there is relatively little in common between pre- and post-Meiji geography, or at least little has as yet been identified and the modern geographer finds little interest in the earlier period, and (2) modern geography is so new that there is little history involved. There are some good studies of early Japanese map makers such as Fujita Motoharu's Kaitei zōho Nihon chirigakushi (History of geography in Japan, revised and enlarged) and thoughtful interpretations of data contained in such works as the Engishiki have been made. In general it has been the practice of those few concerned with early geography to completely ignore the modern period. The task of identifying traditional attitudes and inherited knowledge in modern geography is still to be done.

A. History of Japanese Geographical Thought

154. Arakawa Hidetoshi 荒川秀俊, Nihon kishōgakushi 日本氣象學史 (History of Japanese meteorology), Tōkyō, Kawade Shōbō, 1941, 2+8+192 pp.

An outline history of the modern science of weather and climate in Japan. It traces the organization and growth of the Chūō Kishōdai (Central Meteorological Observatory) and its nation-wide network of observation and forecasting stations, seismological observatories, and other scientific installations. Also, the history and scientific contributions of the principal Japanese meteorological associations are summarized. With a short bibliography of reference works on meteorology and a chronological table of noteworthy developments in Japanese meteorology.

155. Ayuzawa Shintarō 鮎澤信太郎, Chirigakushi no kenkyū 地理學史の研究 (Studies in the history of geography), Tōkyō, Ainichi Shoin, 1948, 1+9+9+429 pp.

A study of the development of Japanese geographical thought beginning with ancient writings in the Kojiki, Nihon shoki, the various fudoki, etc., and extending through the work of Japanese scholars in the Edo period, including the first influence of Western thought, particularly as introduced by the Dutch.

156. Iimoto Nobuyuki 飯本信之, Chirigaku hattatsushi 地理學發達史 (A history of the development of geography), Tōkyō, Chūkōkan, 1940, 2+4+258+10 pp., 41 illus.

This work surveys the history of geography throughout the world. The last chapter is devoted to the development of modern geography in Japan during the present century.

157. Iizuka Kōji 飯塚浩二, Jimbun chiri gakusetsushi 人文地理學說史 (A history of the theories of human geography), Tōkyō, Nihon Hyōronsha, 1949, 9+2+223 pp.

This book is primarily a critical survey of the theories of Ritter, Peschel, and Ratzel. The writer further considers the relationship between geography and world history and the basic character of geo-politics.

158. Fujita Motoharu 藤田元春, Kaitei zōho Nihon chirigakushi 改訂增補 日本地理學史 (History of geography in Japan, revised and enlarged), Tōkyō, Tokō Shoin, 1942, 4+16+8+7+667 pp., 6 photos, 89 illus.

Fundamentally, a history of Japanese cartography to the Meiji era. However, a portion of the work deals with geographic thought and writing. The author regards the fudoki, compilations of provincial information made by local governments in the 8th to 10th centuries, as the beginning of geographic writing. He traces the development of mapping from the ancient Gyōgi map to the surveys of Inō Tadataka.

159. Mochizuki Katsumi 望月勝海, Nihon chigakushi 日本地學史 (History of Japanese earth sciences), Tōkyō, Heibonsha, 1948, 10+184 pp.

A concise historical outline of the development of certain of the earth sciences in Japan from ancient times into the modern period. Mineralogy, mining geology, geomorphology, and other phases of geology are the principal fields considered. However, the inclusion of cartography makes this short work of more specific interest to the geographer. Also, it should be noted that the writer's coverage of the post-Meiji era, although brief, is more ably presented than in most surveys of this nature.

160. Muramatsu Shigeki 村松繁樹, Nihon chirigakushi 日本地理學史 (History of geography in Japan), in Iwanami kōza: chiri (Iwanami lectures in geography), Tōkyō, Iwanami Shoten, 1931-34, 95 pp.

A history of the development of geographic thought in Japan. The growth of the Japanese concept of regionalism is treated in some detail, as traced from early writings in the fudoki, Engishiki, and Wamyōshō. Final chapters treat the enlargement of geographic knowledge under the stimulation of Japan's expansion and the influence of Western geography.

161. Shiga Jūkō 志賀重昂, Nihon fūkeiron 日本風景論 (A review of the scenery of Japan), Tōkyō, Hakubunkan and Tōkyōdō, 1894, 233+8 pp.

A classic work on the physical geography of Japan from the regional approach. As a pioneer study of its type, it is of historical as well as purely geographical value. This important volume should be a basic Japanese-language reference work to the geographer or student concerned with Japan's physical features.

162. Tanaka Keiji 田中啓爾, Chirigaku no honshitsu to genri 地理學の本質と原理 (The nature and principles of geography), Tōkyō, Kokon Shoin, 1949, 2+10+400+10 pp., 64 illus.

An interpretation of regional geography and the approaches to it through an analysis of regional characteristics The author cites abundant specific material derived from his lifetime of regional study in Japan.

163. Watsuji Tetsurō 和辻哲郎, Fūdo 風土 ("Acclimatization"), Tōkyō, Iwanami Shoten, 2nd rev. ed., 1943, 4+2+408 pp.

A monumental work by an outstanding Japanese philosopher. It has had a considerable influence upon modern Japanese thought, geographical and otherwise. The term fūdo represents a complex metaphysical and semi-scientific concept, in part similar to the Western idea of "environmental determinism," pertaining to the regionality of human personality as a reflection of various conditions of the natural environment. The first edition of this book appeared in 1935.

B. History of Japanese Cartography

164. Akioka Takejirō 秋岡武次郎, Chizugakushi 地圖學史 (A history of cartography), in Iwanami kōza: chiri (Iwanami lectures on geography), Tōkyō, Iwanami Shoten, 1931-34, 41 pp., 12 illus.
Following a summary of the historical development of maps and map-making in the world, Akioka presents a brief but comprehensive survey of the history of map-making in Japan. He discusses the early Japanese conception of the map of the world, cites various examples, and comments on the derivation of world maps in Japan together with their routes of introduction. Early Chinese maps also are discussed in some detail since they greatly influenced Japanese cartography.

165. Ōkawa Michihisa 大川通久, "Nihon kokon sokuchi ippan 日本古今測地一斑 (Topographical surveys in Japan, in former times and now: a partial view)," Chigaku zasshi, vol. 1 (1889), 144-147, 187-190, 294-297, 354-358, 419-425, 2 tables, 6 illus.
Ōkawa was a member of the Nōshōmushō, Chishitsukyoku, Chikeibu 農商務省地質局地形部 (Topographic Section, Geological Survey Bureau, Ministry of Agriculture and Commerce) at the time he wrote his article. He describes in technical terms the survey methods then employed, explaining the development of surveying techniques and organizations from the time when Dutch surveying was introduced into Japan during the Tokugawa period. Of definite historical significance.

166. Ōmori Matakichi 大森又吉, Rikuchi sokuryō seika riyōhō 陸地測量成果利用法 (Guide to the use of the results of the land survey), Tōkyō, Kōseisha Kōseikaku, 1951, 130 pp.
Many accurate measurements and detailed surveys have been and are being made preparatory to the production of the 1:50,000 and 1:25,000 topographic maps published by the Chiri Chōsajo (Geographical Survey Institute). All the basic data, however, do not usually appear on the printed map sheets and, in accordance with the newly enacted Land Survey Law, these materials are now available to the general public. This guide to securing and utilizing such data should be an invaluable aid to the field mapper or cartographic researcher.

167. Takagi Kikusaburō 高木菊三郎, Nihon chizu sokuryō shōshi 日本地圖測量小史 (Short history of map surveying in Japan), Tōkyō, Kokon Shoin, 1931, 2+13+170+15 pp., 25 illus.
The historical development of map surveying is outlined in five sections which treat: 1) the beginning of surveying after the 7th century, 2) introduction of Western methods, 3) improvement of surveying techniques, 4) surveys under the Shogunate, and 5) the modern period and the use of photo methods.

C. History of Exploration by Japanese

168. Tōa Kōko Gakkai Mōko Chōsahan 東亞考古學會蒙古調查班 (Archeological Association of East Asia, Survey Group for Mongolia), Mōko kōgen ōdanki 蒙古高原橫斷記 (Record of trips across the Mongolian plateau), Tōkyō, Nikkō Shoin, 1941, 12+318 pp., 69 photos, 70 illus.
Reports on the accomplishments of two expeditions to Inner Mongolia sponsored by the Archeological Association of East Asia in 1931 and 1935, respectively. One geographical paper on the physical geography of the Mongolian plateau by I. Matsuzawa is included, with six anthropological and archeological articles by E. Akabori, N. Egami, Y. Yokoo, and K. Takeuchi.

169. Uehara Yoshitarō 上原芳太郎, ed., Shin-seiikiki 新西域記 (Records of Central Asia), Tōkyō, Yūkōsha, 1937, 2 v.
A collection of the records, travel diaries, and subsequent lectures resulting from the three extensive expeditions of Ōtani Kōzui to Sinkiang and adjacent regions in 1902-04, 1908-09, and 1910-14, respectively. Ōtani was primarily concerned with the historical diffusion of Buddhism. Included are articles by Hada Tōru and Naitō Konan, specialists on the history of Central Asia, and a variety of miscellaneous material pertaining to Ōtani's archeological finds, his museum exhibitions, etc. Ōtani's travel observations and cultural findings remain of definite geographical interest.

Chapter VIII

Physical Geography

It is in the fields of physical geography that Japanese geographers have made the greatest and best contributions. In Japan, as has been commonly true elsewhere, modern geography developed out of two sources, geology and history. The former exerted much the stronger influence, and "converted" geologists played the leading part in establishing geography. Their heirs still comprise the dominant majority in Japan. The proportion of studies in geomorphology is very high indeed. The quality of these studies is also high. Studies in volcanism and in seismology are also comparatively important, both in volume and quality, as might be expected in a land where these forces are so significant. Studies in climatology and in ground water, although a later development, are likewise relatively important — particularly as they pertain to agriculture. Studies in cultural geography are now rapidly increasing in number and in value, but it is still in the area of physical geography that Japanese geography is most eminent.

A. Geomorphology and Geognostics

170. Adachi Toshio 足立俊男, "Kashima-hantō nambu no suna no kenkyū 鹿島半島南部の砂の研究 (A study of the sand in the southern part of the Kashima peninsula)," Chirigaku hyōron, v. 13 (1937), no. 7, 37-48.

Concerns an area along the Pacific coast in southern Ibaragi prefecture. Sand types are classified according to texture and their distribution is shown on a map. The origin of each type is discussed, mainly in connection with the influence of the wind on deposition. This is the fourth in a series of studies on raised benches and sand areas.

171. Akioka Takejirō 秋岡武次郎 "Manshūkoku no chikeiku 滿洲國の地形區 (The physiography of Manchuria)," Chirigaku hyōron, v. 11 (1935), no. 6, 49-63.
The author divides Manchuria into eight physiographic regions and describes each. This paper is presented as a preliminary outline posing future research, yet the geomorphological descriptions are fairly detailed and useful.

172. Anzai Tōru 安齋徹, "Ōtori-ike no seiin ni tsuite 大鳥池の成因に就て (On the origin of Ōtori pond)," Chirigaku hyōron, v. 4 (1928), no. 4, 36-51.
The origin of this small mountain lake had hitherto been attributed to the filling of a depression. Here the author identifies it as resulting from the damming up of a valley. The peculiar characteristics of the lake water in this granite region of Yamagata prefecture are also described.

173. Asai Jihei 淺井治平, "Tama kyūryō no chikei 多摩丘陵の地形 (Topography of the Tama hill region)," Chirigaku hyōron, v. 1 (1925), no. 6, 20-40.
A pioneer study of the hills and terraces near Tōkyō. Although in need of some revision, it is a basic reference work for the geographer interested in the land forms of southwestern Kantō.

174. Asami Shigeo 淺海重夫, "Chikei bunrui ni kansuru shian 地形分類に關する試案 (Some attempts at classifying land forms)," Chirigaku hyōron, v. 24 (1951), no. 12, 20-30, 1 map.
An article on the methodology of land form delimitation including a short discussion of the classification of geomorphological features and the morphometry on topographic maps. An example of the author's preferred system is included in the form of a map of the physiographic divisions of the Bōsō peninsula.

175. Ata Saneo 阿多實雄, "Minami Kyūshū no nisan no onsemmyaku to onsen 南九州の二三の温泉脈と温泉 (Several hot springs and hot springs groups in southern Kyūshū)," Chigaku zasshi, v. 43, no. 511 (1931), 504-512.
Discusses the relationship between hot springs and volcanoes in the Kirishima district, especially in regard to their patterns of alignment.

176. Awaji Shōzō 淡路正三, "Kasen kyokuryū ni tsuite 河川曲流に就いて (On river meandering)," Chirigaku hyōron, v. 9 (1933), no. 10, 52-63.
A morphological study of river meanders in Hokkaidō.

177. Awaji Shōzō 淡路正三, "Kirigamine ni mirareru dansō chikei no ichirei 霧ヶ峯に見られる斷層地形の一例 (An example of displacement along a fault-line seen at Kirigamine)," Chirigaku hyōron, v. 14 (1938), no. 1, 42-56.
Reports on a detailed investigation of the Kirigamine fault and associated dissected topography to the northwest of the Suwa basin in Nagano-ken. The significance of this fault is considered in relation to the general structure of the Suwa graben.

178. Awaji Shōzō 淡路正三, "Nagano-shi o chūshin to shita Saikawa kyūryō tōempen no chikeigakuteki kenkyū (1) 長野市を中心とした犀川丘陵東縁邊の地形學的研究(1) (Geomorphological study of the eastern margin of the Saikawa hills surrounding Nagano city (1))," Chirigaku hyōron, v. 13 (1937), no. 1, 41-66.
The distribution of river terraces in this region is studied and past shiftings of the river channels are traced by a correlation of terrace surfaces.

179. Chishitsu Chosajō 地質調査所 (Geological Institute), Nihon chishitsu kōsanshi 日本地質鑛產誌 (Geological and mineral products of Japan), Tōkyō, Chigaku Kyōkai, 1932, 4+453 pp.
An encyclopedia-type volume consolidating general information on this subject which was published in fragmented form prior to 1932. A 1 : 4,000,000 geological map of Japan is appended.

180. Ehara Shingo 江原眞伍, Taiheiyō no chikaku undō 太平洋の地殼運動 (Crustal movement in the Pacific), Ōsaka, Sekizenkan, 1945, 2+4+188 pp., 2 charts, 89 illus.
A discourse on the geo-tectonics of the western Pacific basin and the influence of crustal pressures and stresses in that region upon the structural geology of the Japanese islands, particularly as reflected in the geomorphology of Shikoku and the fossa-magna sector of central Honshū. Ehara is a noted authority in this field.

181. Fujimoto Haruyoshi 藤本治義, "Kobotoke sō ni tsuite 小佛層に就て (On the Kobotoke series)," Chigaku zasshi, v. 43, no. 509 (1931), 13-19.
A technical stratigraphic study of certain formations widely distributed in the southern Kantō mountains. Geologists had believed the Kobotoke series to be Paleozoic strata, but, in the light of new evidnece, they have come to regard the series as Cretaceous. A bibliography is added.

182. Fujiwara Sakuhei 藤原咲平, "Kazammyaku to santai ressen to ni tsuite 火山脈と山體裂線とに就いて (On the main volcanic ranges and volcanic fissures [in Japan])," Chirigaku hyōron, v. 1 (1925), no. 8, 1-24.
In this study Fujiwara investigated the echelon arrangement of the principal volcanic ranges in relation to volcanic fissure lines in an effort to cast light on the cause and nature of block crustal movements in Japan as a whole.

183. Fujiwara Sakuhei 藤原咲平 , "Norikuradake Kamegaike fukin no kikkōgata rekitai ni tsuite 乗鞍岳亀ヶ池附近の亀甲形礫堆に就て (On the turtle-shell patterned gravel bank in the vicinity of Kamegaike at Mt. Norikura)," Chirigaku hyōron, v. 4 (1928), no. 8, 77-81.
The first report by Fujiwara regarding his discovery in the Japanese Alps of a peculiar type of soil deposit which he believes results from the effects of low temperatures and freezing and thawing conditions.

184. Fukui Eiichirō 福井英一郎 , "Yokote bonchi no chikeigakuteki narabi ni kikōgakuteki kenkyū 横手盆地の地形學的竝びに氣候學的研究 (A geomorphological and climatological study of the Yokote basin)," Chirigaku hyōron, v. 4 (1928), no. 1, 15-45.
The Yokote basin is in Akita prefecture, Tōhoku. Fukui studies the effect of distance to the sea and the influence of surrounding mountains upon local climate. Correlations between climatic variations and rice yields are also made in this paper.

185. Funakoshi Motoichi 船越素一 , "Wakasa-wan to Kita Tajima oyobi Kyōto bonchi o nasu dansōsen no kankei 若狭湾と北但馬及京都盆地を成す断層線の關係に就いて (On the relationship of the fault-lines which comprise Wakasa Bay and the North Tajima and Kyoto basins)," Chikyū, v. 5 (1926), no. 2, 22-31.
Discussing the structural geology of the Tamba plateau, Funakoshi points out the effect faulting in that area has had upon the geomorphology of the Wakasa Bay - Lake Biwa depression.

186. Hanai Jūji 花井重次 , "Katsura-gawa engan no chikei oyobi kagan dankyū no kenkyū 桂川沿岸の地形及河岸段丘の研究 (A study of the topography and river terraces along the Katsura-gawa)," Chirigaku hyōron, v. 3 (1927), no. 3, 1-17; no. 4, 46-64; no. 5, 50-60.
The Katsura River flows out of Lake Yamanaka to the east of Mt. Fuji and enters Sagami Bay, after circling aound the Tanzawa mountain block. In this article, Hanai is mainly concerned with the identification of fault features along that river valley and investigation of the relationship between the formation of terraces and volcanic and geo-tectonic disturbances.

187. Hanai Jūji 花井重次 , "Taiwan Tōen daichi no katsudansō 臺灣桃園臺地の活斷層 (Active faults in the Tōen hills, Formosa)," Chirigaku hyōron, v. 6 (1930), no. 7, 226-237.
Alluvial layers forming the surface of the Tōen hills in northwestern Formosa are terraced and show scars of recent faulting. Hanai investigated this topography and determined the nature of structural displacement in the underlying Tertiary strata.

188. Hanai Jūji 花井重次 , "Tōbu Manshū no kazan chikei 東部満洲の火山地形 (Volcanic topography in eastern Manchuria)," Chirigaku hyōron, v. 15 (1939), no. 12, 1-14.
After a general description of the geology and geomorphology of eastern Manchuria, Hanai explains the structure of basaltic lava plateaus found there in an effort to clarify the age of erosion surfaces in that region.

189. Hanai Jūji 花井重次 , "Ryōtō hantō ni hattatsu-suru iwayuru teiichi junheigen to sono shoseishitsu ni tsuite 遼東半島に發達する所謂低位置準平原と其の諸性質に就いて (The so-called 'low-location peneplain' developed in Liao-tung peninsula and its various characteristics)," Chirigaku hyōron, v. 4 (1928), no. 10, 1-12.
A study of the low-level, gently-sloping erosion surfaces of the major Manchurian peninsula. Hanai's "low-location peneplain" corresponds to the terrain classification of "bevelled hills" identified by L. Reinecke.

190. Hanzawa Seishirō 半澤正四郎 , "Ryūkyū Yaeyama guntō chishitsu oyobi chishi gaikan 琉球八重山群嶋地質及地史概觀 (An outline of the geology and the geological history of the Yaeyama island group in the Ryūkyūs)," Chirigaku hyōron, v. 8 (1932), no. 2, 39-51.
A brief discourse on the post-Paleozoic stratigraphy of the southwestern-most island group of the Ryūkyū chain. Notations on the geological relationship between Formosa and Yaeyama guntō are included.

191. Hanzawa Seishirō 半澤正四郎 , "Taiwan ni okeru ryūki sangoshō ni tsuite 臺灣に於ける隆起珊瑚礁に就いて (On the elevated coral reefs in Formosa)," Chirigaku hyōron, v. 7 (1931), no. 2, 1-14.
Hanzawa reports on the distribution and estimated geologic age of the elevated coral reefs forming coastal terraces in southern Formosa. Correlations with similar formations in Japan are made, e.g., the Numa terrace in Chiba prefecture.

192. Hanzawa Seishirō 半澤正四郎 , "Taiwan ni okeru Ryūkyū sekkaigan no sōi oyobi chikei ni tsuite 臺灣に於ける琉球石灰岩の層位及地形に就て (On the stratigraphy and geomorphology of the Ryūkyū limestone in Formosa)," Chirigaku hyōron, v. 7 (1931), no. 3, 38-55.
A detailed analysis of the geologic origin and characteristics of the Ryūkyū limestone formations in Formosa.

193. Harada Kiyoshi 原田清 , "Shōwa kunen Shōgawa kōzui ryūnyūkō fukin ni okeru chikei chishitsu no keiryōteki chōsa · kekka ni tsuite 昭和九年庄川洪水流入口附近に於ける地形地質の計量的調査・結果に就いて (A statistical survey of the geomorphology and geology of the outer belt of the Shōgawa before and after the flood of 1934)," Chirigaku hyōron, v. 14 (1938), no. 10, 24-48.
A detailed, local investigation of flood deposits to indicate the effect of flooding on downstream areas. The Shōgawa is a major stream in Toyama prefecture.

194. Harada Toyokichi 原田豊吉 , "Nihon chishitsu kōzōron 日本地質構造論 (On the geological construction of Japan)," Chigaku zasshi, v. 1 (1889), no. 2, 46-51; no. 3, 90-98; no. 4, 132-137; no. 5, 190-193, 5 maps.
Though dated, this is a classic interpretation of the topography of Japan with emphasis upon structural evolution. Geological structure is considered in terms of three broad zones: the northern arc, the Fuji or central sector, and the southern arc of Japan.

195. Hayasaka Ichirō 早坂一郎, Nihon chishi no kenkyū 日本地史の研究 (Paleontological geology of Japan), Tōkyō, Uchida Rōkakuho, 1926, 8+5+4+254 pp., 20 photo plates, 13 illus.

The historical geology of the Japanese islands from the paleontological point of view. Twenty pictorial plates show fossil finds.

196. Homma Fujio 本間不二男, Shinano chūbu chishitsushi 信濃中部地質誌 (Regional geology of central Shinano), Tōkyō, Kokon Shoin, 1931, 10+14+331+209+26+6+16+80 pp., numerous maps, photos, and illus.

A highly detailed and technical regional geology of the central Shinano or Nagano district of north-central Honshū, resulting from the author's 16 years of field work and research on that area. Included are general geological maps, structural maps, and geological cross-sections, and a number of schematic diagrams. Part of the work concerns fossil finds and a discussion of their significance. Some of the land form interpretation and summaries of the principal sections will be of interest to the physical geographer. With 75 photographs and more than 100 maps and illustrations, including several separate map and photo sheets.

197. Homma Fujio 本間不二男, "Shinano chūō takadai no chishitsu oyobi chikeigakuteki igi 信濃中央高臺の地質及地形學的意義 (The geologic and geomorphic meaning of the central highland in Shinano)," Chikyū, v. 8 (1927), no. 2, 40-51.

Discusses the erosion cycles and crustal movements which affected the highlands in the old province of Shinano, now a part of Nagano prefecture. The article is chiefly concerned with geological developments in pre-Quarternary periods.

198. Honda Takeo 本田武夫, "Hakubadake fukin no kōzan chikei 白馬岳附近の高山地形 (Geomorphological observations around Mt. Hakuba)," Chirigaku hyōron, v. 13 (1937), no. 9, 26-43; no. 10, 22-39.

Primarily a consideration of faulting and glaciation in the Hakuba area of the Japan Alps. The writer is critical of the geomorphologic conclusions of other scholars as regards this region.

199. Honda Takeo 本田武夫, "Taiwan no chikeigakuteki kenkyū 臺灣の地形學的研究 (A geomorphological study of Formosa)," Chirigaku hyōron, v. 15 (1939), no. 1, 17-34.

A generalized treatment of Formosan landforms. Geomorphic surfaces are classified on the basis of summit-level data and crustal movement interpreted from river profiles. The author's summit-level map and profile cross-sections of the island are included.

200. Hoyanagi Mutsumi 保柳睦美, "Naibu Ajia ni okeru sakyū no shūrai ni tsuite 内部アジアに於ける砂丘の襲来に就て (On the progress of sanddunes in inner Asia)," Chirigaku hyōron, v. 18 (1942), no. 10, 1-16.

A consideration of the enlargement of sanddune areas and related arid topography in the Ordos and Taklamakan deserts, based on observations made in the field in 1940 and 1941. Hoyanagi argues that the expansion of desert conditions in north China and Mongolia is the result of dune migration or sand invasion rather than climatic change.

201. Ichikawa Wataru 市川渡, "Etchū Jōganji senjōchi kibu no chikei hattatsushi 越中常願寺扇狀地基部の地形發達史 (The history of the geomorphological development of the upper part of the alluvial fan of the river Jōganji, Etchū province)," Chirigaku hyōron, v. 7 (1931), no. 1, 26-41.

The geological formation of this alluvial fan in Toyama prefecture is investigated in relation to evidence of past tectonic disturbances as revealed by a study of river terraces and changes in the course of the Jōganji river.

202. Ichikawa Wataru 市川渡, "Echizen Kinosaki-mura chihō no chikei to sono hattatsu ni tsuite 越前城崎村地方の地形と其の發達に就いて (On the landforms and their development in the district of Kinosaki-mura, Echizen province)," Chirigaku hyōron, v. 5 (1929), no. 12, 10-24.

An examination of the forms and origin of elevated marine terraces in the Kinosaki area of Fukui prefecture.

203. Ichikawa Wataru 市川渡, "Etchū heiya ni okeru chikei kubun no sonraku keikan 越中平野に於ける地形區分と村落景觀 (Views of physiographic divisions and settlement on the Etchū plain)," Chirigaku hyōron, v. 17 (1941), no. 2, 1-19.

The Etchū or Toyama lowland is divided into four general physical divisions and a number of subdivisions, based on terrain types; the settlement and land use pattern of each is discussed with reference to historical influences.

204. Ichikawa Wataru 市川渡, "Jinzūgawa oyobi sono shiryū Takaharagawa ryūiki no chikei 神通川及その支流高原川流域の地形 (On the topography of the watershed of the Jinzū river and its tributary, the Takahara river)," Chirigaku hyōron, v. 9 (1933), no. 2, 55-71.

Fundamentally, a geomorphological description of the Toyama plain with comments on the orography of the adjacent Hida highlands.

205. Ichinose Yayokichi 市瀬八代吉, "Iida bonchi oyobi dankyū 飯田盆地及段丘 (The Iida basin and river terraces)," Chirigaku hyōron, v. 2 (1926), no. 3, 35-47; no. 4, 47-60; no. 5, 25-31.

A summary of information obtained by geologic field parties on the origin and development of river terraces along the course of the Tenryūgawa in southern Nagano prefecture. A table is appended, listing the relative heights and surface inclinations of terraces in that area.

206. Igakura Tadatoshi 猪鹿倉忠俊, "Kaigan chikei ga shimesu suijun henka (Chūgoku hokugan tōbu chihō) 海岸地形が示す水準變化（中國北岸東部地方） (Changes of water-level indicated by coastal topography in the eastern part of northern Chūgoku)," Chigaku zasshi, v. 46, no. 547 (1934), 4-12; no. 548, 39-47; 3 photos, 6 illus.

A detailed study of coastal geology along parts of the shorline of Tottori and Shimane prefectures on the Japan Sea.

207. Igakura Tadatoshi 猪鹿倉忠俊, "Tsugaru hantō ni hattatsu-seru kaigan dankyū ni tsuite 津軽半島に發達せる海岸段丘について (Concerning the coastal terraces developed on the Tsugaru peninsula)," Chirigaku hyōron, v. 7 (1931), no. 9, 1-17.

Founded upon Igakura's field observations, this is a discussion of the distribution and geomorphologic development of coastal terraces along the Tsugaru peninsula on the Japan Sea in northeast Chūgoku.

208. Ihara Keinosuke 伊原敬之助 and Ishii Kiyohiko 石井清彦, "Kita Izu jishin chiiki no chihen 北伊豆地震地域の地變 (Land deformations in the North Izu earthquake area)," Chigaku zasshi, v. 43, no. 514 (1931), 41-52; v. 44, no. 516 (1932), 17-23; 1 map and 13 photos.

The article follows upon a second one by the same authors which also appeared in Chigaku zasshi (see next entry). The present work contains a description of faulting, earth cracks, land slides, and changes in man-made features resulting from the Izu earthquake. Observations made on the effects of the quake upon the flow of some of the famous hot springs in the Izu area are particularly interesting. With one geological map in color and 13 photographs.

209. Ihara Keinosuke 伊原敬之助 and Ishii Kiyohiko 石井清彦, "Kita Izu chishin chiiki no chikei oyobi chishitsu 北伊豆地震地域の地形及地質 (On the topography and geology of the North Izu earthquake area)," Chigaku zasshi, v. 43, no. 513 (1931), 41-52, 7 photos.

The North Izu earthquake of 1930 caused heavy damage and was of further seismological interest because of unusual geologic phenomena resulting from it.

210. Iki Tsunenaka 伊木常誠, "Aso kazan 阿蘇火山 (The volcano Aso)," Chigaku zasshi, v. 13, no. 148 (1901), 187-200; no. 155, 671-690; no. 156, 739-757, 1 map, 3 illus.

A thorough geomorphological study of Mount Aso, the Kyūshū volcano, reprinted from a report made by the Association for Earthquake Defence (Shinsai Yobō Chōsakai).

211. Imamura Akitsune 今村明恒, "Taishō jūninen Kantō dai-jishin shingen no tagensei to kono jishin ni kanren-seru dansōkei to no kankei ni tsuite 大正十二年關東大地震震源の多元性と此地震に關聯せる斷層系との關係に就いて (The multiple nature of the sources of the Kantō earthquake and their relationship to the fault systems)," Chirigaku hyōron, v. 6 (1930), no. 7, 27-38.

Imamura discusses the nature of the 1923 earthquale, assuming that there were three epicenters and several zones of secondary faulting.

212. Imamura Gakurō 今村學郎, "Jiyū dakō o suru kawa no dakōtai no ōkisa ni tsuite 自由蛇行をする川の蛇行帶の大いさに就いて (On the width of freely developed river meanders)," Chirigaku nempō, v. 2 (1934), 3-10.

A quantitative investigation of river meanders. Imamura discusses the relationship between the width of the free meander zone, stream gradient, and volume of flow.

213. Imamura Gakurō 今村學郎, "Kaigansen no suichoku idō to suihei idō oyobi sono yūsutatikku undō to no kankei 海岸線の垂直移動と水平移動及びそのユースタティック運動との關係 (Vertical and horizontal displacements of strand lines and their relation to eustatic movements)," Chirigaku hyōron, v. 4 (1928), no. 4, 1-14.

After pointing out that past studies of crustal movement in coastal areas have been largely restricted to detecting horizontal rather than vertical displacement, Imamura cites evidence of positive correlation between upheaval and horizontal displacement, discovered during field investigations in the Kitakami district and on Yaku island.

214. Imamura Gakurō 今村學郎, "Kaishoku daichi no hen'i-jō yori mitaru tōkyoku undō to keisha undō 海蝕臺地の変位上より見たる撓曲運動と傾斜運動 (Warping and tilting seen from the displacement of abrasion platforms)," Chirigaku hyōron, v. 4, no. 3 (1928), 1-11.

Two examples of displacement of the strand line on well developed abrasion platforms are studied, namely: at the Kitakami coast of Tōhoku, and on Yaku island to the south of Kyūshū. The extent of, and the reasons for, crustal movement are analyzed and compared.

215. Imamura Gakurō 今村學郎 and Fukui Eiichirō 福井英一郎, Yakushima no kenkyū sono ichi kaigan dankyū ni kansuru nisan no jijitsu (yohō) — chikeigakuteki kenkyū 屋久島の研究 その1 海岸段丘に關する二三の事實(豫報) — 地形學的研究 (Study of Yaku Island, 1 : two or three facts concerning the coastal terraces (preliminary report) — a physiographical study), Chirigaku hyōron, v. 8 (1932), no. 1, 12-23.

The marine abrasion platform encircling Yakushima, to the south of Kyūshū, is explained geologically. The writers argue that this coastal formation is not a fluvial Piedmontfläche as suggested by some Japanese geologists.

216. Imamura Gakurō 今村學郎 and Inoue Haruo 井上春雄, "Chichibu sanchi no chikeigakuteki kenkyū 秩父山地の地形學的研究 (A geomorphological study of the Chichibu mountains)," Chirigaku nempō, v. 2 (1934), 11-24.

Conjectures on peneplanation in the Chichibu mountains following a general topographic analysis of summit levels in that range.

217. Inami Etsuji 稻見悦治, "Hompō ni okeru kasen sōdatsu no gen'in to katei 本邦に於ける河川爭奪の原因と過程 (The causes and processes of river piracy in Japan)," Chirigaku hyōron, v. 24 (1951), no. 10, 1-7.

A paper on an interesting physiographic process which has developed rather frequently on the elevated peneplains of Japan. Examples are cited for the Chūgoku district and the Abukuma plateau in Tōhoku.

218. Inoue Haruo 井上春雄, "Fujigawa karyū ryūiki no chishitsu gaikan 富士川下流流域の地質概観 (A geological view of the district along the lower course of the River Fuji)," Chigaku zasshi, v. 46, no. 539 (1934), 18-34, 6 photos, 3 illus.

A concise geological description of the lower Fujigawa area in Shizuoka prefecture.

219. Ishihara Hatsutarō 石原初太郎, Fujisan no shizenkai 富士山の自然界 (Natural history of Mount Fuji), Tōkyō, Hakubunkan, 1925, 216 pp., 41 photos.

A popular guidebook to the Mt. Fuji district. Emphasis is on the general geology, petrology, and vegetation of that area.

220. Ishihara Iwao 石原巌, "Dammenzu ni yoru Tanzawa sanchi no chikeiteki kōsatsu 斷面圖に依る丹澤山地の地形的考察 (Morphological study of the Tanzawa mountains, with special reference to the profile curve)," Chirigaku hyōron, v. 8 (1932), no. 7, 49-69.

The border between Yamanashi and Kanagawa prefectures runs along the coastline of the Tanzawa range. Profile and cross-section diagrams are used to illustrate the discussion of the general morphologic characteristics of these mountains.

221. Ishihara Iwao 石原巌, "Hodakadake oyobi Yarigatake no hyōshoku chikei ni tsuite 穂高岳及槍ヶ岳の氷蝕地形に就いて (On the glacial topography of Hodaka and Yari mountains)," Chirigaku hyōron, v. 9 (1933), no. 6, 45-51.

Map identification, using 1 : 50,000 topographic sheets, of glacial terrain features in the highest portion of the Japanese Alps or Hida mountain range.

222. Ishii Itsutarō 石井逸太郎, "Toyama-wan teisen no idō ni tsuite 富山灣汀線の移動に就て (On the shoreline migration of Toyama Bay)," Chirigaku hyōron, v. 9 (1933), no. 7, 1-6 and v. 12 (1936), no. 10, 1-17.

The submergence of the Japan Sea coast in this area has been a much discussed topic. In this paper the author points out the existence of drowned coastal valleys and cites evidence such as submerged tree stumps, peat deposits, etc., to validate his conclusions.

223. Ishikawa Seishō 石川成章, "Kyūshū narabi ni Yamaguchi-ken no jūsekikō 九州並に山口縣の重石鑛 (On wolframite in Kyūshū and Yamaguchi prefecture)," Chikyū, v. 11 (1929), no. 1, 30-43.

Regarding wolframite mines, their geology, extraction techniques, and the quality and amount of their output.

224. Ishikawa Seishō 石川成章, "Nanki Yusaki onsen 南紀湯崎温泉 (The hot springs of Yusaki in southern Wakayama)," Chikyū, v. 5 (1926), no. 5, 33-50, 1 map.

A study of a group of hot springs on the Kii peninsula, accompanied by a map showing the geological structure of the area and indicating the location of each spring.

225. Ishimoto Mishio 石本巳四雄, "Chikaku no kammansei undō ni tsuite 地殻の緩慢性運動に就いて (On chronic movements of the earth's crust), Chirigaku hyōron, v. 6 (1930), no. 7, 13-26.

A discussion of the orogenic formation of basic landforms.

226. Itō Ryūkichi 伊藤隆吉, "Geibi chiiki no suikei keitai to dansō chikei 藝備地域の水系形態と斷層地形 (The drainage pattern and fault topography of the Geibi area of Chūgoku)," Chirigaku hyōron, v. 17 (1941), no. 6, 14-35; no. 7, 1-14.

The Geibi district constitutes a part of central Chūgoku, lying within Hiroshima prefecture. Itō considers it to be a region of fault-block mountains, an interpretation based largely on his analysis of the drainage pattern.

227. Itō Ryūkichi 伊藤隆吉, "Kaigan oyobi kagan no ōketsu shinshoku 海岸及河岸の甌穴侵蝕 (Pothole erosion along the seashore and at river bank sites)," Chirigaku hyōron, v. 16 (1940), no. 2, 1-17.

A discussion of potholes formed by both wave and stream action. The author cites some 30 illustrations of this peculiar type of erosion but is particularly interested in two examples, one caused by marine erosion at Cape Inubo in Chiba prefecture, and the other resulting from river abrasion in the Hidagawa valley in Gifu prefecture.

228. Itō Ryūkichi 伊藤隆吉, "Kashokukei no bunrui 河蝕系の分類 (Classification of river erosion)," Chirigaku hyōron, v. 19 (1943), no. 3, 1-10.

Conclusions of writer's intensive studies on the aspects and mechanics of river erosion.

229. Iwatsuka Shukō 岩塚守公, "Uenohara-machi o chūshin to shita Katsuragawa engan no chikei ni tsuite 上野原町を中心とした桂川沿岸の地形に就いて (On the geomorphology of the Katsura river basih at Uenohara-machi)," Tōkyō Daigaku chirigaku kenkyū, no. 2 (1952), 137-146.

An analytical survey of river terraces and pediment-like erosion surfaces to throw light on post-Tertiary crustal movements which have affected the development of land forms in this basin.

230. Jimbo Kotora 神保小虎, "Hokkaidō chisei sōron 北海道地勢總論 (Outline of the topography of Hokkaidō)," Chigaku zasshi, v. 4 (1892), no. 40, 155-164; no. 41, 209-214; no. 42, 252-259; no. 43, 318-322; no. 45, 410-416.

The first two of seven sections in this article are not devoted to landforms exclusively but rather give a general cultural background of the island prior to the outline of physical geography that follows. The other sections are on: topography and geology as keys to regional division; coasts; mountains, plateaus, and plains; hydrography; and terraces. Although one of the earliest studies of this kind, it remains as a good geomorphologic description of Hokkaidō.

231. Juen Shingo 壽圓晋吾, "Tama kyūryō no chikeigakuteki kenkyū 多摩丘陵の地形學的研究 (Geomorphological research on the Tama hills)," Chirigaku hyōron, v. 24 (1951), no. 11, 15-27.
The Tama hills, located west of Tōkyō, are notable examples of high, maturely-dissected diluvial terraces. In this study the writer is especially interested in the distribution of certain asymmetrical ridges developed in the area.

232. Kagawa Kan'ichi 香川幹一, Nihon no chikei 日本の地形 (Geomorphology of Japan), Tōkyō, Kokon Shoin, 1933, 10+230 pp., 224 maps and illus.
A geomorphological study of Japan, beginning with a general approach, followed by a description of landforms on a regional basis. The author's basis for determining geomorphologic regions may be of interest to the physical geographer. With many maps and schematic diagrams.

233. Kagawa-ken Kyōikukai 香川縣教育會 (Kagawa Prefectural Educational Society), Sanuki no gansō to chisō 讃岐の岩層と地層 (Rocks and geological layers of Sanuki province), Takamatsu, Kagawa-ken Kyōikukai, 1939, 92 photos.
A general guidebook to field work in geology and paleontology in Kagawa prefecture, based on the collected materials and experience of T. Sugiyama.

234. Kaizuka Sōhei 貝塚爽平, "Chūgoku chihō seibu no chikei 中國地方西部の地形 (Geomorphology of the western part of Chūgoku)," Tōkyō Daigaku chirigaku kenkyū, no. 1 (1950), 87-98.
Problems of interpreting marginal peneplanation and fault topography in western Chūgoku are investigated.

235. Kambara Shin'ichirō 神原信一郎, Fujisan no chiri to suiri 富士山の地理と水理 (Geography and hydrology of Mount Fuji), Tōkyō, Hakushinkan, 1929, 500 pp.
The summing up of 19 years of field study in the Fuji area, the majority of the text is devoted to geological and hydrological observations on the Five Lakes of Fuji. The historical significance of Fuji from a geographical point of view is briefly treated. Descriptions are detailed and Kambara's data have a reputation for accuracy among Japanese scientists. Included are both early and recent topographic maps of the Fuji area, geologic cross-sections, maps on the distribution of lava flows, and tables on water levels in the lakes and on local precipitation. Measurements of the rate of flow and volume of the Katsuragawa are appended at the end of the volume.

236. Kano Tadao 鹿野忠雄, "Chōsen tōhokubu sanchi no hyōga chikei ni tsuite 朝鮮東北部山地の氷河地形に就いて (On glacial topography in the mountains of northeastern Korea)," Chirigaku hyōron, v. 13 (1937), no. 12, 82-101.
A descriptive analysis of glaciated topography in the mountains of northeastern Korea where a variety of unusual landforms, believed to be of glacial origin, are found in association with volcanic structure.

237. Kano Tadao 鹿野忠雄, "Hompō kazantai no hyōshoku ni kansuru yosatsu 本邦火山體の氷蝕に關する豫察 (Preliminary notes on the glaciation of volcanoes in Japan)," Chirigaku hyōron, v. 13 (1937), no. 2, 1-20.
Glaciation in Japan has been a rather controversial topic among Japanese scientists. In this article on three volcanoes (two in Japan, and the third in the Kuriles), Kano describes glacial topography which he believes he has clearly identified. Ontake and Norikuradake in central Honshū are the two Japanese peaks.

238. Kano Tadao 鹿野忠雄, "Taiwan kōzan chiiki ni okeru nisan no chikeigakuteki kansatsu 臺灣高山地域に於ける二三の地形學的觀察 (Two or three preliminary notes on the morphology of the high mountain lands of Formosa)," Chirigaku hyōron, v. 8 (1932), no. 3, 26-32; no. 6, 45-60.
General physiographic observations on the Fomosan highlands, with particular reference to the existence of glacial topography in the Tsugitakayama area.

239. Kano Tadao 鹿野忠雄, "Taiwan Tsugitaka san'i ni okeru hyōga chikei kenkyū (dai-ippō) 臺灣次高山彙に於ける氷河地形研究(第一報) (A study of glacial topography in the Tsugitaka mountains, first report)," Chirigaku hyōron, v. 10 (1934), no. 7, 54-71; no. 8, 74-93; no. 9, 62-81; no. 11, 52-79; v. 11 (1935), no. 3, 30-49.
The initial part of the first and classic study of glacial landforms in the high mountains of Formosa founded on field research undertaken by the author's zoological and geological expedition of 1933. Many types of glacial terrain features are described in detail. Kano believes that they resulted from the last glaciation at the end of the Pleistocene. With numerous excellent photographs.

240. Katabira Jirō 帷子二郎, "Hokkaidō Akaigawa karudera ni tsuite 北海道赤井川カルデラに就いて (On the Akaigawa caldera, Hokkaidō)," Chirigaku hyōron, v. 2 (1926), no. 11, 1-26; no. 12, 38-66.
A detailed account of the formation and geological aspects of the Akaigawa caldera, located west of Otaru city.

241. Katō Takeo 加藤武夫, "Jishin dansō to gankō rekka 地震斷層と雁行裂罅 (Earthquake faults and echelon fissures)," Chirigaku hyōron, v. 3 (1927), no. 7, 1-5.
Katō discusses differences in the origin and form of the two types of fault, as well as variations and interrelationships of these features. He cites the complex Gōnomura fault in the northern Tango region, Kyōto prefecture, as an example.

242. Kawada Saburō 川田三郎, "Azusagawa no kagan dankyū 梓川の河岸段丘 (River terraces of the Azusagawa)," Chirigaku hyōron, v. 18 (1942), no. 2, 31-44.
The Azusagawa rises in the northern part of the Japan Alps and flows into the Saigawa in the Matsumoto basin. This is detailed geomorphological study of river terraces along some 60 kilometers of the course of that river.

243. Kawada Saburō 川田三郎, "Jinzūgawa no kagan dankyū 神通川の河岸段丘 (River terraces of the Jinzū-gawa)," Chirigaku hyōron, v. 18 (1942), no. 10, 44-55.

This river flows north from the Japan Alps into Toyama Bay. In this article Kawada attempts to correlate the development of terraces and volcanic activity.

244. Kawada Saburō 川田三郎, "Shin tōei dammenzu to Hida sammyaku 新投影斷面圖と飛驒山脈 (A new method of projected profile and the Hida mountain range)," Chirigaku hyōron, v. 16 (1940), no. 12, 42-58.

Geomorphological analysis of the Hida mountains, or Japan Alps, using the projected profile method introduced by D. W. Johnson in his studies of sea-bottom topography.

245. Kawada Saburō 川田三郎, "Taiwan no chikei ni kansuru jakkan no kiroku 臺灣の地形に關する若干の記錄 (Some records on the topography of Formosa)," Chirigaku hyōron, v. 19 (1943), no. 2, 1-21.

Brief notations on a variety of landforms and geomorphological features such as mud-volcanoes, landslide lakes, and karst topography.

246. Kawada Saburō 川田三郎, "Taiwan Takkirikei no kagan dankyū 臺灣タッキリ溪の河岸段丘 (Terraces along the river Takkiri, Formosa)," Chirigaku hyōron, v. 17 (1941), no. 8, 40-62; no. 9, 17-41.

Correlation of river terraces on Formosa, particularly along the Takkiri, founded largely on a map study of valley profiles. On this basis, primarily, the author discusses the development of the island's landforms as a whole.

247. Kawada Shirō 河田四郎, "Junheigan no chikei keisokuteki ichikōsatsu: Ryōtō junheigen ni tsuite 準平原の地形計測的一考察：遼東準平原に就いて (A morphometric study of peneplains -- On the Liao-tung peneplain)," Chirigaku hyōron, v. 6 (1930), no. 6, 42-47.

A quantitative map analysis of the peneplain topography of the Manchurian peninsula with mathematical notations on the slope relationship between peneplain and monadnocks.

248. Kimizuka Kōjirō 君塚康治郎, Nagano-ken Shimominochi-gun chishitsushi 長野縣下水内郡地質誌 (Regional geology of Shimominochi district, Nagano prefecture), Nagano, Nagano-ken Shimominochi Kyōikukai, 1929, 2+74 pp., 1 map and 29 illus.

A summary of the observations and data derived from geological field surveys in Shimominochi district, located in the extreme northern portion of Nagano prefecture. Considerable attention is given to the formations of Tertiary and igneous rock in that district, an area of relatively low mountains when compared with the other highlands of the same prefecture. Includes a general geomorphological description of the area, notes on mineral deposits, and a bibliography.

249. Kiuchi Shinzō 木内信藏, "Kazan no hipugurafu kyokusen 火山のヒプソグラフ曲線 (Hypsographic curves of volcanoes)," Chirigaku hyōron, v. 11 (1935), no. 8, 10-24.

An experimental analysis of the correlation between the height above sea-level and the area of the cross section at every 100-meter interval for three types of volcano: the conical, lava dome, and shield forms. Of interest to the geomorphologist.

250. Kiuchi Shinzō 木内信藏, "Yōganryūkei 熔岩流形 (Morphological study of lava flows)," Chirigaku hyōron, v. 9 (1933), no. 8, 54-63.

The profiles and morphology of the volcanoes Sakurajima, Asama, and Fuji are analyzed according to the physical characteristics of their lava flows as determined from topographic map analysis.

251. Kobayashi Giichirō 小林儀一郎, "Hompō onsen no chishitsugakuteki bunrui narabi ni onsen yūshutsu to chishitsu kōzōsen no kankei ni tsuite 本邦溫泉の地質學的分類並びに溫泉湧出と地質構造線の關係に就いて (Geological classification of the hot springs of Japan and the relation between amount of flow and certain geo-tectonic lines)," Chigaku zasshi, v. 51, no. 608 (1939), 19-23; v. 52, no. 613 (1940), 33-36; no. 616, 28-37; no. 619, 36-44; 6 photos.

Although there are many geological studies of Japanese hot springs, none of them attempts a correlation between the distribution and types of hot spring and the overall geological structure of Japan, as this article does. Kobayashi sees a definite relationship between the location of hot springs and the pattern of certain geotectonic lines extending throughout Japan. With a bibliography in English of source materials pertaining to geotectonics and to a tectonic map of Japan.

252. Kobayashi Teiichi 小林貞一, "Chōsen hantō chikei hattatsushi to kinseidai chishi to no kankei ni tsuite no ichi kōsatsu 朝鮮半島地形發達史と近生代地史との關係に就いての一考察 (On the relationship between development of topography and Cenozoic geology of the Korean peninsula)," Chirigaku hyōron, v. 7 (1931), no. 7, 1-28; no. 8, 16-36; no. 9, 18-43.

The geomorphology of peninsular Korea and its stratigraphy are discussed and comparison made with the Cenozoic geology of Japan and south Manchuria.

253. Komaki Saneshige 小牧實繁, "Awaji no kaigan 淡路の海岸 (The Awaji coast)," Chikyū, v. 5 (1926), no. 1, 44-49.

A brief description of the physical geography of the southern shore of Awaji island with conjectures on the subsidence of that coast.

254. Komaki Saneshige 小牧實繁, "Satsuma no kuni Fukiage sakyū no kenkyū 薩摩國吹上砂丘の研究 (A study of the Fukiage sanddune in Satsuma province)," Chirigaku nempō, v. 2 (1934), 1-91, 29 illus. and 98 photos.

Komaki specialized in the study of sandy terrain in Japan. This monograph constitutes a report of his detailed field observations on the geomorphology, vegetation, associated land use, etc. of the Fukiage dune located on the western coast of Satsuma peninsula in Kagoshima prefecture. A supplement of 20 pages of photographs on gravure paper is added.

255. Komaki Saneshige 小牧實繁, "Shōnai no sakyū 庄内の砂丘 (The Shōnai sanddunes)," Chikyū, v. 5 (1926), no. 3, 16-37.
Notes on the Shōnai sand dunes in Yamagata prefecture and travel observations on that region. This is a sequel to an earlier article by Komaki on the coastal dunes of northern Echigo, which appeared in the publication Rekishi to chiri.

256. Kumagai Naoichi 熊谷直一, "Tanhoku jishin no shingen no fukasa to shingen no seishitsu ni tsuite 但北地震の震源の深さと震源の性質に就いて (On the depth and nature of the epicenter of the northern Tajima earthquake)," Chikyū, v. 4 (1925), no. 3, 1-14.
A seismological analysis of the nature of the northern Tajima earthquake of 1925 determined by data derived from mechanical observations of the after-shocks which followed that quake.

257. Kuno Hisashi 久野久, "Hakone-kazan o ōdan-suru Kintokisan-Makuyama kōzōsen ni tsuite 箱根火山を横断する金時山 — 幕山構造線に就いて (On the Kintokisan-Makuyama tectonic line traversing Hakone volcano)," Chirigaku hyōron, v. 13 (1937), no. 9, 58-67.
A geological and geomorphological study of the extinct volcano, Mt. Hakone. Kuno discusses in this paper the nature of several assumed tectonic lines in the area and explains the distribution of parasitic cones. Also briefly treated is the relationship between earthquake occurrence and faulting in the Hakone vicinity.

258. Kuno Hisashi 久野久, "Saikin no chishitsu jidai ni okeru Tanna dansō no undō ni tsuite 最近の地質時代に於ける丹那斷層の運動に就いて (On the amount of displacement of the Tanna fault since the Pleistocene period)," Chirigaku hyōron, v. 12 (1936), no. 1, 18-32.
The author investigated the Tanna fault following the great northern Izu earthquake of 1930 and measured the amount of displacement which occurred at that time. In this article he discusses the movements along this fault during the recent geological period and presents his opinions on its formation and general significance.

259. Machida Tadashi 町田貞, "Numata bonchi ni okeru kosui no keisei ni tsuite 沼田盆地に於ける湖水の形成に就いて (On the formation of the lake in the Numata basin)," Chirigaku hyōron, v. 22 (1949), no. 2, 11-16.
The formation and disappearance of an ancient lake which occupied the Numata basin in Gumma prefecture is explained on the basis of a study of terrace deposits in the upstream drainage areas of the Tonegawa and Katakaigawa and of mud and lava flows from the Akagi volcano.

260. Machida Tadashi 町田貞, "Usuigawa nangan chiiki no chikei 碓氷川南岸地域の地形 (Geomorphology of the region on the south side of the Usui river)," Chirigaku hyōron, v. 23 (1950), nos. 9-10, 11-18.
The geomorphology of the area south of the Usuigawa, between Usui pass and Takasaki city. The author is interested in erosion surfaces and made detailed stratigraphic studies of the high terraces in that neighborhood.

261. Matsumoto Tadaichi 松本唯一, "Aira kazan ni tsuite 姶良火山について (On Aira volcano)," Chirigaku hyōron, v. 9 (1933), no. 7, 62-74.
From geomorphological and petrological evidence, Matsumoto concludes that Sakurajima volcano, in Kagoshima bay, is a remnant central cone of a huge caldera, which he terms "Aira" and relates seismologically to the Kirishima volcanic cone.

262. Matsumoto Tadaichi 松本唯一, "Aso yōgan no bumpu 阿蘇熔岩の分布 (The distribution of the Aso lava)," Chirigaku hyōron, v. 11 (1935), no. 2, 1-12; no. 3, 15-29.
A detailed description of the distribution of lava from Mount Aso in central Kyūshū, one of the world's major volcanoes. For more than ten years Matsumoto studied the Aso lava fields, which have a diameter of over 170 kilometers.

263. Matsuzawa Isao 松澤勲, "Manshū Nekka chihō no dansō keitō to sono Chūseidai chikaku hendō ni tsuite 滿洲熱河地方の斷層系統とその中生代地殼變動に就いて (Fault systems in the Jehol, Manchuria area and their crustal movements during the Mesozoic)," Chirigaku hyōron, v. 11 (1935), no. 1, 24-47.
An interpretation of the geological history of the Jehol region during the Mesozoic era in terms of faulting and folding zones identified from topographic map analysis.

264. Matsuzawa Isao 松澤勲, "Mōko kōgen chishi to sono chikei hattatsushi ni tsuite 蒙古高原地史とその地形發達史に就いて (On the geographical history of the Mongolian upland and its geomorphological development)," Chirigaku hyōron, v. 9 (1933), no. 1, 50-60; no. 3, 36-53; no. 4, 46-57; no. 5, 42-56; no. 6, 38-44.
The geological structure of the Mongolian plateau is outlined, and the historical geology of that region since the late Cretaceous period is summarized. A correlation of the structure and geomorphology of north China and Mongolia is made.

265. Mino Yokichi 三野與吉, "Abukuma kōgen no shōkifuku chikei to dojō (chūi heitammen ni kansuru chirigakuteki kenkyū, dai gohō) 阿武隈高原の小起伏地形と土壤（中位平坦面に關する地理學的研究 第五報） (Low-relief topography and soil layers of the Abukuma plateau: A geographical study of a flat plain of medium altitude, report no. 5)," Chigaku zasshi, v. 50, no. 518 (1938), 5-18; 2 maps, 12 photos, and 12 illus.
Correlation of soils, topography, and erosion in an effort to clarify the stage of peneplanation of the Abukuma plateau in southeastern Tōhoku.

266. Mino Yokichi 三野與吉, Chikei genron 地形原論 (Principles of geomorphology), Tōkyō, Kokon Shoin, 1942, 8+517+44 pp., 202 illus.

A monograph on a special phase of geomorphology concerning the standard theories relating to "rock floors" and the relationship between these theories and the known aspects of peneplanation. The author reviews the ideas of other geologists on this subject and advances his own conclusions. A supplement of nearly 100 pages contains a summary of the field of geomorphology designed for beginning students. With a 44 page index in both Japanese and English.

267. Mino Yokichi 三野與吉, "Chūgoku chihō ni okeru heitammen ni tsuite 中國地方に於ける平坦面に就いて (On some flat plains of Chūgoku district)," Chirigaku hyōron, v. 9 (1933), no. 8, 1-26.

A study of peneplanation and erosion surfaces in western Honshū. Three levels of "flat" land are classified, mainly on the basis of summit-level analysis.

268. Mino Yokichi 三野與吉, "Kōchi-shi shikin ni okeru shinshokumen no taihi to Tosa-wan tōhambu no kaigan chikeishi (yohō) 高知市附近に於ける侵蝕面の對比と土佐灣東半部の海岸地形誌(豫報) (Correlation of erosion surfaces in the neighborhood of Kōchi city and the topography of the eastern coast range of Tosa Bay — a preliminary report)," Chigaku zasshi, v. 43, no. 507 (1931), 10-21, 10 maps.

A geomorphological study of the southern coast of Shikoku along Tosa bay near Kochi city. As a report of field reconnaissance, the article is concerned chiefly with a brief description of coastal terraces, sand dunes and bars, marine erosion surfaces, and inshore sea-bottom topography. Of considerable interest to the physical geographer.

269. Mino Yokichi 三野與吉, "Inadani no chikei 伊那谷の地形 (Geomorphology of the Inadani intermontane basins)," Chirigaku hyōron, v. 24 (1951), no. 7, 1-16.

Observations on the problems involved in the correlation of erosion cycles and erosion surfaces, with particular reference to the Ina fault valley in Nagano prefecture. Other geomorphologic aspects of that area are also briefly considered.

270. Mino Yokichi 三野與吉, "Okayama-ken Takahashi-machi fukin no chikei 岡山縣高梁町附近の地形 (Geomorphology in the vicinity of Takahashi in Okayama prefecture)," Chirigaku hyōron, v. 12 (1936), no. 4, 34-65.

The writer divided the Takahashi district into four landform divisions and here describes the geomorphology of each with special reference to the formation of gorges in areas of gravel strata.

271. Miyazaki Kenzō 宮崎健三, "Kii sanchi ni okeru kōdo bumpu no tōkeiteki kenkyū 紀伊山地に於ける高度分布の統計的研究 (A statistical study of the distribution of heights in the Kii mountains)," Chirigaku hyōron, v. 6 (1930), no. 9, 17-30.

A map analysis of the mountainous topography of the Kii peninsula in an effort to establish the positive existence of low-relief surfaces in that region. Using a grid technique, Miyazaki identifies five terrace-like surfaces and explains their geomorphological origin.

272. Miyazaki Kenzō 宮崎健三, "Kyūshū sanchi no sanrokumen 九州山地の山麓面 ("Piedmontfläche" in the Kyūshū mountains)," Chirigaku hyōron, v. 8 (1932), no. 8, 1-20.

In this paper, southern Kyūshū is delimited geomorphologically as that region between the Yatsushiro-Usuki fault line and the Kirishima volcanic group. This is a study of six areas of low-relief, often named baru or bara in Japanese, but known more technically as Piedmontfläche.

273. Mochizuki Katsumi 望月勝海, "Abukuma sanchi no kaseki-junheigen ni tsuite 阿武隈山地の化石準平原に就いて (On the fossil peneplain of the Abukuma mountains)," Chikyū, v. 11 (1929), no. 6, 1-13.

A technical monograph on the historical geology and geomorphology of the Abukuma mountains, and the neighboring Yamizo range, in Tōhoku.

274. Mochizuki Katsumi 望月勝海, "Etchū Futakamiyama burokku ni tsuite 越中二上山ブロックに就いて (On the Futakamiyama block, Etchū province)," Chirigaku hyōron, v. 7 (1931), no. 2, 23-33.

On the basis of his study of terraces and fossil finds, Mochizuki identifies the Futakamiyama area as a structural unit independent of the surrounding Tertiary strata. He points out the similarity of past tectonic movement in this area and in the neighboring Noto peninsula.

275. Mochizuki Katsumi 望月勝海, "Kojō sammyaku ni tsuite no ichishōken 弧狀山脈に就いての一小見 (An hypothesis on the Japanese arcuate mountains)," Chirigaku hyōron, v. 13 (1937), no. 2, 21-30.

No clearly-defined theory had been advanced at the time of this article to show the relationship between the Honshū and Shichitō-Mariana arcs. Mochizuki here presents the hypothesis that these arcuate mountain chains resulted from pressures generated under the Asiatic mainland, and, in this light, remarks on the significance of the giant Yamato bank recently discovered in the Japan Sea.

276. Mochizuki Katsumi 望月勝海, "Kōzōron-teki ni mitaru Noto to Sado 構造論的に見たる能登と佐渡 (Noto and Sado tectonically considered)," Chirigaku hyōron, v. 8 (1932), no. 3, 1-25.

A comparative study of the geo-tectonic structure of the Noto peninsula and neighboring Sado island. The writer considers the origin of these areas and their structural relationship to the main island arc of Honshū.

277. Mochizuki Katsumi 望月勝海, "Noto hantō kibu o chūshin to seru kochiri oyobi chikei hattatsushi 能登半島基部を中心とせる古地理及び地形發達史 (The paleogeography and historical geomorphology of the district around the southern part of Noto peninsula)," Chirigaku hyōron, v. 4 (1928), no. 11, 22-42.

The post-Tertiary structural development of the Noto peninsula region in Ishikawa and Toyama prefectures is studied in detail. An analysis of fossil types and distribution is included.

278. Mochizuki Katsumi 望月勝海, "Shichitō-Mariana-ko no seiin 七島マリアナ弧の成因 (The origin of the Shichitō-Mariana arc)," Chirigaku hyōron, v. 16 (1940), no. 4, 1-11.
An attempt at improving on the theories explaining the origin of the Shichitō-Mariana arc, such as those advanced by L. Kober and E. Suess. Mochizuki assumes circulation or movement in the "Sima" zone of the Pacific floor which he believes would account for the arcuate form of the island chain. He dates the beginning of the arc's formation as middle Tertiary and correlates it with Alpine and Himalayan orogeny.

279. Mochizuki Katsumi 望月勝海 "Tōa daichikei no ichikōsatsu — toku ni tairikukan to shite no kansatsu 東亞大地形の一考察 — 特に大陸環としての觀察 (A view of mountain systems in eastern Asia — especially as the peripheral ring of a continent)," Chirigaku hyōron, v. 15 (1939), no. 5, 25-54.
A short treatise on some aspects of the interrelationship of deep-seated earthquakes, volcanism, and arcuate mountain formation in the western Pacific.

280. Mochizuki Katsumi 望月勝海 "Tonamiyama kyūryō oyobi Kanda kyūryō no chikei 礪波山丘陵及蟹谷丘陵の地形 (The geomorphology of the Tonami and Kanda hills)," Chirigaku hyōron, v. 10 (1934), no. 10, 1-35.
These hill areas in Ishikawa and Toyama prefectures are part of the mountains of the Noto peninsula but their topography is more like that of areas in northeast Japan. This article compares the geology and topography of the two similar areas and includes a discussion of fault scarps and low relief surfaces. A geo-historical outline is appended.

281. Murata Teizō 村田貞藏, "Eboshi-Mitsumata-Yari rempō ni okeru jakkan no kiroku 烏帽子 — 三俣 — 槍連峰に於ける若干の記錄 (Some observations on the Yari-Mitsumata-Eboshi range)," Chirigaku hyōron, v. 14 (1938), no. 4, 65-76.
An investigation of possible glacial landforms in the highest range of the Japan Alps, the Hida range. Aspects of nevation and moraine-like topography are particularly considered and classified but the author is unable to reach positive determination of glacial features.

282. Murata Teizō 村田貞藏, "Isawa senjōchi no keitaigakuteki kenkyū 膽澤扇狀地の形態學的研究 (Geomorphological study of the Isawa alluvial fan, Iwate prefecture)," Chirigaku hyōron, v. 15 (1939), no. 2, 52-65.
The Isawa fan is at the southern end of a series of alluvial outwash deposits along the west side of the Kitakami lowland. Murata discusses the form and development of this fan, the nature of other topographic features in the vicinity, and their relationship.

283. Murata Teizō 村田貞藏, "Senjōchi keitai ni kansuru rironteki kōsatsu 扇狀地形態に關する理論的考察 (Theoretical research on the form of alluvial fans)," Chirigaku hyōron, v. 7 (1931), no. 7, 47-64.
The geometric forms of alluvial fans as determined by a study of topographic maps.

284. Murata Teizō 村田貞藏, "Senjōchi no keitai to shūi no chikei to no kankei 扇狀地の形態と周圍の地形との關係 (Relationship between the form of alluvial fans and their marginal topography)," Chirigaku hyōron, v. 7 (1931), no. 8, 37-51.
Another in the author's series of papers on the subject of alluvial fans, several of which appeared in this journal.

285. Murata Teizō 村田貞藏, "Senjōchi to sono jōryū kakoku to no kōbai kankei ni tsuite 扇狀地とその上流河谷との勾配關係に就て (On the relationship of the slopes of alluvial fans and their up-river valleys)," Chirigaku hyōron, v. 9 (1933), no. 10, 39-51.
A study of the slope and form of alluvial fans in terms of the gradient of the parent valley.

286. Murata Teizō 村田貞藏, "Shinshokkoku ni kansuru ichikōsatsu 侵蝕谷に關する一考察 (A study of the nature of erosion valleys)," Chirigaku hyōron, v. 6 (1930), no. 6, 48-62.
The characteristics of V-shaped valleys are studied by Murata, using his slope formula on the relationship between the inclination of valley walls and valley bottoms.

287. Murata Teizō 村田貞藏, "Yamagata-ken Midaregawa senjōchi no chikeigakuteki kenkyū 山形縣亂川扇狀地の地形學的研究 (Geomorphological investigations of the Midaregawa fan, Yamagata prefecture)," Chirigaku hyōron, v. 12 (1936), no. 11, 65-88.
The Midaregawa fan is the largest of the alluvial fans which have developed in the Yamagata basin, a north-south graben in Tōhoku. Murata uses this fan as an example supporting the validity of his theory concerning the metamorphosis of the contours of alluvial fans caused by displacement.

288. Murata Teizō 村田貞藏, "Yamagata-ken Shinjō bonchi no keitaigakuteki kenkyū 山形縣新庄盆地の形態學的研究 (A geomorphologic study of the Shinjō basin in Yamagata prefecture)," Chirigaku hyōron, v. 17 (1941), no. 6, 36-53.
Primarily conjectures on local crustal movement. Based on the study of alluvial fans in the Shinjō basin.

289. Nagatani Minoru 長谷實, "Omaezaki kaishokudai 御前崎海蝕臺 (The Omaezaki marine abrasion platform)," Chirigaku hyōron, v. 20 (1944), no. 3, 6-20.
A geomorphological investigation of marine terraces and other coastal features at Cape Omae, Shizuoka prefecture.

290. Nakamura Keisaburō 中村慶三郎, "Echigo Ōmi no karusuto chikei 越後青海のカルスト地形 (Karst topography in the neighborhood of Ōmi in Echigo province)," Chirigaku hyōron, v. 10 (1934), no. 12, 33-38.
A paper dealing with the nature of karst landforms in the Mt. Kurohime area of Niigata prefecture.

291. Nakamura Keisaburō 中村慶三郎, "Yamakuzure to kōsuiryō narabi ni sono jikanteki bumpu 山崩れと降水量並びにその時間的分布 (Landslides, amount of rainfall, and their distribution in time)," Chirigaku hyōron, v. 8 (1932), no. 7, 32-48.

The first portion of this article deals with the relationship between landslides, geological factors, and precipitation. The last part concerns the distribution of major slides in the Kubiki district of Niigata prefecture and the chronological occurrence of the slides.

292. Nakano Takamasa 中野尊正, "Hokkaidō no kaigan teichi 北海道の海岸低地 (The coastal lowlands of Hokkaidō)," Chirigaku hyōron, v. 24 (1951), no. 8, 11-19.

The topography of the Ishikari, Notsuke, Yufutsu, and Embetsu lowlands is described, and from his comparison Nakano draws conclusions regarding differential crustal movements which occurred during the Pleistocene epoch.

293. Nakano Takamasa 中野尊正, "Nara bonchi tōen no sanrokugai 奈良盆地東縁の山麓崖 (Piedmont scarps on the eastern border of the Nara basin)," Chirigaku hyōron, v. 19 (1943), no. 4, 34-43.

Stratigraphical clarification of the geologic nature of piedmont scarps at the margins of the Nara basin.

294. Nakano Takamasa 中野尊正 and Yoshikawa Torao 吉川虎雄, Chikei chosahō 地形調査法 (A guide to the field survey of geomophology), Tōkyō, Kokon Shoin, 1951, 1+4+169+3+4 pp., 82 illus.

A booklet of interest to physical geographers in general as well as to the geomorphologist and topographer. The main chapters deal with preparations before the field survey, morphometric techniques, field survey techniques, and the handling of data and materials obtained in the field. Map interpretation of 22 landform types is explained and problems in geomorphology and recent Japanese trends in this field are discussed. A selected bibliography is included at the end.

295. Nasu Shinji 那須信治, Tango jishin no yoshin no rittaiteki hairetsu yori suitei-saruru chikai no keijō ni tsuite 丹後地震の餘震の立体的配列より推察さるゝ地塊の形状に就いて (On the form of blocks presumed from the cubic disposition of after-shocks of the Tango earthquake)," Chirigaku hyōron, v. 5 (1929), no. 6, 1-17.

The arrangement of structural units of the earth's crust in the Tango region of the Kinki district is speculated upon in relation to seismological evidence. The major earthquake referred to occurred in March, 1927.

296. Niino Hiroshi 新野弘, "Nihonkai Musashitai no teishitsu ni tsuite 日本海武藏堆の底質に就て (On soundings at Musashi shoal)," Chirigaku hyōron, v. 12 (1936), no. 2, 1-11.

A report on oceanographic investigations conducted in the area of the Musashi bank or shoal, located in the Sea of Japan to the west of Hokkaidō. Mainly a description of soundings and bottom deposit.

297. Niino Hiroshi 新野弘, "Nihon kinkai kaitei ni okeru reki no bumpu 日本近海海底に於ける礫の分布 (Distribution of sea-bottom gravels around Japan)," Chigaku zasshi, v. 45, no. 533 (1933), 22-35, 1 map.

The occurrence of gravels on the continental shelf surrounding Japan is, in some areas, difficult to account for, particularly when rather great depth and extreme distance from the coast are involved. This is a detailed report on the nature of the bottom samples taken while investigations of the distribution of these sea-floor gravels were being made. Eight specimen tables included.

298. Nishimura Kasuke 西村嘉助, "Tonegawa jōryū no kagan dankyū 利根川上流の河岸段丘 (River terraces on the upper Tonegawa)," Chirigaku hyōron, v. 16 (1940), no. 5, 1-8.

A general description of river terraces developed along the upper Tonegawa in Gumma prefecture, comparing them with terraces in other districts.

299. Nishio Keijirō 西尾銈次郎, "Tōkyō chihō ni okeru jiban chinka ni taisuru chishitsugakuteki kōsatsu 東京地方に於ける地盤沈下に対する地質學的考察 (Geological considerations on the sinking of the earth's surface in the Tōkyō area)," Chigaku zasshi, v. 48, no. 572 (1936), 9-28, 5 illus.

Subsidence of land in the down-town area of Tōkyō is a rather serious problem which the author investigates in this study from both the engineering and geological viewpoints. Of interest to the general student of Japanese urban geography, as well as to the technical specialist, since local settling and large-scale sinking of land are problems in several other metropolitan districts. Contains three lists of observational data.

300. Noda Seijirō 野田勢次郎, "Hompō ni okeru sekiboku kōshō 本邦に於ける石墨鑛床 (Graphite deposits in Japan)," Chigaku zasshi, v. 25, no. 290 (1913), 10-23; no. 291, 32-48; 10 maps, 1 graph.

An old report on Japanese deposits of graphite ore, based on work done by the Imperial Geological Survey. Presents a description of the operations at the 14 graphite mines then existing in Japan. Although outdated, much of the technical information remains valid.

301. Nōtomi Shigeo 納富重雄, "Ariakekai oyobi Yatsushirokai no seiseiki to hensen narabi ni Aso yōgan ni tsuite 有明海及八代海の生成期と變遷並に阿蘇熔岩に就いて (The geological age and history of the Ariake and Yatsushiro seas and the Aso lavas)," Chigaku zasshi, v. 42, no. 493 (1930), 5-13; no. 494, 29-39.

The Ariake and Yatsushiro "seas" constitute the major embayments of western Kyūshū. This article presents a general explanation of the formation of the landforms in that region.

302. Obara Nobuhiko 尾原信彦, "Izu Shimoda fukin no kaishoku dankyū chikei to ryūki teisen 伊豆下田附近の海蝕段丘地形と隆起汀線 (Elevated shorelines and marine abrasion platforms found in the vicinity of Shimoda, Izu peninsula)," Chirigaku hyōron, v. 13 (1937), no. 3, 61-73; no. 4, 46-68.

A former coastline in the southeastern sector of the Izu peninsula is traced and associated crustal disturbances interpreted as the result of geomorphological analysis of marine terraces and other elevated, low-relief erosion surfaces.

303. Obara Nobuhiko 尾原信彦, "Shimane-hantō Hinomisaki kimbō no ryūki-teisen to sono chikaku undō 島根半島日御碕近傍の隆起汀線と其地殻運動 (The elevated shoreline near Cape Hino on the Shimane peninsula and crustal movements there)," Chirigaku hyōron, v. 15 (1939), no. 2, 1-10.

One of a series of papers by Ōbara concerning the elevated shorelines of Japan. This is a study of coastal terraces to the west of Cape Hino, in Shimane prefecture, and of their significance as indicators of crustal uplift.

304. Ogawa Takuji 小川琢治, Nihon guntō 日本群島 (The Japanese archipelago), Tōkyō, Kōbundō, 1944, 9+372 pp., 47 illus.

Written by a former geologist, this is a regional geography of the Japanese islands. It is designed for popular appeal and is of little value to the advanced student. Emphasis in the first portion is upon the physical elements. Eleven regions are identified in the second part. The author's basis for determining them may be of interest.

305. Ogawa Takuji 小川琢治, "Nihon guntō chishitsu kōzōron 日本群島地質構造論 (The geological construction of the Japanese archipelago)," Chigaku zasshi, v. 11, no. 126 (1899), 413-423; no. 127, 475-505; no. 128, 537-560; no. 130, 685-695; v. 14, no. 157 (1902), 1-8.

Subsequent to the early classical studies on this subject by Dr. Harada and Dr. Kotō, a mass of new observations and data became available from the work of several comprehensive geological surveys. On the basis of this new information, particularly the completion of a one-millionth geological map of Japan, Ogawa in this study reviews the salient concepts of the past.

306. Ogawa Takuji 小川琢治, "Seinan Nihon chishitsu kōzōron 西南日本地質構造論 (On the geological construction of southwest Japan)," Chigaku zasshi, v. 19, no. 218 (1907), 92-118; no. 219, 167-188; no. 220, 227-241.

Concerning the geotectonic geology of southwest Japan, this is an excellent technical article on structural theory. Written following the author's field surveys, accomplished during the compilation of a one-millionth geological map of the Japanese empire, it deals with the delimitation of inner and outer structural zones in southwest Japan and discusses the relationship between Kyūshū and the Ryūkyū arc and between southwest Japan and the Ogasawara arc.

307. Ogawa Takuji 小川琢治, "Tōa chishitsu kōzōron kara mita jishin genshō no setsumei 東亜地質構造論から見た地震現象の説明 (The explanation of the phenomenon of earthquakes seen from theories on the geological structure of east Asia)," Chikyū, v. 4 (1925), no. 2, 1-19; no. 3, 15-27.

The seismic and volcanic activity which affects Japan is accounted for in terms of the generally accepted concepts pertaining to the geotectonic structure of East Asia. The author's interpretations are in concordance with the theories expressed in E. Argand's work, La tectonique de l'Asie.

308. Ogura Tsumotu 小倉勉, "Hoku-Man kazan no keitai 北満火山の形態 (Types of volcano in northern Manchuria)," Chirigaku hyōron, v. 14 (1938), no. 11, 1-15.

The distribution and types of volcano in northern Manchuria are described in their relation to geological structure, and the volcanoes are classified. This article is a general introduction to the study of volcanoes in this region.

309. Ōhashi Ryōichi 大橋良一, "Akita-shigai Takashimizu kyūryō ni okeru nijō no yokosuberi dansō to Tenchō jishin to no kankei 秋田市外高清水丘陵に於ける二條の横辷断層と天長地震との關係 (Two parallel fissures at Takashimizu hill near Akita and their bearing on the earthquake of the Tenchō era)," Chirigaku hyōron, v. 12 (1936), no. 9, 34-43.

This is a revision of an article by the same author concerning the Akita fault, which appeared in Chirigaku hyōron, v. 3 (1927), no. 8. It is based upon later field work.

310. Ōhashi Ryōichi 大橋良一, "Bunka shichinen no Oga jishin to Magurogawa dansō 文化七年の男鹿地震と鮪川斷層 (The 1810 Oga earthquake and Magurogawa fault)," Chirigaku hyōron, v. 4 (1928), no. 2, 56-73.

In part, a recapitualtion of information on the nature of the 1810 earthquake which affected the Oga peninsula in Akita prefecture. It is based on an examination of old documents. Ōhashi explains that faulting in the Mt. Kampu area caused the 1810 seismic disturbance.

311. Ōhashi Ryōichi 大橋良一, "Dewa chihō ni okeru Kōshinsei igo no chikei no henka ni tsuite 出羽地方に於ける更新世以後の地形の變化について (On Post-Miocene topographical changes in the Dewa district), "Chishitsugaku zasshi, v. 39, no. 465 (1932), 393-400.

The general structural geology and geomorphology of the Tōhoku coastal district fronting on the Sea of Japan are outlined and regional units explained and identified with maps. The major changes of the land and sea since mid-Tertiary times are summarized with four diagrammatic maps.

312. Ōhashi Ryōichi 大橋良一, "Dewa sammyaku no hatsuikushi to rikukai jōshōsetsu 出羽山脈の發育史と陸塊上昇説 (The geological history of the Dewa mountains and the theory of land-mass upheaval)," Chirigaku hyōron, v. 3 (1927), no. 2, 48-70.

A theoretical discussion of geotectonics accompanied by a brief topographic description. The Dewa mountains constitute a portion of the Tōhoku region.

313. Ōhashi Ryōichi 大橋良一, "Kita-Nihon no chishitsu-ku ni tsuite 北日本の地質區について (On the geological division of northern Japan), Chirigaku hyōron, v. 3 (1927), no. 4, 37-45.

A modification of von Richthofen's geological divisions.

314. Ōhashi Ryōichi 大橋良一, Oga hantō no chitai kōzō 男鹿半島の地帯構造 (Geological structure of the Oga peninsula) [= Nihon Gakujutsu Kyōkai hōkoku (Reports of the Japan Science Federation), v. 4], Tōkyō, 1928, 343-350, 2 illus.

The peculiar volcanic topography and landforms of coastal erosion are described in conjunction with a brief outline of the geological structure of this peninsular area in Akita prefecture.

315. Ōhashi Ryōichi 大橋良一, "Onsen-shiki kazan to karudera-shiki kazan 温泉式火山とカルデラ式火山 (Hotspring-type volcanoes and caldera-type volcanoes)," Chirigaku hyōron, v. 3 (1927), no. 12, 47-52.

After a general description of the geological characteristics of volcanoes, Ōhashi classifies Japan's volcanoes into the two broad categories indicated in the title.

316. Ōinoue Yoshichika 大井上義近, "Hompō no shuyō naru kōsen no bumpu ni tsuite 本邦の主要なる鑛泉の分布に就いて (On the distribution of the main mineral springs in Japan)," Chigaku zasshi, v. 43, no. 512 (1931), 559-562.

A geological classification of 316 mineral springs. Distributional factors are considered together with chemical and petrological criteria.

317. Okada Osamu 岡田修, "Kokukei no heimenkei ni tsuite 谷形の平面形に就いて (On valley topography and drainage patterns)," Chirigaku hyōron, v. 10 (1934), no. 10, 46-70.

A discussion of the relationship between valley topography and drainage patterns in an area of relatively gradual slopes in the southeastern portion of the Kantō district. Okada bases his article on observations made during topographic mapping surveys.

318. Okayama Toshio 岡山俊雄, "Akaishi sanchi no seppōmen 赤石山地の切峯面 (Summit levels of the Akaishi mountains)," Chirigaku hyōron, v. 4(1928), no. 8, 1-25.

In the first half of this article, the general geomorphological characteristics of the Akaishi range are described. The latter half comprises an analysis of the mountain landforms in conjunction with an interpretation of recent crustal movement; in this regard the author refers to his accompanying summit-level maps.

319. Okayama Toshio 岡山俊雄, "Chūō Nihon no shōkifukumen ibutsu to sore ni kanren-seru nisan no mondai 中央日本の小起伏面遺物とそれに關聯せる二,三の問題 (Remnants of low relief surfaces in central Japan and some problems related to them)," Chirigaku hyōron, v. 6 (1930), no. 11, 91-113.

Examples of low-relief surfaces common to the mountain districts of central Japan are shown and the origin and development of such topography is discussed.

320. Okayama Toshio 岡山俊雄, "Chūō Nihon sanchi no seppōmen oyobi kubun 中央日本山地の切峯面及び區分 (Summit levels and mountain division in central Japan)," Chirigaku hyōron, v. 6 (1930), no. 6, 1-21.

For the purposes of this geomorphological discussion, Okayama delimits central Japan as the region lying between the Tsuruga bay and Ise bay depression and the fossa magna. The writer determines physiographic divisions on the basis of a summit-level map he prepared. The question of peneplanation in central Japan is also explored in this article.

321. Okayama Toshio 岡山俊雄, "Kai no kuni Jizō Hōō-san shita no gyakudansō 甲斐國地藏鳳凰山下の逆斷層 (The reversed fault at the foot of Jizō and Hōō mountains in Kai province)," Chirigaku hyōron, v. 5 (1929), no. 11, 1-12.

The writer appraises map evidence and field observations indicating the possible existence of a reversed fault at the margin of a granite region in the mountains of Yamanashi prefecture.

322. Okayama Toshio 岡山俊雄, "Mino Echizen sanchi naibu no dansō chikei 美濃越前山地内部の斷層地形 (Fault topography of the interior part of the Mino-Echizen mountain-land)," Chirigaku hyōron, v. 7 (1931), no. 11, 20-42; no. 12, 31-58.

On the relationship between peneplanation and faulting in the Mino-Echizen highlands of west-central Honshū. Okayama concludes his detailed study with a general summary of the geological history of that region.

323. Okayama Toshio 岡山俊雄, "Suijun sokuryō kaisoku no kekka to chikei to no kankei 水準測量改測と地形との關係 (The relationship between the result of precise levellings and topography)," Chirigaku hyōron, v. 6 (1930), no. 7, 440-452.

A correlation of crustal movement and landform type based on detailed topographical measurements of vertical displacement.

324. Ōnishi Manjirō 大西萬次郎, "Chikaron yori mitaru Nihon rettō no kikō gaisetsu 地渦論より觀たる日本列島の機構概説 (Outline of the physical construction of the Japanese islands from the standpoint of the earth vortex theory)," Chigaku zasshi, v. 48, no. 564 (1936), 36-44; no. 565, 22-34, 10 maps.

A geotectonic hypothesis, known as the "earth vortex theory," was advanced by Dr. Fujiwara Sakuhei to explain the existence of arcuate landforms and island chains distributed on certain parts of the earth's surface, such as those occurring in the circum-Pacific region. In this paper the construction of the Japanese island arc is considered in terms of Fujiwara's theory.

325. Ōtani Hisao 大谷壽雄, "Bonchisetsu no gaiyō 盆地説の概要 (An outline of the basin theory)," Chirigaku hyōron v. 11 (1935), no. 9, 1-19.

An article founded on Ōtani's analysis of the geological structure of the Liao-tung peninsula in Manchuria, which he interprets as an agglomeration of structural basins. Ōtani presents a general theory which he believes applicable in explaining the nature of many landform types. He classifies the various basin forms and cites examples of major earth features which, he concludes, support his concept.

326. Ōtsuka Yanosuke 大塚彌之助, "Bōsō hantō ni okeru ichishō Dekken kōzō 房總半島に於ける一小デッケン構造. (A minor Decken structure on the Bōsō peninsula)," Chirigaku hyōron, v. 13 (1937), no. 3, 19-25.

A survey of abrasion platforms on the east side of the Awa mountains in reference to their correlation with the general fault system of the Bōsō peninsula, Chiba prefecture.

327. Ōtsuka Yanosuke 大塚彌之助, "Hyūga kaigan heiya no senshinsei shoki kara no chikaku undō to sore ni kansuru gimon 日向海岸平野の鮮新世初期からの地殻運動とそれに關する疑問 (Crustal movements from the early Pliocene period of the southeastern coastal plain in Kyūshū and doubts concerning them)," Chirigaku hyōron, v. 8 (1932), no. 2, 1-15.

A study of the landforms and geologic strata of the Hyūga coastal plain and an interpretation of the crustal movements which affected the neo-Tertiary series and terraces of that region.

328. Ōtsuka Yanosuke 大塚彌之助, "Ise-Shima chihō no chishitsu oyobi chikei kara suitei-sareru shuyō naru dansō 伊勢志摩地方の地質及び地形から推定される主要なる斷層 (Main fault-lines inferred from the geology and geomorphology of Ise and Shima regions)," Chirigaku hyōron, v. 4 (1928), no. 7, 1-22.

Ōtsuka discovered and studied many fault lines during his field surveys and, in this paper, he relates them to the pattern of major graben in the Ise region.

329. Ōtsuka Yanosuke 大塚彌之助, "Ko Yamazaki Hakase no Miura hantō Busan dansō no chishitsugakuteki kōsatsu 故山崎博士の三浦半島武山斷層の地質學的考察 (Geological consideration of the Busan fault of the late Dr. N. Yamazaki in the Miura peninsula)," Chirigaku hyōron, v. 11 (1935), no. 5, 49-56.

Ōtsuka's field observations supporting the arguments of the late N. Yamazaki claiming the existence of a fault scarp in the Busan area of Kanagawa prefecture.

330. Ōtsuka Yanosuke 大塚彌之助, "Miyazaki-ken Takanabe-machi fukin no chishitsugakuteki mondai 宮崎縣高鍋町附近の地質學的問題 (Geological problems in the vicinity of Takanabe village in Miyazaki prefecture)," Chirigaku hyōron, v. 6 (1930), no. 7, 496-522.

A survey of coastal stratigraphy and tectonics in southeastern Kyūshū.

331. Ōtsuka Yanosuke 大塚彌之助, "Nihon guntō ni mirareru chinseki rinne no aru mono 日本群島に見られる沈積輪廻の或るもの (A certain cycle of sedimentation in the Japanese archipelago)," Chirigaku hyōron, v. 8 (1932), no. 12, 1-28.

A correlation of orogenic movement and the cycle of sedimentation in neo-Tertiary times with remarks on the necessity of extending this form of research to cover the Mesozoic and Paleozoic eras.

332. Ōtsuka Yanosuke 大塚彌之助, Nihon no chishitsu kōzō 日本の地質構造 (Geological structure of Japan), Tōkyō, Dōbun Shoin, 1942, 3+7+272+8 pp.

This book is not specially concerned with the geology of Japan, as the title suggests, but rather is a general text on structural geology. However, references to the structure of the Japanese islands appear frequently. Ōtsuka is a leading Japanese geologist and this work merits the interest of the physical geographer as well as the student of geology.

333. Ōtsuka Yanosuke 大塚彌之助, "Nihon no kaigansen no hattatsu ni kansuru aru kangae 日本の海岸線の發達に關する或る考へ (Certain thoughts on the development of the Japanese coastline)," Chirigaku hyōron, v. 9 (1933), no. 10, 1-25.

The geomorphology of the Japanese coastline. Ōtsuka indicates some relationships between geological structure and marine erosion. He cites examples of coastal subsidence and discusses the Rias coast in Tōhoku. The study concludes with observations on the present tendencies of coastal development based on seismological and other evidence.

334. Ōtsuka Yanosuke 大塚彌之助, "Nihon-tō no chūsekiki shoki no kaigansen no henka to sono engan rikudana ni hattatsu-suru chindekikoku ni kansuru jidaiteki kōsatsu sonota 日本島の沖積期初期の海岸線の變化とその沿岸陸棚に發達する沈溺谷に關する時代的考察其の他 (Changes in the coastlines of the Japanese islands in the early alluvial epoch, some chronological considerations on the drowned valleys on the continental shelves surrounding them, and other matters)," Chirigaku hyōron, v. 7 (1931), no. 6, 1-12.

A consideration of the coastal geomorphology and inshore submarine geology of the Japanese archipelago. Particular reference is made to the distribution and characteristics of marine alluvial deposits and shell mounds as clues to the chronology of coastal development.

335. Ōtsuka Yanosuke 大塚彌之助, "Noto hantō Ōchi chikō 能登半島邑知地溝 (The Ōchi graben of the Noto Peninsula)," Chirigaku hyōron, v. 10 (1934), no. 8, 1-26.

The Ōchi graben cuts across the neck of the Noto peninsula, Ishikawa prefecture, in north-central Japan. This is a geological study of this graben and its surrounding area. It includes a fairly detailed description of marine terraces along the southeastern coast of the Noto peninsula.

336. Ōtsuka Yanosuke 大塚彌之助, "Ōchi chikō tōhokubu no chishitsu kōzō 邑知地溝東北部の地質構造 (Geological structure of the northeastern part of the Ōchi graben)," Chirigaku hyōron, v. 11 (1935), no. 11, 1-25.

The last of three articles by Ōtsuka on the Ōchi graben and its vicinity.

337. Ōtsuka Yanosuke 大塚彌之助, "Ōiso chikai o chūshin to shita saishin chishitsu jidai no chishi 大磯地塊を中心とした最新地質時代の地史 (Recent geologic history of the Ōiso block and its adjacent area)," Chirigaku hyōron, v. 6 (1930), no. 1, 1-20; no. 2, 11-41.

The Ōiso block is a structural unit forming a portion of the southern Kantō region. In this paper the author is chiefly concerned with interpreting the geotectonic disturbances which have affected this area in Kanagawa prefecture during the recent geological period.

338. Ōtsuka Yanosuke 大塚彌之助, "Shima Isobe-mura fukin no shizen-chirigakuteki chishi no ichibu no kenkyū 志摩磯部村附近の自然地理學的地史の一部の研究 (A study of one part of the natural geological history of the region around Isobe village, Shima district)," Chirigaku hyōron, v. 4 (1928), no. 2, 41-55; no. 3, 48-61.
Mainly a discussion of the distribution and nature of diluvial deposits in this district. From the characteristics of the layers, the period of deposition and/or crustal activity is determined. Strata from diluvium to alluvium are termed in this paper "pre-Shima layers."

339. Ōtsuka Yanosuke 大塚彌之助, "Shimantogawa no ryūiki ni okeru kyokuryū no kenkyū 四萬十川の流域に於ける曲流の研究 (A study of meanders in the drainage area of the Shimanto river)," Chirigaku hyōron, v. 3 (1927), no. 5, 27-49.
Marked development of incised meanders in the course of the Shimantogawa and its tributary streams is described and the cause of their formation considered geologically.

340. Ōtsuka Yanosuke 大塚彌之助, "Taiheiyō-gan ni okeru kōsekisei shoki no kaishin to chikeimen no bumpu (yohō) 太平洋岸に於ける洪積世初期の海進と地形面の分布(豫報) (Diluvial marine invasion on the Pacific coast and the distribution of terraces — a preliminary report)," Chirigaku hyōron, v. 9 (1933), no. 12, 33-39.
In this study Ōtsuka divided the Quarternary period into six phases and used that time breakdown to classify the various erosion surfaces of Japan's east coast. The study is based upon fossil identification.

341. Ōtsuka Yanosuke 大塚彌之助, "Tōkyō no shitamachi 東京の下町 (Downtown Tōkyō)," Chiri kyōiku, v. 20 (1934), no. 1, 16-21.
A simplified explanation of the geological structure underlying downtown Tōkyō with brief notes on the topography of that area.

342. Ōtsuka Yanosuke 大塚彌之助, "Yamanashi-ken nansei-bu Fujimiyama dansōgai no chishitsugakuteki kōsatsu 山梨縣南西部富士見山斷層崖の地質學的考察 (A geologic study of the Fujimiyama fault scarp in the southwestern part of Yamanashi prefecture)," Chirigaku hyōron, v. 14 (1938), no. 12, 1-16.
A geological study of step-fault topography along the southern sector of the Itoigawa-Shizuoka fault line which marks the western margin of Naumann's fossa magna.

343. Ōtsuka Yanosuke 大塚彌之助 and Mochizuki Katsumi 望月勝海, Chikei hattatsushi 地形發達史 (History of the development of landforms), in Iwanami kōza: chiri (Iwanami lectures on geography), Tōkyō, 1931-34, 69 pp., 24 maps and illus.
Following an introductory outline of the structural geology and geomorphology of Japan as a whole, specific landforms and physiographic regions are considered. The discussions are not systematically presented, although one lecture deals with the entire series of major structural units forming the Japanese archipelago. Other lectures treat briefly such subjects as: the Bōsō peninsula, the outer zone of southwest Japan, the fossa magna and its adjacent area, etc.

344. Ozawa Yoshiaki 小澤儀明, "Akiyoshidai no chishi to chikei to chikasui 秋吉臺の地史と地形と地下水 (Historical geology, geomorphology, and the under-ground water in Akiyoshidai)," Chirigaku hyōron, v. 1 (1925), no. 1, 32-49; no. 2, 34-44; no. 3, 25-39.
Akiyoshidai is a plateau of characteristic karst topography in Yamaguchi prefecture and constitutes the largest area of its type in Japan. All stages in the cycle of karst development observed there are discussed and brief references are made to the historical geology of the entire Chūgoku district.

345. Ozawa Yoshiaki 小澤儀明, "Seinan Nihon naitai ni okeru daisanki izen no chikaku hendō 西南日本内帶に於ける第三紀以前の地殼變動 (Pre-Tertiary crustal movements in the inner zone of Southwest Japan)," Chirigaku hyōron, v. 2 (1926), no. 2, 43-57.

346. Ozegahara Sōgō Gakujutsu Chōsadan 尾瀬ヶ原總合學術調査團 (The Ozegahara Joint Scientific Research Group), Ozegahara 尾瀬ヶ原 (The Ozegahara moor), Tōkyō, Nihon Gakujutsu Shinkōkai, 1953, 16+841 pp., 25 photos.
The Ozegahara moor, situated at an elevation of 1,400 meters in the mountains on the boundary of Fukushima Gumma, and Niigata prefectures, is a unique remnant of an extensive prehistoric moorland in central Japan. The proposed flooding of this area in conjunction with dam construction raised a strong protest among Japanese scientists and nature lovers and resulted in this research conducted under the sponsorship of the Ministry of Education in the period 1950-1952. Of 55 articles, 15 are in English, and 12 were written by geographers, treating such topics as: landforms, drainage, plant and animal distribution, limnological conditions, etc.

347. Rikuchi Sokuryōbu 陸地測量部 (Imperial Land Survey), "Fujisan no hyōkō ni tsuite 富士山の標高に就て (Regarding the altitude of Mt. Fuji)," Chigaku zasshi, v. 42, no. 493 (1930), 35-39.
The exact height of Mt. Fuji was long a subject of argument. This brief statement was published by the Imperial Land Survey to explain the status of information on that subject and to give the official measurements.

348. Sakata Isao 坂田勲, Kagawa-ken Chishitsu gaikan 香川縣地質概觀 (Outline of the geology of Kagawa prefecture), Takamatsu, Kagawa-ken Kyōiku Tosho Kabushiki Kaisha, 1932, 2+3+77+3 pp., 6 photos and 44 illus.
This survey of the regional geology of Kagawa prefecture is the second in a series of provincial studies conducted by staff members of Kagawa Normal School. The treatment is relatively simple, with stress upon the historical geology of the area. With bibliography.

349. Sakita Ryūji 崎田龍二, "Tateyama Gozenzawa no kokutō ni yokotawareru hyōkai ni tsuite 立山御前澤の谷頭に横はれる氷塊に就て (On the ice deposits at the head of Gozenzawa valley on Mt. Tateyama)," Chirigaku hyōron, v. 7 (1931), no. 9, 44-57.
A description of the Gozenzawa Kar — an ice accumulation some 1,700 meters in length on the upper slopes of Mt. Tateyama in Toyama prefecture. Sakita points out the features of that deposit which favor its classification as a rudimentary glacier.

350. Sasa Yasuo 佐々保雄, "Iwate-ken Kuji chihō no chishitsu ni tsuite 岩手縣久慈地方の地質について (On the geology of the Kuji district, Iwate prefecture)," Chishitsugaku zasshi, v. 39, no. 466 (1932), 401-430; no. 467, 481-501; no. 468, 552-580.
A detailed report on the stratigraphy of this district. Sasa earlier published a summary of this material in collaboration with Saeki.

351. Sasakura Masao 笹倉正夫, "Himekawa jōryū engan no chishitsu gaihō 姫川上流沿岸の地質概報 (An outline report on the geology of the upper Himekawa district)," Chikyū, v. 17 (1933), no. 1, 33-44; no. 2, 11-26, 1 map.
Report on the findings of detailed geological field surveys made in the upper drainage area of the Himekawa, Shizuoka prefecture, along an important geo-tectonic line of weakness.

352. Satō Denzō 佐藤傳藏, "Akiyoshidai karusuto (sekkaigan keikan) 秋吉臺カルスト(石灰岩景觀) (The Akiyoshidai karst: limestone survey)," Chigaku zasshi, v. 40, no. 474 (1928), 7-25; no. 475, 36-44; no. 476, 48-52; no. 477, 48-50; no. 478, 50-54; v. 41, no. 479 (1929), 39-45, 8 illus.
Studies of the Akiyoshidai karst landscape in Japan and a general explanation of the development of karst topography.

353. Satō Denzō 佐藤傳藏, "Shimokita hantō 下北半島 (The Shimokita peninsula)," Chigaku zasshi, v. 28, no. 335 (1916), 1-9; no. 336, 29-42, 9 illus.
A geological and geomorphological study of the Shimokita peninsula at the extreme northern end of Honshū. Emphasis is upon petrology and stratigraphy.

354. Satō Denzō 佐藤傳藏, "Tsushima no chihō oyobi chishitsu 對馬の地貌及地質 (The topography and geology of Tsushima)," Chigaku zasshi, v. 25, no. 300 (1913), 33-40; v. 26, no. 302 (1914), 18-27.
A capable description of the geology and topography of the Tsushima islands.

355. Satō Hisashi 佐藤久, "Kazantō Torishima 火山島鳥島 (The volcanic island, Torishima)," Chirigaku hyōron, v. 21 (1949), no. 12, 19-27.
An outline of the geomorphology of Torishima, a double volcanic cone, in the Kazan archipelago south of Japan.

356. Satō Hisashi 佐藤久, "Nishi Nyū Ginia hokugan no chikei (1) — toku ni dansō to sangoshō ni tsuite 西ニューギニア北岸の地形(上) 特に斷層と珊瑚礁に就いて (The topography of the northern coast of Western New Guinea (1) — especially regarding faults and coral reefs)," Tōkyō Daigaku chirigaku kenkyū, no. 2 (1952), 67-91.
Field observations and a study of aerial photography led to this general geomorphological classification of the major landforms of the northern coast of western New Guinea. The dominant topographic features as well as the unusual geomorphological forms in five physiographic regions are described.

357. Satō Hisashi 佐藤久, "Yōganryū no chikei bunrui — hompō kazantai no chikeigakuteki kenkyū (1) — 熔岩流の地形分類—本邦火山體の地形學的研究(一) (A geomorphological classification of lava flows — a study of volcanic topography in Japan (1))," Tōkyō Daigaku chirigaku kenkyū, no. 1, 1950, 114-132.
Satō identifies three basic and nine metamorphosed types of lava in terms of their viscosity. The surface forms of lava flows are also classified. Many illustrations are presented; they are based upon the writer's field study and stereoscopic analysis of air photographs.

358. Shikama Tokio 鹿間時夫, Nihon kaseki zufu 日本化石圖譜 (Illustrated dictionary of Japanese fossils), Kyōto, Nihon Kōbutsu Shumi no Kai, 1943, 45+279 pp.
An outline paleontography of Japan, including a description of selected fossil finds from Korea, Manchuria, and the South Seas. Geological and paleontological time scales are appended. With Latin index.

359. Shimomura Hikoichi 下村彦一, "Hiroshima-ken Takada-gun Kamine fukin no chibō 廣島縣高田郡上根附近の地貌 (Landforms of the district around Kamine, Takada district, Hiroshima prefecture)," Chirigaku hyōron, v. 4 (1928), no. 11, 55-65.
A study of fault topography, the influence of fault features upon stream capture, etc., in this coastal district of southern Chūgoku.

360. Shimomura Hikoichi 下村彦一, "Hompō no kaigan dankyū ni kansuru ichini no shikō 本邦の海岸段丘に關する一二の私考 (Some considerations concerning coastal terraces in Japan)," Chirigaku hyōron, v. 6 (1930), no. 7, 298-322.
Observations on the nature and structural origin of marine terraces located along the west coast of the Kii peninsula and at Cape Muroto in Shikoku.

361. Shinano Kyōikukai Shimo-Ina Bukai 信濃教育會下伊那部會 (The Shinano Education Association, Shimo-Ina Branch), Shimo-Ina-gun chishitsushi 下伊那郡地質誌 (Regional geology of Shimo-Ina district), Tōkyō, Kokon Shoin, 1925, 208+20 pp., 27 illus.
A geomorphological survey of a district in Nagano prefecture, with one section on petrology.

362. Shimomura Hikoichi 下村彦一, "Nihon guntō no chikeiku 日本群島の地形區 (Geomorphological divisions of the Japanese Islands)," Chirigaku hyōron, v. 2 (1926), no. 12, 1-13; v. 3 (1927), no. 4, 65-74; no. 9, 45-55.
Shimomura bases his physiographic divisions largely on the delineation of the major fault structures in Japan. The geomorphological provinces delimited are discussed comparatively.

363. Sugimura Arata 杉村新, "Kantō chihō shūhen no kaitei dankyū sonota ni tsuite 關東地方周邊の海底段丘その他に就いて (On the submarine terraces along the coast of the Kantō region and other matters)," Chirigaku hyōron, v. 23 (1950), no. 1, 10-16.
Observations on submarine terraces along the Pacific coast of central Japan which are believed to have been formed at the time of the lowered sea level caused by the maximum glaciation during the fourth glacial period.

364. Sugiyama Masahide 椙山正英, "Kibi kōgen no seppōmen to sekkokumen 吉備高原の切峯面と切谷面 (Summit heights and river levels on the Kibi plateau)," Chirigaku hyōron, v. 7 (1931), no. 4, 25-39.
The Kibi plateau in Chūgoku is perhaps the best example of an elevated peneplain to be found in Japan. This is a geomorphological study of that upland with particular attention being paid to stream action. With maps compiled by the author.

365. Sugiyama Masahide 椙山正英, "Mikawa kōgen no heitando ni tsuite 三河高原の平坦度に就いて (Concerning the levelness of the Mikawa highland)," Chirigaku hyōron, v. 6 (1930), no. 8, 27-44.
The Mikawa highland, located in south-central Honshū, is identified geologically as an uplifted surface of peneplanation. This is a morphometric survey of that upland as interpreted from 1:50,000 topographic maps.

366. Sugiyama Masahide 椙山正英, "Mimasaka Naginosen nanroku no colluvium to sono fukin no chikei ni tsuite 美作那岐山南麓のcolluviumとその附近の地形に就て (On the colluvium along the southern foot of Mt. Nagi in Mimasaka and the topography in its vicinity)," Chirigaku hyōron, v. 9 (1933), no. 8, 64-76.
A geomorphological discussion of the colluvial fans developed along the base of a mountain of subdued topography which forms part of the water divide in Chūgoku.

367. Sugiyama Masahide 椙山正英, "Ryōtō hantō junheigen no heitando 遼東半島準平原の平坦度 (Evenness of the peneplain in the Liaotung peninsula)," Chirigaku hyōron, v. 8 (1932), no. 1, 24-37.
A morphometric study of the Kwantung territory. 1:25,000 topographical maps are analysed. Sugiyama refers to similar map methods he utilized when interpreting the form of the Mikawa plateau in Aichi prefecture, Japan.

368. Suzuki Jun 鈴木醇, "Ōboke kyōkoku ni tsuite 大崩壊峡谷に就いて (On the Ōboke gorge)," Chirigaku hyōron, v. 2 (1926), no. 4, 1-17.
A reconsideration of the geomorphological origin of the gorge occupied by the Yoshino river in the mountains of central Shikoku. From an analysis of river terraces, Suzuki explains that the Ōboke canyon resulted from the deepening of an antecedent valley rather than being formed by faulting as had been generally theorized.

369. Suzuki Jun 鈴木醇, "Tajima Gembudō fukin no chikei ni tsuite 但馬玄武洞附近の地形に就いて (On the topography around the Gembudō cave in Tajima province)," Chirigaku hyōron, v. 1 (1925), no. 4, 42-51.
The Gembudō is a basaltic cave in Kyōto prefecture noted for beautifully developed columnar structure. Suzuki outlines the geomorphology of the coastal area in the vicinity of this cave with emphasis upon an interpretation of past crustal disturbances there.

370. Suzuki Kōichi 鈴木好一 and Inagaki Seiji 稲垣誠二, "Kumano kaigan sanchi oyobi sono shūenchiiki no chikei to chikei hattatsushi 熊野海岸山地及其周緣地域の地形と地形發達史 (The morphology and its development in and around the Kumano coastal mountain region)," Chirigaku hyōron, v. 9 (1933), no. 11, 19-46; no. 12, 40-49; v. 10 (1934), no. 3, 45-71; no. 4, 59-89.
Written by two geologists, this article is a geomorphological study of the Kumano coastal mountains, Tertiary in origin, which extend along the southeastern coast of the Kii peninsula, mostly in Mie prefecture.

371. Tada Fumio 多田文男, "Enshūnada engan Yokosuka fukin no kaigan chikei 遠州灘沿岸横須賀附近の海岸地形 (The morphology of the Enshūnada coast near Yokosuka)," Chirigaku hyōron, v. 19 (1943), no. 12, 1-7.
Enshūnada is the local name given the inshore Pacific waters off Aichi and Shizuoka prefectures. In this paper Tada presents a survey of the coastal topography in the vicinity of Yokosuka, east of Hamamatsu, where features are characteristic of a shoreline of emergence.

372. Tada Fumio 多田文男, "Gyakudansō ni yoru dansōgai no ichirei 逆斷層による斷層崖の一例 (An example of fault scarps formed by reverse faulting)," Chirigaku hyōron, v. 6 (1930), no. 7, 44-59.
A trigonometric analysis of horizontal displacement along the Gōnomura fault undertaken in order to correlate its theoretical and actual forms.

373. Tada Fumio 多田文男, "Heikō kyokusen no kenkyū ni yotte shirieta Kitsugawa no ryūro no hensen 平衡曲線の研究によって知り得た木津川の流路の變遷 (Changes in the course of the Kitsugawa river as discovered through a study of its profiles and shiftings)," Chirigaku hyōron, v. 5 (1929), no. 3, 1-29.
Primarily concerned with geotectonics, this is a study of the profile and course of the Kitsugawa, a braided stream flowing from the Iga highlands in Mie prefecture. This article is an outgrowth of Tada's earlier investigations of Mt. Kasagi, a "block mountain" in that region.

374. Tada Fumio 多田文男, "Kagoshima-ken Shirasu daichi no gakekuzure -- futatabi horidasareta maisekikoku no ichirei 鹿児島縣シラス台地の崖崩れ — 再び掘り出された埋積谷の一例 (Headward erosion along the Shirasu valley in Kagoshima prefecture -- an example of an underground valley newly uncovered)," Tōkyō Daigaku

chirigaku kenkyū, no. 2 (1952), 27-34.
Examines the effects of unusually rapid erosion in the Shirasu section of the ash plateau of southern Kagoshima prefecture, caused by heavy typhoon rains in 1949.

375. Tada Fumio 多田文男, "Kita-Manshū heiya no chikei -- kikō shitsujunka ni tomonau chikei henka no rei -- 北満洲平野の地形 — 氣候濕潤化に伴う地形變化の例 --(The topography of the north Manchurian plain -- an example of topographic transformation caused by climatic change)," Tōkyō Daigaku chirigaku kenkyū, no. 1 (1950), 133-145.
Tada led expeditions in northern Manchuria in 1943 and 1944. His study of the climate, landforms, soils, and vegetation of that region led him to conclude that the now humid interior basin was once an arid plain. Here he presents considerable evidence to support the theory of cyclical climatic change.

376. Tada Fumio 多田文男, "Mammō kōkai chihō no chikei ni tsuite 満蒙交界地方の地形に就いて (On the landforms of the border area between Manchuria and Mongolia)," Chirigaku hyōron, v. 1 (1925), no. 5, 45-72.
A scholarly monograph on the general geomorphology and topography of the Manchurian and Mongolian border zone. Includes the author's observations based on an expedition to this region in 1924. With numerous references to research on this subject by other persons.

377. Tada Fumio 多田文男, "Nan'yō shotō no kaishoku dankyū 南洋諸島の海蝕段丘 (Coastal terraces of the Micronesian islands)," Chirigaku hyōron, v. 2 (1926), no. 5, 1-19.
Elevated marine terraces on Saipan and Yap are investigated and their relation to present coral reefs discussed.

378. Tada Fumio 多田文男, "Sanchō no kōdo to kifukuryō to no kankei narabi ni kore yori mitaru sanchi no kaisekido ni tsuite 山頂の高度と起伏量との関係並に之より見たる山地の開析度に就いて (On the relationship between the altitude and relief energy of mountains and on degrees of dissection in mountain areas as seen in this relationship)," Chirigaku hyōron, v. 10 (1934), no. 11, 1-29.
A theoretical consideration of the erosion cycle in reference to mountain topography. The degree of dissection in twelve highland regions in Japan and Manchuria is considered. Diagrammatic maps show the relationship between the degree of relief and altitudinal differences in summit and valley elevations.

379. Takizawa Teikichi 瀧澤鼎吉, "Asamayama sono ta no kazan sokusen 淺間山其他の火山側線 (Slope curvi of the volcano Asama and some other examples)," Chirigaku hyōron, v. 7 (1931), no. 10, 1-13.
This paper is a sequel to the article next listed in this bibliography. Takizawa calculates formulae giving the slope curves for the following volcanoes: Asama, Yatsugadake, Komagatake, Iwatesan, Daisen, Iwakisan, and Ezo Fuji.

380. Takizawa Teikichi 瀧澤鼎吉, "Fuji-san no sokusen 富士山の側線 (Slope curvi of Mt. Fuji)," Chirigaku hyōron, v. 4 (1928), no. 7, 49-55.
The profile contours of Mt. Fuji are measured on topographic maps and it is pointed out that altitude may be regarded as an exponential function of horizontal distance.

381. Tanabe Ken'ichi 田邊健一, "Nihon kaigan sakyū no keitaiteki bunrui narabi ni tochi riyō 日本海岸砂丘の形態的分類並びに土地利用 (A morphological classification of coastal sand dunes in Japan and their utilization)," Chirigaku hyōron, v. 17 (1941), no. 5, 41-69.
The core of this article is concerned with the form and distribution of coastal dunes in Japan, and the types of land use characteristic of dune areas, as determined by an examination of topographic maps. Appended is the writer's small-scale distribution map of sand dunes which includes delimitation of areas where associated forms of land-tied islands occur. Attention is given to the problem of dune forestation.

382. Tanabe Ken'ichi 田邊健一, "Noto hantō no dansō chikei narabi ni kaitei chikei 能登半島の斷層地形並に海底地形 (Fault topography and submarine topography at the Noto peninsula)," Chirigaku hyōron, v. 16 (1940), no. 7, 41-54; no. 8, 34-44.
Geomorphology of the fault-block topography of the Noto peninsula. The final portion of this article is devoted to a discussion of submarine topography off that coast with particular regard to bottom deposits.

383. Tanaka Kaoru 田中薫, "Yari-Hotaka rempō no hyōshoku chikei ni tsuite 鎗穂高連峯の氷蝕地形に就いて (On the Rundhocker in the Yari-Hotaka range)," Chirigaku hyōron, v. 17 (1941), no. 8, 1-18.
A survey of glacial topography in the Yari-Hotaka area of the north Japanese Alps. Tanaka attempts to determine the type, distribution, and extent of glaciation in that region of central Japan. With thirty excellent photographs.

384. Tanaka Kaoru 田中薫 and Kano Tadao 鹿野忠雄, "Taiwan Nankotaisan san-i ni okeru hyōshoku chikei ni tsuite 臺灣南湖大山山彙に於ける氷蝕地形に就いて (On the glacial topography of the Nankotaisan mountains in Formosa)," Chirigaku hyōron, v. 10 (1934), no. 3, 1-22.
The glacial landforms in the northern mountains of Formosa are described and the authors make general comparisons with the glaciated topography of other areas in Formosa and in Japan proper. The influence of rock type upon the formation of cirques is emphasized and speculation on the height of the snowline at the time of glaciation is presented.

385. Tanaka Toyoji 田中豊治, "Oki Dōgo no kazan chikei 隠岐島後の火山地形 (The volcanic landforms of Dōgo, Oki islands)," Chirigaku hyōron, v. 22 (1950), no. 11, 16-24.
A paper on the geomorphology of Dōgoshima, a volcanic island in the Oki group north of Chūgoku.

386. Tanakadate Hidezō 田中館秀三, "Komagatake no fusekiryū ni kansuru ichikōsatsu 駒ヶ嶽の浮石流に關する一考察 (Some considerations on the pumice flow at Komagatake)," Chirigaku hyōron, v. 6 (1930), no. 7, 387-396.
Treats the cause and nature of a pumice flow at the major volcano in southern Hokkaidō.

387. Tanakadate Hidezō 田中館秀三, "Okutango jishin no sai kaigan no shōkō undō 奥丹後地震の際海岸の昇降運動 (The vertical movement of the coastal region at the time of the Okutango earthquake)," Chigaku zasshi, v. 39, no. 465 (1927), 1-11; no. 466, 22-35.
In this paper Tanakadate presents evidence which he hopes will support his theory that gradual coastal subsidence is offset eventually by a sudden seismic upheaval.

388. Tayama Risaburō 田山利三郎, Bōsō hantō no chikei, toku ni shinshokumen no taihi ni tsuite 房總半島の地形特に侵蝕面の對比に就いて (Regarding the topography of the Bōsō peninsula, especially the correlation of erosion surfaces) [=Saitō Hōonkai Gakujutsu Kenkyū Sōmubu 齋藤報恩會學術研究總務部 (Saitō Requital Foundation, General Affairs Section for Scientific Research), Gakujutsu kenkyū hōkoku 學術研究報告 (Scientific research reports), no. 9], Sendai, 1930, 64+16 pp., 21 figures and 12 tables.
An attempt to correlate erosion surfaces quantitatively on the basis of the degree of dissection. Concerns an area of terraces along the eastern margin of the Kantō structural basin on the Bōsō peninsula, east of Tōkyō.

389. Tayama Risaburō 田山利三郎, "Chikeizujō ni miru saikin no chikaku undō no nisanrei ni tsuite 地形圖上に見る最近の地殼運動の二・三例に就いて (On some examples of recent crustal movements interpreted from topographic maps)," Ōtsuka Chiri Gakkai rombunshū, no. 1 (1933), 1-18.
A geomorphological study of river terraces in the Toyooka basin (Kyōto prefecture), on the Bōsō peninsula (Chiba prefecture), and along the Kitakami coast in Tōhoku, in an effort to cast light on post-Tertiary crustal movement in Japan.

390. Tayama Risaburō 田山利三郎, Kita-Izu jishin to chishitsu kōzō to no kankei 北伊豆地震と地質構造との關係 (The relation between the north Izu earthquake and geological structure) [=Saitō Hōonkai Gakujutsu Kenkyū Sōmubu 齋藤報恩會學術研究總務部 (Saitō Requital Foundation, General Affairs Section for Scientific Research), Gakujutsu kenkyū hōkoku 學術研究報告 (Scientific research reports), no. 11], Sendai, 1931, 1+54 pp., 32 photos., 5 illus.
Faulting, landslides, and other phenomena of crustal dislocation which accompanied the north Izu earthquake of 1930 are studied in regard to the general geological structure of the affected area in Shizuoka and Kanagawa prefectures.

391. Tayama Risaburō 田山利三郎, "Kitakami sanchi higashi shamen no kaigan dankyū ni tsuite 北上山地東斜面の海岸段丘に就いて (On the coastal terraces on the eastern margin of the Kitakami mountains)," Chirigaku hyōron, v. 7 (1931), no. 5, 1-26.
A study of well-developed coastal terraces on the eastern foot-slopes of the Kitakami mountains in Iwate prefecture as an aid in the analysis of the complicated crustal movements which have affected that region.

392. Tayama Risaburō 田山利三郎, Kitakami sanchi no chikeigakuteki kenkyū 北上山地の地形學的研究 (Geomorphological study of the Kitakami mountains) [=Saitō Hōonkai Gakujutsu Kenkyū Sōmubu 齋藤報恩會學術研究總務部 (Saitō Requital Foundation, General Affairs Section for Scientific Research), Gakujutsu kenkyū hōkoku 學術研究報告 (Scientific research reports), no. 20], Sendai, 1935, 30 pp., 21 illus., 9 tables.
The degree of dissection in the Kitakami and Abukuma mountains of Tōhoku was determined from map analysis (1:200,000) and a correlation made between these two regions.

393. Tayama Risaburō 田山利三郎, Oguni jishin oyobi sono shin'iki chihō no chikei chishitsu ryakuhō 小國地震及びその震域地方の地形地質略報 (The Oguni earthquake and a brief report on the landforms and geology of the earthquake area) [=Saitō Hōonkai Gakujutsu Kenkyū Sōmubu 齋藤報恩會學術研究總務部 (Saitō Requital Foundation, General Affairs Section for Scientific Research), Gakujutsu kenkyū hōkoku 學術研究報告 (Scientific research reports), no. 16], Sendai, 1932, 1+26 pp., 3 maps, 12 photos., 9 illus.
A seismological discussion of the nature and effects of a severe earthquake which occurred in the Shimohei district of Iwate prefecture.

394. Tayama Risaburō 田山利三郎, "Shiobara kazan higashi shamen oyobi Shiobara bonchi ni hattatsu-suru dankyū ni tsuite 鹽原火山東斜面及び鹽原盆地に發達する段丘に就て (On the terraces developed on the eastern slope of Shiobara volcano and in the Shiobara basin)," Chirigaku hyōron, v. 5 (1929), no. 7, 1-19.
General discussion of the development of landforms in this region. A table giving the heights of terraces is shown in connection with a correlation of tectonic lines and their influence upon present topographic forms.

395. Tayama Risaburō 田山利三郎, Tango Tajima jishin chiiki no chitai kōzō to saikin chishi 丹後但馬地域の地帯構造と最近地史 (Structure and recent geological history of the region of the Tango and Tajima earthquakes) [=Saitō Hōonkai Gakujutsu Kenkyū Sōmubu 齋藤報恩會學術研究總務部 (Saitō Requital Foundation, General Affairs Section for Scientific Research), Gakujutsu kenkyū hōkoku 學術研究報告 (Scientific research reports), no. 6], Sendai, 1928, 32+44 pp., 7 illus.
Mainly based on topographic map analysis, this is an interpretation of the geological characteristics of the Kinosaki (Tajima) and Tango regions of the Kinki district, affected by major earthquakes in May, 1925, and March, 1927.

396. Tayama Risaburō 田山利三郎, "Waga Nan'yō ni okeru sangoshō no bumpu (yohō) 我が南洋に於ける珊瑚礁の分布（豫報） (The distribution of coral reefs in the South Seas -- a preliminary report)," Chirigaku hyōron, v. 10 (1934), no. 7, 35-53.
Ten types of coral formation are classified and described and their distribution in the Marianas, Caroline, Marshall, and other Micronesian islands discussed. Tayama led an expedition to these regions of the Pacific in 1933-34.

397. Tayama Risaburō 田山利三郎 and Aoki Renjirō 青木廉二郎, Kantō bonchi toku ni seihembu no chikei oyobi chishitsu ni tsuite 關東盆地特に西邊部の地形及地質に就て (On the landforms and geology of the

Kantō basin, especially of its western margin) [= Saitō Hōonkai Gakujutsu Kenkyū Sōmubu 齊藤報恩會學術研究總務部 (Saitō Requital Foundation, General Affairs Section for Scientific Research), Gakujutsu kenkyū hōkoku 學術研究報告 (Scientific research report), no. 8], Sendai, 1930, 13 pp., 9 illus.

The aim of this paper is primarily to correct errors made in an earlier article on this subject published by Tayama and R. Yabe in the Chirigaku hyōron in 1927. New information led to a revised concept of certain aspects of the geomorphology of western Kantō.

398. Tōki Ryūshichi 東木龍七, "Chikei to kaizuka bumpu yori mitaru Kantō teichi no kyūkaigansen 地形と貝塚分布より見たる關東低地の舊海岸線 (The former shoreline of the Kantō lowland, studied in terms of its topography and the distribution of shell-mounds)," Chirigaku hyōron, v. 2 (1926), no. 7, 45-55; no. 8, 27-46; no. 9, 24-52.

Tōki reconstructs the old shoreline of south Kantō following a survey of the micro-topography and an investigation of shell-mounds in that area. Indications are that the present lower valleys were embayments in the near geologic past.

399. Tōki Ryūshichi 東木龍七, "Kagan dankyū no hitaishōteki haichi to sono seiin 河岸段丘の非對象的配置と其の成因 (Asymmetric arrangement of river terraces and their origin)," Chirigaku hyōron, v. 5 (1929), no. 5, 44-80; no. 6, 31-54; no. 7, 20-38; no. 8, 15-41; no. 9, 23-35; no. 10, 50-58; no. 11, 37-54; no. 12, 25-38; v. 6 (1930), no. 1, 49-63; no. 2, 42-60; no. 3, 42-54.

This series of papers comprises the results of an elaborate investigation of stream terraces conducted over a period of several years. Tōki presents many examples of asymmetric terrace arrangement and analyzes the causes of their formation on the basis of their peculiar characteristics and site influences. The author's aim is to show the significance of terrace interpretation in detecting crustal disturbances too minute to be determined in the study of other relief features. This method of analysis will be of value to the geomorphologist particularly concerned with plains and diluvial uplands.

400. Tōki Ryūshichi 東木龍七, "Kantō heiya no bichikeigakuteki kenkyū 關東平野の微地形學的研究 (Micromorphologic studies in the Kantō plain)," Chirigaku hyōron, v. 6 (1930), no. 9, 31-68; no. 10, 49-83.

Tectonic movement in the Kantō lowland was investigated through microtopographic observations. Displacement diagrams were drawn using data derived from a detailed study of erosion surfaces. The general geomorphologic development of Kantō landforms was considered on the basis of this specific data.

401. Tōki Ryūshichi 東木龍七, "Kantō heiya no keitai keitō, undōmen oyobi undō jikusen — shinshokumen kenkyū no ōyō, heiya keitaigaku— 關東平野の形態系統，運動面及運動軸線 — 侵蝕面研究の應用，平野形態學 — (Morphology of the Kantō plain, its movement surfaces, and movement axis -- application of the study of erosion surface; morphology of plains --)," Chigaku zasshi, v. 42, no. 491 (1930), 6-17, 7 maps.

A unique morphological study of the Kantō or Tōkyō plain. The first portion of the article is devoted to classifying the Kantō into morphological regions. The core of the study is an explanation of the development of these regions based on the author's theory of erosion and deposition surfaces, with special attention being given to the relationship between erosion surfaces and earth movement.

402. Tōki Ryūshichi 東木龍七, "Nihon guntō no kōsekisō kyūryōchi ni okeru shinshokumen no keitai no sanshu no keishiki 日本群島の洪積層丘陵地に於ける侵蝕面の形態の三種の型式 (Three types of erosion surface in the diluvial hills of the Japanese archipelago)," Chirigaku hyōron, v. 4 (1928), no. 10, 60-70.

The diluvial uplands of Japan are classified according to their topographical characteristics and with reference to their processes of formation.

403. Tōki Ryūshichi 東木龍七, "Nihon Naikai seiiki Suōnada nambu no seiinron 日本内海西域周防灘南部の成因論 (On the origin of the southern part of Suōnada, in the western part of the Inland Sea)," Chirigaku hyōron, v. 5 (1928), no. 1, 16-41.

Tōki wrote other articles on faulting and structural warping in this region. Here he discusses the geomorphological origin of the Suōnada district on the basis of an analysis of post-diluvial erosion surfaces.

404. Tōki Ryūshichi 東木龍七, "Setonaikai to Chikuhō chikai no kyōkaitai no kigenron 瀬戸内海と筑豊地塊の境界帶の起源論 (A discussion of the origin of the boundary-zone between the Inland Sea and the Chikuhō block)," Chirigaku hyōron, v. 4 (1928), no. 12, 1-19.

This article is one part of the author's broad study of post-diluvial erosion surfaces and deals mainly with the development of fault scarps and related landforms in northern Kyūshū.

405. Tōki Ryūshichi 東木龍七, "Shinshokumen no hattatsushi yori mitaru Kasumigaura chihō no chikaku undō 侵蝕面の發達史より見たる霞ヶ浦地方の地殼運動 (Crustal movement in the Kasumigaura district seen from the point of view of the development of erosion surfaces)," Chirigaku hyōron, v. 4 (1928), no. 2, 23-40.

The origin of lowland swamps in the Kantō district is discussed and the nature of crustal movements in this region analyzed in relation to the development of erosion surfaces.

406. Tōki Ryūshichi 東木龍七, "Tama kyūryō no bichikeigakuteki kenkyū 多摩丘陵の微地形學的研究 (Micromorphological studies of the Tama hills)," Chigaku zasshi, v. 42, no. 497 (1930), 22-33; no. 498, 32-44, 8 maps.

Tōki was the first scientist to apply "micro" field techniques to the study of topography and geomorphology in Japan. This report of his detailed field survey of the Tama terraces, west of Tōkyō, is representative of his other field projects. He has attempted to work out field survey methods and an approach to the study of landforms in limited areas as a contribution not only to geomorphology but also to the fields of agriculture, soil science, and regional geography.

407. Tōki Ryūshichi 東木龍七, "*Tōbu Chikuhō chikai nambu no chikei hattatsushi* 東部筑豊地塊南部の地形發達史 (Historical geomorphology of the southern part of the east Chikuhō block)," *Chirigaku hyōron*, v. 4 (1928), no. 11, 43-54.

Tōki interprets the fault system of the Chikuhō region as a basis for describing the overall structural development of landforms in that district Emphasis is upon the geomorphology of fault basins and *Graben*.

408. Tōki Ryūshichi 東木龍七, "*Tōkyō shitamachi chiiki narabi ni sono fukin ni okeru kōsekisei igo no chikei hattatsushi no kenkyū* 東京下町地域並びに其の附近に於ける洪積世以後の地形發達史の研究 (A study of the development of landforms of downtown Tōkyō and its vicinity since the diluvial era)," *Chirigaku hyōron*, v. 4, (1928), no. 4, 15-35; no. 5, 34-49; no. 6, 12-30.

A detailed survey of the geomorphology of the Tōkyō area. The geological age of local landforms is estimated, based on an appraisal of their relationship to the development of erosion surfaces and swamps in the Kantō region.

409. Tōki Ryūshichi 東木龍七, "*Teibō no bichikeiteki kenkyū* 堤防の微地形的研究 (Micromorphological studies of embankments)," *Chirigaku hyōron*, v. 6 (1930), no. 5, 43-56.

An argument on the importance of making detailed investigations and interpretations of the immediate topography prior to undertaking construction of road embankments and river dykes.

410. Tokuda Teiichi 德田貞一, "*Shichitō-Marianna zōzantai ni tsuite* 七島マリアンナ造山帶について (On the Shichitō-Marianas orogenic zone)," *Chirigaku hyōron*, v. 3 (1927), no. 4, 1-18.

Tokuda's theory of symmetrical echelon-folding in the development of the Izu Shichitō and Marianas islands and in the Japanese archipelago is presented with numerous references to other orogenic concepts.

411. Tokuda Teiichi 德田貞一, "*Shūkyokuku to burokkuku no katei oyobi kukaku* 皺曲區とブロック區の假定及區劃 (An assumption concerning folded regions and block regions, and a classification of them)," *Chirigaku hyōron*, v. 1 (1925), no. 4, 27-41.

Theoretical arguments on the orogeny of the Japanese archipelago.

412. Tokushige Eisuke 德重英助, "*Echigo ni okeru gichinkō genshō to saikin no kaigansen idō* 越後に於ける僞沈降現象と最近の海岸線移動 (The phenomenon of pseudo-submergence in Echigo and the recent changes of shoreline)," *Chirigaku hyōron*, v. 3 (1927), no. 9, 26-44; no. 10, 11-30; no. 11, 53-74; v. 4 (1928), no. 1, 46-75.

An attempt to clarify the causes of recent local changes in the shoreline along certain sectors of the Niigata coast. Tokushige discusses the general nature of coastal warping and tries to determine the trends of crustal movement in that region by an investigation of marine terraces.

413. Tokushige Eisuke 德重英助, "*Gensei ni okeru Sado Ogi kaigan no hattatsushi* 現世に於ける佐渡小木海岸の發達史 (Shoreline development in modern times along the Ogi coast of Sado Island)," *Chirigaku hyōron*, v. 15 (1939), no. 11, 1-12.

As the result of several field surveys at the southern extremity of Sado island, the author summarizes the geological history of that area. His article is based on detailed study of the coastal topography and the evidences of crustal movement he observed there.

414. Tokushige Eisuke 德重英助, "*Saikin ni okeru Shinanogawa kakō fukin no kaigansen no ijō henka ni tsuite* 最近に於ける信濃川河口附近の海岸線の異常變化に就いて (On recent abnormal changes of the shoreline near the mouth of the Shinano river)," *Chirigaku hyōron*, v. 6 (1930), no. 7, 90-99.

A study of the coastline retreat caused by marine erosion and river action at the mouth of the Shinanogawa.

415. Tomita Yoshirō 富田芳郎, "*Taiwan no keikokunai ni aru senjōchi ni tsuite* 臺灣の溪谷内に在る扇狀地に就いて (On the alluvial fans in the valleys of Formosa)," *Chirigaku hyōron*, v. 14 (1938), no. 10, 1-23.

Emphasis is put on the geomorphological significance of alluvial fans formed at stream confluences which are characteristic of the main transverse valleys on Formosa. The relationship between this type of fan, terraces, and fault systems is discussed.

416. Toriyama Takeo 鳥山武雄, "*Tsukushi heiya oyobi sono shikin no chishitsu ni kansuru kōzōshiteki kōsatsu* 筑紫平野及其四近の地質に關する構造史的考察 (Tectonic and historical considerations on the geology of the Tsukushi plain and its surroundings)," *Chikyū*, v. 18 (1933), no. 5, 1-12.

The geology and geomorphology of the lowland region extending south from Fukuoka city to the hills on the south of Kurume city in northern Kyūshū.

417. Toyoda Hideyoshi 豊田英義, "*Jōetsu-sen Shimizu tonneru fukin no chigaku kansatsu gaihō* 上越線清水トンネル附近の地學觀察概報 (Survey and general report of the geology of Shimizu tunnel and its environs)," *Chigaku zasshi*, v. 43, no. 511 (1931), 38-50; no. 512, 19-28, 1 map and 2 photos.

A study of the geology and petrology in the area of the Shimizu pass where the loop tunnel system on the Jōetsu railway line constitutes the longest tunnel system in Japan. During the construction stages of this engineering project the author investigated the underlying rock formations and was able to make vertical and horizontal geological cross sections of the locality.

418. Toyama Shirō 外山四郎, "*Tōhoku ura-Nihon kaigan chihō no iwayuru kaishoku daichi* 東北裏日本海岸地方の所謂海蝕臺地 (On the so-called marine terraces on the Tōhoku coast of the Japan Sea)," *Chigaku zasshi*, v. 39, no. 460 (1927), 27-33; no. 461, 25-32.

An argument on the origin of the marine abrasion platforms widely distributed along the Japan Sea coast of Honshū from the Tōhoku region southward to the Tsugaru area. For convenience in description the author subdivides the coast into four districts: Akita, Echigo, Noto, and Tsugaru.

419. Tsuboi Chūji 坪井忠二, "Jishin to chikaku undō 地震と地殻運動 (Earthquakes and crustal movements)," Chirigaku hyōron, v. 5 (1929), no. 1, 42-52.
Tsuboi supports the block theory, explaining the structure of the earth's crust by citing seismological evidence determined by accurate measurements.

420. Tsuboi Chūji 坪井忠二, "Tango chihō no sankaku sokuryō kaisoku no kekka kara mita chikai undō 丹後地方の三角測量改測の結果から見た地塊運動 (Block movements in the Tango district as revealed by triangulations)," Chirigaku hyōron, v. 6 (1930), no. 7, 60-72.
A report on field analysis of the differing displacement of structural blocks, the cause of recent crustal movement in this district of the Kinki region.

421. Tsujimura Tarō 辻村太郎, Banshūki 晩秋記 (Essays on late fall), Tōkyō, Kokon Shoin, 1941, 6+793 pp.
A collection of the author's travel records and articles on travel and sight-seeing in Japan, written from a geographic viewpoint. The book contains some 69 papers, reprinted mainly from bulletins published by the Nihon Sangakkai 日本山岳會 (Japanese Alpinists' Society) and the Kokuritsu Kōenkai 國立公園會 (National Park Service), and from trade journals of the sight-seeing and recreation industry.

422. Tsujimura Tarō 辻村太郎, Chikeigaku 地形學 (Geomorphology), Tōkyō, Kokon Shoin, 1923, 12+610 pp., 85 illus.
A monumental general text which marked the establishment of geomorphology as a distinct field in Japanese science. There are only occasional references to Japanese landforms, however, and the work is primarily of historical significance as far as the purpose of the present bibliography is concerned. The two-volume revision of this text, published in 1932-33, is devoted entirely to Western concepts and examples.

423. Tsujimura Tarō 辻村太郎, Chikei no hanashi 地形の話 (Talks on geomorphology), Tōkyō, Kokon Shoin, 1949, 4+4+450+6 pp., 20 photos.
An improved general text on Japanese geomorphology, expanded from Tsujimura's earlier work Chikeigaku (Geomorphology), published in 1923 (see previous entry). Illustrative material in this book, however, is not so abundant as in Chikeigaku because of post-war publishing difficulties. However, the volume is more readable and more commonly available.

424. Tsujimura Tarō 辻村太郎, Dansōchikeironkō 斷層地形論考 (Studies on fault topography), Tōkyō, Kokon Shoin, 1942-43, 2v., 40 maps.
A series of geomorphological studies of faulting and fault topography by an authority on this subject. This is a collection of Tsujimura's published articles and previously unpublished monographs, the majority of which are concerned with specific fault features and zones in Japan. Discussions of fault basins and block mountains predominate. Part of the second volume is devoted to a selection of topographic maps illustrating examples cited. A post-war edition of this work appeared in 1948.

425. Tsujimura Tarō 辻村太郎, "Dansōkoku no seishitsu narabi ni Nihontō ichibu no chikeigakuteki dansō kōzō (yohō) 斷層谷の性質並に日本島一部の地形學的斷層構造 (豫報) (The nature of fault valleys and geomorphic fault structures in some parts of the Japanese Islands — a preliminary report)," Chirigaku hyōron, v. 2 (1926), no. 2, 20-42; no. 3, 8-34.
Following a geomorphologic analysis of fault valleys and other fault structures, Tsujimura describes the major fault systems in Japan.

426. Tsujimura Tarō 辻村太郎, "Hida sammyaku no hokutan ni okeru dansōgai no ichikeishiki 飛騨山脈の北端に於ける斷層崖の一型式 (A type of fault-scarp at the northern end of the Hida range)," Chirigaku hyōron, v. 2 (1926), no. 8, 47-63.
An authoritative discussion of the origin and nature of relatively recent fault scarps found along the northern margins of the Japanese Alps.

427. Tsujimura Tarō 辻村太郎, "Kai no kuni Komagatake narabi ni Senjōgatake no chikeigakuteki kansatsu 甲斐國駒ヶ岳並に仙丈ヶ岳の地形學的觀察 (Geomorphological observations of Mt. Komagatake and Mt. Senjōgatake, in Kai province)," Chirigaku hyōron, v. 10 (1934), no. 1, 43-57; no. 4, 25-40.
Both Komagatake and Senjōgatake at the northern end of the Akaishi range are about 3,000 meters high and the author's interest in this article is concentrated on determining evidence of nevation and/or glaciation on those peaks.

428. Tsujimura Tarō 辻村太郎, Kaigan no chiri 海岸の地理 (Marine coast geography), Tōkyō, Heibonsha, 1948, 4+264 pp.
There are few systematic Japanese works confined to coastal geomorphology and geography. This is a general textbook but much of the content is specifically concerned with the Japanese coast. Tsujimura discusses factors of both physical and human geography in the introductory chapters but the book is primarily a geomorphology text.

429. Tsujimura Tarō 辻村太郎, "Kashima-Yarigatake fukin no chikei ni tsuite 鹿島槍岳附近の地形に就て (Geomorphology in the vicinity of Kashima-Yarigatake)," Chirigaku hyōron, v. 11 (1935), no. 10, 20-33.
The fault topography of this section of the Japanese Alps, where a series of Graben exist along the margin of the fossa magna, is explained in terms of geological origin and characteristics.

430. Tsujimura Tarō 辻村太郎, Nihon chikeishi 日本地形誌 (Geomorphology of Japan), Tōkyō, Kokon Shoin, 1929, 5+2+455 pp., 134 maps.

An outstanding pioneer work, both theoretical and descriptive. The first portion concerns the geomorphology of Japan as a whole. The second is devoted to a discussion of type-examples, the majority being landforms located in southwestern Japan.

431. Tsujimura Tarō 辻村太郎, "Nihon no kōzankei 日本の高山形 (Orography of the highlands in Japan)," Chigaku zasshi, v. 35, no. 416 (1923), 1-13; no. 418, 10-24.

An article on the high mountain topography of central Japan, a topic previously discussed in this same publication by Yamazaki Naomasa and Ōzeki Kyūgorō. Tsujimura advances new theories which had developed in Japan and abroad concerning the distinctions between nevation and glaciation and the development of Kar.

432. Tsujimoto Yoshirō 辻本芳郎, "Sagamihara no chishitsu kōzō 相模原の地質構造 (Geological structure of the Sagami plain)," Chirigaku hyōron, v. 13 (1937), no. 9, 44-57.

Tsujimoto reports on a survey made of the depth of the water table on the Sagami alluvial fan and concludes that observed variations in ground water levels are not attributable to geological causes.

433. Tsujimura Tarō 辻村太郎, "Setonaikai no chikei seiinron 瀬戸内海の地形成因論 (A theory on the morphological origin of the Inland Sea)," Wakayama, Wakayama-ken Shihan Gakkō Chirekibu, 1929, 47 pp.

A monograph on the historical and structural geology of the Setouchi or Inland Sea region.

434. Tsujimura Tarō 辻村太郎, "Tōhoku Nihon no dansō bonchi 東北日本の断層盆地 (Fault basins in northeastern Japan)," Chirigaku hyōron, v. 8 (1932), no. 9, 1-18; no. 10, 21-34; no. 12, 73-88.

A monograph summing up the knowledge on this subject accumulated by the research and field work of geographers and geologists. Each fault basin of geomorphologic consequence is described in detail together with its related fault system. Two maps show the distribution of the various fault series of Tōhoku.

435. Tsujimura Tarō 辻村太郎, Yama 山 (Mountains), [=Iwanami shinsho (New Iwanami works, no. 63], Tōkyō, Iwanami Shoten, 1940, 2+2+177 pp.

A pocket-sized outline of mountain geomorphology, written in popular style for the climber and camper.

436. Tsujimura Tarō 辻村太郎, Zoku banshūki 續晩秋記 (Essays on late fall, continued), Tōkyō, Kokon Shoin, 1940, 2+628+3 pp.

This is a collection of the author's geographical essays and papers reprinted from publication other than in Chirigaku hyōron, Chigaku zasshi, and Nihon chishitsugaku shūhō. It contains 21 articles covering a variety of topics but most of them deal entirely or in part with landforms. Tsujimura is a keen observer and most of this writings are highly regarded in Japan and should be of considerable interest to anyone concerned with the natural landscape and its influence upon man in that country.

437. Tsujimura Tarō 辻村太郎 and Awaji Shōzō 淡路正三, "Shinki no undō ni yoru dansō chikei 新期の運動に依る断層地形 (Fault topography due to young crustal movement)," Chirigaku hyōron, v. 10 (1934), no. 12, 64-84.

Some examples of topographic features caused by recent faulting in Japan are presented, and special reference is made to such landforms observable on the western slopes of the north Izu volcanoes, particularly on Mt. Yugawara.

438. Tsujimura Tarō 辻村太郎 and Kawada Shirō 河田四郎, "Kii hantō ni okeru kaigan dankyū no kōdo bumpu yori mitaru jiban undō no seishitsu 紀伊半島に於ける海岸段丘の高度分布より見たる地盤運動の性質 (The nature of crustal movements seen from the distribution of coastal terraces in the Kii peninsula)," Chirigaku hyōron, v. 6 (1930), no. 12, 58-71.

Structural warping is investigated through an analysis of the height and width of marine terraces. In conclusion the authors attempt to correlate coastal and interior crustal changes.

439. Tsuya Hiromichi 津屋弘逵, "Izu Kōzushima kazantō no chikei to chishitsu kōzō to ni tsuite 伊豆神津島火山島の地形と地質構造とに就いて (On the topography and geology of the volcanic island Kōzushima in the Izu group)," Chirigaku hyōron, v. 6 (1930), no. 7, 453-469.

The geological formation of this volcanic island in the Izu-shichitō chain is explained.

440. Ueda Shinzō 上田信三, "Kokutei no jūdammen kyokusen ni shimesareta Hida sammyaku hokubu no jiban undō 谷底の縦断面曲線に示された飛騨山脈北部の地盤運動 (Earth movements in the northern part of the Hida mountain range as revealed by the longitudinal profiles of rivers)," Chirigaku hyōron, v. 11 (1935), no. 7, 34-48.

Analyzing the profiles of two main rivers running through the northern part of the Hida mountain range, Ueda comes to the conclusion that there had been at least two upheavals of the land since the formation of low relief surfaces in that part of the range.

441. Ueda Shinzō 上田信三, "Sanryō no takasa to tani no heimenkei kara mita Akaishi sanchi 山稜の高さと谷の平面形から見た赤石山地 (The Akaishi mountains as seen from the height of the profiles and peneplanation of the valleys)," Chirigaku hyōron, v. 12 (1936), no. 6, 28-50.

An article on the geotectonics and geomorphology of the Akaishi mountains of south-central Honshū, one of Japan's highest ranges. The emphasis is this paper is upon piedmont plains and the relationship between peneplanation and faulting.

442. Ushimaru Shūtarō 牛丸周太郎, Shinshū Takatō chihō no chishitsu 信州高遠地方の地質 (Geology of the Takatō district in Shinano province), Ina, Shinano Kyōikukai Kami-Ina Bukai, 5+2+59 pp., 1 map and 10 illus.

A paper on the geology and petrology of the Takatō area and its vicinity in Shinano province (Nagano prefecture) in the northern sector of the Akaishi mountain block. Emphasis is upon petrology.

443. Wada Yaezō 和田八重造 and Awatsu Hideyuki 粟津秀幸, Genshoku Nihon ganseki zufu 原色日本岩石圖譜 (Iconography of Japanese rocks in natural colors), Tōkyō, Meguro Shoten, 1950, 2+3+56+4+3+33 pp., 56 photo plates.

An illustrated petrography of the principal rock types. Indices in English and in the order of the Japanese syllabary.

444. Watanabe Akira 渡邊光, "Akaishi sankei nambu no chikei to chikei hattatsu 赤石山系南部の地形と地形發達 (Morphology of the Akaishi mountain region and its physiographic history)," Chirigaku hyōron, v. 6 (1930), no. 7, 175-198.

The drainage basin of the Ōi river is discussed in terms of its geomorphology. This is followed by an analysis of crustal movement in relation to peidmont surfaces based on a geological survey of the elevated deltas and river terraces of this region in Shizuoka prefecture.

445. Watanabe Akira 渡邊光, "Bōsō hantō nambu no kaigan dankyū 房總半島南部の海岸段丘 (The coastal terraces of the southern part of the Bōsō peninsula)," Chirigaku hyōron, v. 5 (1929), no. 2, 21-28.

Conjecture on the formation of low marine terraces along the southern coast of the Bōsō peninsula in Chiba prefecture.

446. Watanabe Akira 渡邊光, "Hokkaidō chikeishi gaihō 北海道地形誌概報 (Outline of the landform divisions of Hokkaidō)," Chirigaku hyōron, v. 7 (1931), no. 11, 1-19.

An elaboration of Watanabe's paper read at the Michigan Academy of Science, Arts and Letters and published in the Michigan papers in geography. Here the author is principally concerned with a new geomorphological classification of the Ōshima peninsula in relation to the landform divisions of Hokkaidō proper.

447. Watanabe Akira 渡邊光, "Hompō no ryūki sankakusu ni kansuru kōsatsu 本邦の隆起三角洲に關する考察 (Some considerations on elevated deltas in Japan)," Chirigaku hyōron, v. 5 (1929), no. 1, 1-15.

A comparative geomorphological analysis of elevated deltas in Toyama and Shizuoka prefectures and in the Tōkaichi district of Hokkaidō.

448. Watanabe Akira 渡邊光, "Nihon no chikeiku 日本の地形區 (Landform divisions of Japan)," Chigaku zasshi, v. 61 (1952), no. 1, 1-7.

A basic geomorphological subdivision of Japan into landform regions by name and location. Four primary regions are delimited: 1) Hokkaidō, 2) northeast Japan, 3) central Japan, and 4) southwest Japan. These in turn are differentiated into eight sub-regions, 51 districts, and 240 sections.

449. Watanabe Akira 渡邊光, Nihon no sugata -- Kokudosei no tankyū -- 日本の姿 -- 國土性の探究 -- (Features of Japan — research for regional character), Tōkyō, Nihon Chizu Kabushiki Kaisha, 1948, 52 pp.

A comprehensive description and explanation of the physical features of Japan. Also the historical development of Japanese maps and map-making is summarized in the introduction to the booklet. Following the general topographic analysis, Watanabe's classification of the geomorphologic divisions of Japan is presented.

450. Watanabe Akira 渡邊光, "Okushiritō no kaisei dankyū to sono kōtaiteki keisha undō 奥尻島の海成段丘とその交代的傾斜運動 (Coastal terraces of Okushiri island and their alternate tilting)," Chirigaku hyōron, v. 4 (1928), no. 3, 62-73.

A geomorphological investigation of the terraces of Okushiri island in the central part of the Kurile chain in relation to determining the nature of crustal upheaval in that archipelago.

451. Watanabe Akira 渡邊光 and Imaizumi Masakichi 今泉政吉, "Nihon guntō ni okeru kazan no bumpu narabi ni chikeigakuteki kazangun no settei 日本群島に於ける火山の分布並に地形學的火山群の設定 (Distribution of volcanoes in Japan and determination of geomorphological volcanic groups)," Chirigaku hyōron, v. 3 (1927), no. 6, 1-23; no. 7, 28-45; no. 8, 19-32; no. 10, 31-46.

The authors classify the initial forms of volcanoes according to the formula of K. Schneider. A final division into geomorphologic groups was determined by a study of the present degree of dissection. This article is primarily intended to serve as an explanation of the compilation methods followed in making the accompanying distribution map of Japan's volcanoes. Included is a concise table which summarizes the geographic location and morphological characteristics of the volcanic groups identified on the map.

452. Watanabe Akira 渡邊光 and others, Chikeiku 地形區 (Geomorphological regions), in Iwanami kōza: chiri (Iwanami lectures on geography), Tōkyō, Iwanami Shoten, 1931-34, 98 pp., 11 maps.

One of a series of publications on the geography and geology of Japan printed by Iwanami Shoten. The collaboration of five noted specialists makes this a valuable and standard work on the geomorphological divisions of Japan. The contributing geomorphologists are: J. Hanai, T. Okayama, H. Shimomura, M. Sugiyama, and A. Watanabe.

453. Watanabe Kyūkichi 渡邊久吉, "Daisanki jidai ni okeru Nihon guntō no kochiri 第三紀時代に於ける日本群島の古地理 (Tertiary paleography of the Japanese Islands)," Chigaku zasshi, v. 50, no. 594 (1938), 1-22, 9 maps.

A treatise on the historical geology of the Japanese archipelago. The article is chiefly concerned with the distribution of land and sea during the several stages of the Cretaceous and Tertiary periods. Accompanying the text are nine maps which show the land-sea relationship in sequence from the dawn of the Cretaceous to the late Pliocene.

454. Watanabe Kyūkichi 渡邊久吉 , "Jōban enkai chitai no chikei hattatsushi 常磐沿海地帯の地形發達史 (Topographical development of the Jōban coastal belt)," Chigaku zasshi, v. 42, no. 494 (1930), 18-28; no. 495, 42-48.

Geomorphological development of the Jōban district on the north-central Pacific coast of Honshū is studied from the standpoint of geological changes detected mainly in examinations of Tertiary strata. Some reference is made to Cretaceous and Quaternary formations in the area. A detailed stratigraphic field study of interest mainly to those geologists and geographers concerned with the Jōban coal fields.

455. Watanuki Isahiko 綿貫勇彦 , "Tomboro to māru トンボロとマール (Tombolo and Maare)," Chirigaku hyōron, v. 1 (1925), no. 7, 36-49.

A description and explanation of volcanic features on the Oga peninsula in Akita prefecture. The writer is principally interested in the several Maare or crater lakes on that peninsula, which itself is a tombolo.

456. Yabe Hisakatsu 矢部長克 , "Daisanki oyobi sono chokugo ni okeru Kyūshū chishi no taiyō 第三紀及其直後に於ける九州地史の大要 (Outline of the geologic history, Tertiary and post-Tertiary, of Kyūshū)," Chirigaku hyōron, v. 2 (1926), no. 1, 1-16.

Much new information on the geology of north and central Kyūshū resulted from surveys made by T. Nagao in the 1920s. In this paper, Yabe discusses geologically recent crustal disturbances in Kyūshū in the light of Nagao's findings.

457. Yabe Hisakatsu 矢部長克 , "Shiobara kazan to chitai kōzō to no kankei 鹽原火山と地帯構造との關係 (Shiobara volcano and its relation to geological structure)," Chirigaku hyōron, v. 5(1929), no. 2, 44-60.

An investigation of fossiliferous Tertiary strata and general dissection at Mt. Shiobara in Tochigi prefecture in relation to a correlation of the historical geology of that area with that of the Kantō district as a whole.

458. Yabe Hisakatsu 矢部長克 and Aoki Renjirō 青木廉二郎 , "Kantō kōzō-bonchi shūen sanchi ni soeru dankyū no chishitsu jidai 関東構造盆地周縁山地に沿へる段丘の地質時代 (The geological age of terraces found in mountains surrounding the Kantō structural basin)," Chirigaku hyōron, v. 3 (1927), no. 2, 1-9.

General considerations on the characteristics of the high terraces developed along the western and northern margin of the Kantō structural basin and on the nature of the basin itself.

459. Yabe Hisakatsu 矢部長克 and Tayama Risaburō 田山利三郎 , "Hōjō ōchitai 北條凹地帯 (The Hōjō depression zone)," Jishin Kenkyūjo ihō, (The bulletin of the Earthquake Research Institute), v. 10 (1933), no. 1.

Hitherto, the lowland near the south end of the Bōsō peninsula was called the "Hōjō Graben" by geologists. On the basis of their studies of terraces in that area, the authors challenge this terminology, contending that the lowland is not a true graben and should instead be identified as a "depression zone."

460. Yagi Teisuke 八木貞助 , Kami-Takai-gun chishitsushi 上高井郡地質誌 (Regional geology of Kami-Takai district), Suzaka, Kami-Takai Kyōikukai, 1941, 3+6+230+9 pp., 4 maps, 55 illus.

A fairly detailed, but not highly technical regional geology of the Kami-Takai area, a mountain district on the eastern side of the Zenkōji basin in Nagano prefecture. The study is not restricted by prefecture boundaries and references are made to ore deposits, hot springs, and volcanoes in the neighboring districts as far north as Gumma prefecture. Contains one chapter on applied geology.

461. Yagi Teisuke 八木貞助 , Kōshoku chishitsushi 更埴地質誌 (Regional geology of the Sarashina and Hanishina districts), Nagano, Shinano Mainichi Shimbunsha, 1943, 1+2+9+289+4 pp., numerous illus.

A geological survey and geomorphological description of two gun on opposite sides of the Chikumagawa in Nagano prefecture. One section of the study is devoted to various aspects of applied geology.

462. Yagi Teisuke 八木貞助 , "Nagano-ken Jigokudani funsen ni tsuite 長野縣地獄谷噴泉に就いて (On the hot-spring of Jigokudani, Nagano prefecture)," Chikyū, v. 11 (1929), no. 5, 1-11.

An explanation of the geological structure and physical mechanism of a large, constantly following hot spring. Although solfatara are common in Japan, there are few springs comparable to the Nagano Jigokudani and it was declared a national monument because of its unusual nature.

463. Yamagishi Tadao 山岸忠夫 , "Amagi sanroku onsen chitai no kōsatsu (天城山麓温泉地帯の考察 (A study of the hot-spring zone at the foot of Mt. Amagi)," Chirigaku hyōron, v. 9 (1933), no. 9, 1-8.

Observations on the distribution of hot springs and rural settlement in the central portion of the Izu peninsula in respect to the influence of geological structure upon underground water. The raising of wasabi (Eutrema wasabi, a kind of horse-radish) in that area is described briefly in relation to water quality.

464. Yamagishi Tadao 山岸忠夫 ," Hompō onsen bumpu to chishitsu kōzō to no kankei ni tsuite no kōsatsu 本邦温泉分布と地質構造との關係についての考察 (Thoughts on the relationship between the distribution of hot springs and the geological structure)," Chirigaku hyōron, v. 19 (1943), no. 9, 1-10.

An investigation of geological structure in the vicinity of hot springs in the Mikuni mountains and on the Izu peninsula.

465. Yamagishi Tadao 山岸忠夫 , "Izu Kawazu-mura onsen chitai no kōsatsu 伊豆河津村温泉地帯の考察 (Thoughts on the Kawazu hot-spring district in the Izu peninsula)," Chirigaku hyōron, v. 8 (1932), no. 1, 48-58.

The controlling effect of tectonic structure upon human activity is examined in reference to the distribution of hot springs and ordinary springs in south Izu.

466. Yamagishi Tadao 山岸忠夫, "Toi onsen chitai no kōsatsu 土肥温泉地帯の考察 (Thoughts on the Toi hot-spring district)," Chirigaku hyōron, v. 7 (1931), no. 2, 15-22.
Toi is located on the west coast of the Izu peninsula, Shizuoka prefecture. The characteristics of hot springs in that area are noted briefly and the distribution of the springs is discussed in relation to the nature of local landforms.

467. Yamaguchi Sadao 山口貞夫, "Izu shotō no kaishokugai 伊豆諸島の海蝕崖(The sea cliffs of the Izu Islands)," Chirigaku hyōron,v. 9 (1933), no. 2, 1-18.
Many coastal sectors in the volcanic Izu islands south of Tōkyō are marked by impressive sea cliffs. In this paper Yamaguchi reports on his efforts to determine the relationship between the height and distribution of these cliffs and the direction of the prevailing winds.

468. Yamaguchi Sadao 山口貞夫, "Kaigan dammen ni kansuru nisan no kōsatsu 海岸断面に関する二・三の考察 (Some considerations on shore profiles)," Chirigaku hyōron,v. 9 (1933), no. 6, 1-20.
Changes in shore profiles are considered in relation to the various causative types of crustal movement.

469. Yamaguchi Sadao 山口貞夫, "Tōhoku Nihon Taiheiyōgan no kaigan dammen ni tsuite 東北日本太平洋岸の海岸断面に就いて (Shore profiles of the Pacific coast of northeastern Japan)," Chirigaku hyōron, v. 9 (1933), no. 11, 47-76.
Interpretation of crustal movement based upon an analysis of the author's profiles of coastal topography. The cross sections studied cover inshore submarine topography as well as coastal landforms.

470. Yamanari Fujimaro 山成不二麿,"Chōsen Kōgendō no rimpen kōzō 朝鮮江原道の鱗片構造 (Imbricate structure in Kōgendō, Korea)," Chirigaku hyōron, v. 2 (1926), no. 7, 20-38.
A study in the structural geology of an area in eastern Korea where Ordovician limestone strata overthrust middle Mesozoic layers. From an investigation of this formation, the author draws conclusions on the general nature of past crustal movement in that region.

471. Yamazaki Naomasa 山崎直方, "Bōsō hantō ni okeru keidō chikei ni tsuite 房總半島に於ける傾動地塊に就いて(On tilted blocks on the Bōsō peninsula)," Chirigaku hyōron, v. 1 (1925), no. 1, 1-14.
An explanation and description of the complex system of tilted blocks which forms the southern peninsular portion of Chiba prefecture.

472. Yoshikawa Torao 吉川虎雄, "Kaigan jūdammen no hattatsu -- benchi no keisei o chūshin to shite -- 海岸縦断面の発達——ベンチの形成を中心として——(The development of shore profiles, with emphasis on the formation of benches)," Tōkyō Daigaku chirigaku kenkyū, no. 1 (1950), 99-113.
A geomorphological study of the influence of wave-cut benches upon the development of shore profiles, citing investigations made on the coasts of the Akashi and Himi areas in the Kinki regions. The interaction of marine and subaerial agents in the formation of shore profiles is pointed out.

473. Yoshikawa Torao 吉川虎雄, "Kurobegawa senjōchi no chikei to sono kaigan shinshoku 黒部川扇状地の地形とその海岸侵蝕(Shore erosion and its bearing on the topography of the Kurobe alluvial fan)," Tōkyō Daigaku chirigaku kenkyū, no. 2 (1952), 92-109.
The alluvial fan under consideration in this paper is situated at the mouth of the Kurobe river gorge on the Sea of Japan coast in eastern Toyama prefecture. The retreat of the shoreline in that area, caused by wave erosion, and its effect upon landforms and land use are investigated.

474. Yoshino Masumi 吉野益見, "Taishakudai karusuto 帝釈臺カルスト (The karst scenery of Taishakudai)," Chigaku zasshi, v. 42, no. 501, (1930), 5-21; no. 502, 32-39; v. 43, no. 504, (1931), 46-55.
Taishakudai is the name applied to the karst district in the Chūgoku mountains of northeastern Hiroshima prefecture. Next to the Akiyoshidai area, it is the largest region of karst topography in Japan. In this article the karst forms of the Taishakudai are described in detail. The district is divided into four sectors for convenience in treatment. It might be noted here that similar articles on the "karst scenery of the Akiyoshidai" by Satō Denzō appear in volumes 40 and 41 of Chigaku zasshi.

B. Volcanism and seismology

475. Chūō Kishōdai 中央氣象臺 (Central Meteorological Observatory), Tōnankai daijishin chōsa gaihō 東南海大地震調査概報(General report of research on the great Tōnankai earthquake), Tōkyō, Chūō Kishōdai, 1945, 94 pp.
A symposium concerning research conducted on the major earthquake which occurred on December 7, 1944, in the Tōkaidō and Nankaidō regions. Contents are mainly summaries of field surveys made by prominent Japanese geologists in the quake area and an expression of their opinions on the significance of the quakes. Contributors are Homma Shōsaku, Takagi Hijiri, Yamanoue Akikazu, Okumura Hiroji, Sagisaka Kiyonobu, Sakai Otohiko and Musha Kingo.

476. Fujiwara Sakuhei 藤原咲平, Chika, chiretsu oyobi jishin 地渦・地裂及地震 (Earth vortex, earth cracks and earthquakes), Tōkyō, Kokon Shoin, 1932, 17+2+2+161+14 pp., 40 illus.
A collection of Fujiwara's technical writings on the relationship between earthquakes and surface fissures, crustal movements, earth mechanics, and related topics, all previously published in the principal Japanese journals on earth sciences.

477. Imamura Akitsune 今村明恒, "Tajima daijishin no kansoku ni tsuite 但馬大地震の観測に就いて (Concerning observations of the Tajima earthquake)," Chirigaku hyōron, v. 1 (1925), no. 5, 73-78.

The characteristics of the Tajima earthquake, which occurred in Hyōgo prefecture in June, 1925, are discussed on the basis of data recorded at the Tōkyō Daigaku Jishin Kenkyūjo 東京大學地震研究所 (Seismological Institute of Tōkyō University).

478. Ishimoto Mishio 石本巳四雄, Jishin to sono kenkyū 地震とその研究 (Earthquakes and their studies), Tōkyō, Kokon Shoin, 1935, 5+8+336 pp., 92 illus.
A three-part volume on seismology, the first two dealing with the author's lectures on that topic, and the third concerning his theories on the origin of earthquakes with a chronological summary of theoretical developments in that field. Seismological investigations in Japan form the core of the work, and examples cited are accompanied by abundant analytical data. With an index and a bibliography of Japanese publications on seismology.

479. Kobayashi Fusatarō 小林房太郎, Nihon no kazan 日本の火山 (Japanese volcanoes), Tōkyō, Ryūbunkan, 1920, 3+3+13+3+541 pp.
A useful, detailed reference work for the physical geographer or geomorphologist. A general summary of volcanology and an outline of the distribution of Japanese volcanoes and volcanic zones occupy the first part of this book. The second portion provides individual descriptions of each major volcano.

480. Matsumoto Tadaichi 松本唯一, "Ibusuki kazan deiyōgan no bumpu 指宿火山泥溶岩の分布 (Distribution of mud-lava from the Ibusuki volcano)," Chirigaku hyōron, v. 15 (1939), no. 6, 1-15.
A study of the Ibusuki and Aira calderas in southern Kagoshima prefecture and the distribution of the materials erupted by these former volcanoes.

481. Nakamura Saemontarō 中村左衛門太郎, Kantō daishinsai chōsa hōkoku 關東大震災調査報告 (Report of research on the great Kantō earthquake), Tōkyō, Chūō Kishōdai, 1924, 4+61 pp., 11 maps and 21 photos.
Reports of the research made by the Central Meteorological Observatory of Tōkyō on the great Kantō quake of 1923. A basic study of this famous earthquake containing not only seismological observations and data but also a report on geological changes and damage caused by the quake and the disastrous fires which followed it. With several graphs.

482. Nakamura Saemontarō 中村左衛門太郎, Okutango jishin hōkoku 奥丹後地震報告 (A report on the Okutango earthquake) [=Saitō Hōonkai Sōmubu, Saitō Hōonkai scientific reports, no. 5], 1928, 2+56 pp., 9 photos, 26 illus.
General field observations made in the Tango district of Kyōto prefecture following the severe earthquake of March, 1927. Followed by a seismological discussion of zonal damage, crustal displacement, etc.

483. Nihon Kazan Gakkai 日本火山學會, Nihon kazanshi 日本火山誌 (Materials on volcanoes in Japan), Supplements to Kazan, v. 2 (1932), no. 3, 53-121 and v. 3 (1933), no. 1, 74-124.
These two volumes contain detailed studies on Sakurajima and Unzendake, Kyūshū volcanoes. They were published as supplements to Kazan, the regular bulletin of the Volcanological Society of Japan. The articles, by leading volcanologists and geologists in Japan, are generally regarded as outstanding works on these famous volcanoes.

484. Sagisaka Kiyonobu 鷺坂清信, Jishin to tsunami 地震と津波 (Earthquakes and tsunami), Tōkyō, Meguro Shoten, 1949, 2+168 pp., 16 tables, 49 illus.
Designed for college students with elementary training in earth sciences, this is a general discussion of the field of seismology, citing numerous researches and personal experiences of the author during his years devoted to investigating seismic phenomena. References to tsunami are secondary. With chronological tables on Japanese quakes and tidal waves.

485. Suirokyoku 水路局 (The Hydrographical Bureau), "Shōwa nijūichinen Nankai daijishin chōsa hōkoku 昭和21年南海大地震調査報告 (Report on the Nankai earthquake in 1946)," Suiro yohō, special issue, 1948, 117 pp., 49 maps, 11 illus.
This monograph presents the observations and conclusions of a systematic geological field survey made in the Nankaidō district following the great earthquake of 1946. With 12 tables of data on crustal displacement, changes in the coastline, hydrological alterations, etc.

486. Tsuboi Seitarō 坪井誠太郎 and Kuno Hisashi 久野久, "Hompō kazangan 'kiseki anzangan rui' ni tsuite 本邦火山岩 '輝石安山岩類' に就いて (On the so-called 'pyroxene-andesites,' volcanic rocks in Japan)," Kazan, v. 1 (1932), no. 1, 20-37.
An outline of volcanic petrography.

487. Yagi Teisuke 八木貞助, Asamayama 淺間山 (Mt. Asama), [=Nagano kyōdo sōsho (Materials on the Nagano homeland), no. 11], Nagano, Shinano Kyōdo Bunka Fukyūkai, 1929, 102 pp.
No. 11 in a series of pamphlets published by the Shinano Kyōdo Bunka Fukyūkai or Association for the Diffusion of the Provincial Culture of Shinano. This is a popular description of Mt. Asama, including an explanation of its structure, a summary of its past activity, and a discussion of its peculiar characteristics and related volcanic phenomena.

488. Yamaguchi Ikutomo 山口生知, Jishin to tsunami 地震と津波 (Earthquakes and tsunami), Tōkyō, Kokon Shoin, 1951, 79+6 pp.
This pamphlet describes, in non-technical language, the general nature of earthquakes and their concommitant phenomenon, the tidal wave or tsunami. Also discussed are current forecasting techniques, research being done to improve forecasting methods, and other possible defenses against these natural disasters. A six-page table presents a chronology of big earthquakes accompanied by tsunami which affected Japan during the period from 648 to 1946.

489. Yamasaki Naomasa 山崎直方, "Tajima jishin no shingen 但馬地震の震源 (The epicenter of the Tajima earthquake)," Chirigaku hyōron, v. 1 (1925), no. 5, 79-85.

Yamasaki surveyed the Tajima district following the earthquake of June, 1925, and examined fault lines in the vicinity of Tsujiyama bay which he concluded marked the quake's epicenter. This paper summarizes his observations.

C. Drainage

490. Hirata Tokutarō 平田徳太郎 and Mizutani Takashi 水谷鏘, Shinrin no kasui chōsetsu sayō oyobi kasui tōsei no gaikyō 森林の河水調節作用及河水統制の概況 (River-control effects on forests and an outline of river regulation), Tōkyō, Teikoku Chisan Chisui Kyōkai, 1942, 72 pp.

The first section is a general portrayal of the role of the forest in natural watershed protection and stream regulation, and a discussion of the inter-relationships between vegetative cover and precipitation, ground water, run-off, erosion, flooding, etc. Part two concerns the present necessity for artificial river control work in Japan and the status of water research and water regulation projects in the major drainage basins and stream valleys.

491. Inami Etsuji 稲見悦治 and Tochigi Yoshimasa 栃木義正, "Tamagawa kasui to engan chikasui to no kankei 多摩川河水と沿岸地下水との関係 (The relationship between the water of the Tama River and underground water along its course)," Chirigaku hyōron, v. 23 (1950), no. 8, 16-23.

The hydrology of the area along the south bank of the Tamagawa in Kanagawa prefecture. The authors investigate changes in the level of the water table in relation to variations in river level.

492. Konno Yoshinobu 今野義信, "Sukawasui no suiso ion nōdo: sono seibutsu, jimbun ni oyobosu eikyō 須川水の水素イオン濃度，その生物，人文に及ぼす影響 (Hydrogen ion concentration in the river water of the Sukawa: its influence on living things and on human culture)," Chirigaku hyōron, v. 10 (1934), no. 5, 31-43.

The Sukawa, one of the tributaries of the Mogamigawa, rises on the western slopes of Zaō volcano in Yamagata prefecture and its waters are too acidic to support biota or to be used for irrigation. The adverse effect of this stream upon local life and human activities is investigated.

493. Mizutani Takashi 水谷鏘, Nihon kasenron 日本河川論 (Theory on the rivers of Japan), Tōkyō, Tokiwa Shobō, 1941, 2+9+341 pp., 70 illus.

A discussion of river and stream control problems confronting Japan, with emphasis upon social and economic aspects. The first portion deals with theories of stream control, various methods actually employed in Japan, and an outline of some planned projects supposedly of an improved variety. Chapter four presents a more technical explanation of certain phases of hydrology. The book closes with a description of selected river control programs in existence and an analysis of the effect of these efforts on local economy and the social scene. With 98 tables of hydrographic data.

494. Sasaki Haryō 佐々木巴陵, "Kitakamigawa enkaku shirabe 北上川沿革調 (Survey of the Kitakami River)," Sendai sōsho, no. 5 (1924), 241-279.

A comprehensive survey of river control and land reclamation projects, past and present, in the Kitakami valley in Tōhoku, including a brief physical description of the Kitakami watershed. Settlement, land use, and transportation and communication highlights are mentioned. This report is based on a general inventory of the Kitakami basin made by the Engineer Section of the Miyagi prefectural government.

495. Tago Kenkichi 田子健吉, Fukushima-ken kasen no kenkyū 福島縣河川の研究 (A study of rivers in Fukushima prefecture), Fukushima, Fukushima Min'yū Shimbunsha, 1924, 6+164+35 pp., 1 map and 9 photos.

Summarizing flood control measures and the use of river water for hydro-electric power generation and irrigation in Fukushima prefecture, this study is presented in a semi-scientific vein but will be of interest to those geographers concerned with water resources and water control problems in Japan. Rather typical presentation and a forerunner of numerous later planning papers and reports on this topic in other areas. Includes 2 tables of data.

496. Takahashi Jun'ichi 高橋純一, "Kitakami Abukuma ryōga jōryū chiku no ryūsuiritsu 北上阿武隈兩河上流地區の流水率 (Coefficient of water discharge in the upper streams of the Kitakami and Abukuma rivers)," Chirigaku hyōron, v. 6 (1930), no. 7, 397-406.

The amounts of rainfall, run-off, evaporation, and infiltration are calculated for the two principal Tōhoku watersheds and comparisons made which reflect the differing physiographic characteristics of the drainage basins of these rivers.

497. Takeoka Mitsutada 武岡充忠, Yodogawa chisuishi 淀川治水誌 (Yodo River riparian work), Ōsaka, Yodogawa Chisuishi Kankōkai, 1931, 18+14+8+292 pp., 1 map, 2 photos.

This is not a general history of the development of control works on the Yodo river as the title might indicate, but rather it is a description of riparian improvements constructed following the major flood of 1897 on that stream. Of interest to anyone concerned with water control problems and stream reclamation projects.

D. Groundwater

498. Ishikawa Seishō 石川成章, "Kurobe kyōkoku to onsen 黒部峡谷と温泉 (The Kurobe canyon and hot springs)," Chikyū, v. 13 (1929), no. 2, 41-49.

An outline of the geology and geomorphology of the Kurobe canyon, a famous gorge in Toyama prefecture, and a description of the hot springs in its vicinity.

499. Saijō Yatsuka 西條八束, Asama kazan no chikasui 淺間火山の地下水 (Groundwater of the volcano Asama), Tōkyō Daigaku chirigaku kenkyū, no. 1 (1950), 73-86.
A comparative analysis of the chemical content of water from two differing types of mineral spring in the Asama area of Gumma prefecture.

500. Yajima Nikichi 矢嶋仁吉, "Haruna kazan tōnanroku no chikasui 榛名火山東南麓の地下水 (Groundwater at the southeastern foot of Mount Haruna)," Chirigaku hyōron, v. 21 (1948), nos. 7-8, 8-15.
Following field surveys made to investigate the groundwater situation in this gently sloping area on the northwestern margin of the Kantō plain, Yajima mapped the depth of the water table. On the basis of his findings, the author discusses the feasibility of introducing rice culture into this district.

501. Yamamoto Sōki 山本荘毅, Chikasui chōsahō 地下水調査法(Methods of groundwater surveying), Tōkyō, Kokon Shoin, 1953, 240 pp., 101 illus.
A general textbook on hydrology and groundwater surveying. This is one of the most recent and comprehensive works in a field of study which has attracted great interest in recent years among Japanese earth scientists. Chapters are devoted to both the theoretical aspects of the subject and to practical field survey techniques and methods, including the description and explanation of apparatus and equipment.

502. Yamamoto Sōki 山本荘毅 and Katayama Tomooki 片山朝與, "Kakoku no chikasui (daiippō) — Yahagigawa chikasui 河谷の地下水(第1報)矢作川地下水(The groundwater hydrology of river valleys (part 1) — the groundwater of the Yahagigawa valley)," Chirigaku hyōron, v. 23 (1950), nos. 9-10, 1-10.
Technical observations on the hydrology of an area in Aichi prefecture.

503. Yamamoto Sōki 山本荘毅, "Kakoku no chikasui (dainihō)— Kinokawadani no chikasui 河谷の地下水(第2報)紀川谷の地下水 (The groundwater hydrology of river valleys (part 2) — the groundwater of Kinokawadani)," Chirigaku hyōron, v. 24 (1951), no. 1, 8-16.
The second article in this journal by Yamamoto on valley hydrology, in this case concerned with the relationship between groundwater and surface water in the Kinokawa valley, Wakayama prefecture.

504. Yoshida Yoshinobu 吉田義信, "Toyama-ken no sandaisenjōchi no chikasui 富山縣の三大扇狀地の地下水 (The groundwater of the three great alluvial fans in Toyama prefecture)," Chirigaku hyōron, v. 17 (1941), no. 12, 29-42; v. 18 (1942), no. 1, 29-41.
A technical study of the hydrology of large alluvial fans formed by the Kurobe, Shō, and Jōganji rivers.

505. Yoshimura Shinkichi 吉村信吉, Chikasui 地下水(Groundwater), Tōkyō, Kawade Shobō, 1942, 2+3+258 pp., 4 photos, 67 illus.
A popular explanation of Japanese hydrology, including a discussion of natural mineral springs. Yoshimura was a pioneer in the field of limnology. With index.

506. Yoshimura Shinkichi 吉村信吉, "Haramachida fukin Sagamino daichi no chikasui, toku ni chūmizu no seijō 原町田附近相模野臺地の地下水, 特に宙水の性狀 (The groundwater of the Sagamino upland in the neighborhood of Haramachida, with special reference to perched water tables), Chirigaku hyōron, v. 19 (1943), no. 7, 1-17.
This is a report of hydrologic investigations made during the dry season, with special regard to perched water and its significance to upland settlement on the Sagamino terraces in Kanagawa prefecture.

507. Yoshimura Shinkichi 吉村信吉, "Musashino daichi no chikasui ni taisuru Ōizumi chikasui bakufusen no igi 武藏野臺地の地下水に對する大泉地下水瀑布線の意義 (The importance of the Ōizumi groundwater fall-line to the groundwater of the Musashino upland)," Chirigaku hyōron, v. 16 (1940), no. 10, 17-24.
The twelfth paper in a series on the hydrology of the Musashino district.

508. Yoshimura Shinkichi 吉村信吉, "Musashino daichi tōbu Ōizumi chikasui bakufusen oyobi fukin shochikasuitai no chikasui seisa 武藏野臺地東部大泉地下水瀑布線及附近諸地下水堆の地下水精査 (Detailed survey of the groundwater fall-line at Ōizumi in the eastern Musashino upland and measurements on groundwater mounds)," Chirigaku hyōron, v. 19 (1943), no. 11, 32-47; no. 12, 31-40.
Measurements of groundwater levels and water temperature were made at over 200 wells on the Musashino upland in order to determine the nature of perched water masses and underground falls. Yoshimura comments upon the distribution of settlements and wooded areas in terms of his conclusions regarding the water situation.

509. Yoshimura Shinkichi 吉村信吉, "Shōwa jūsannen ni okotta Musashino daichi chikasui no kassui oyobi daizōsui 昭和13年に起った武藏野臺地地下水の渇水及大增水 (Drought and abnormal rise of the groundwater table of the Musashino upland in 1938)," Chirigaku hyōron, v. 15 (1939), no. 3, 1-23.
The first in a series of articles reporting on pioneer studies of hydrologic conditions in the Musashino district and in other areas west of Tōkyō. Although the core of Yoshimura's work deals with technical aspects of hydrology, his frequent observations on the correlation between groundwater and topography and its relationship to land use and settlement patterns make Yoshimura's papers of interest to both the human and physical geographer.

510. Yoshimura Shinkichi 吉村信吉, "Tokorozawa-machi fukin no chikasui to shūraku no hattatsu 所澤町附近の地下水と聚落の發達 (Groundwater in the vicinity of Tokorozawa and its effect upon the development of settlements)," Chirigaku hyōron, v. 17 (1941), no. 1, 1-13; no. 2, 48-62.
Investigations of local differences and seasonal changes in groundwater levels in the area of Tokorozawa, Saitama prefecture.

511.. Yoshimura Shinkichi 吉村信吉, "Tokorozawa-machi tōhō Musashino daichi no chikasui, toku ni chūmizu to asai kubochi no seiin, shūraku ritchi to no kankei 所澤町東方武藏野臺地の地下水,特に宙水と淺い窪地の成因,聚落立地との關係 (Groundwater of the Musashino upland, east of Tokorozawa, with special reference to the origin of perched water and shallow valleys, and rural settlements)," Chirigaku hyōron, v. 16 (1940), no. 3, 1-25.
Yoshimura's sixth paper on groundwater conditions and settlement in the Musashino upland. Field work for this study was done in 1938-39 and in an area to the east of Tokorozawa.

512. Yoshimura Shinkichi 吉村信吉, "Tōkyō-shi seibu Iogi-Amanuma chikasuitai 東京市西部井荻天沼地下水堆 (The Iogi-Amanuma groundwater mounds of western Tōkyō)," Chirigaku hyōron, v. 16 (1940), no. 4, 31-46.
The second article in this journal by Yoshimura on hydrology and topography in the western suburbs of Tōkyō.

513. Yoshimura Shinkichi 吉村信吉, "Tōkyō-shi seibu Kamitakaido-Yodobashi chikasui bakufusen to keikan hattatsu 東京市西部上高井戸,淀橋地下水瀑布線と景觀發達 (The Takaido-Yodobashi groundwater fall-line in western Tōkyō and the development of the cultural landscape)," Chirigaku hyōron, v. 16 (1940), no. 8, 1-15.
A report on hydrologic investigations in the western suburbs of Tōkyō with comment on the relationship between groundwater conditions and the settlement pattern in that area.

514. Yoshimura Shinkichi 吉村信吉, "Tōkyō-shi seikō Ōizumi-machi no chikasui bakufusen to keikan hattatsu no kankei 東京市西郊大泉町の地下水瀑布線と景觀發達の關係 (Groundwater fall-line at Ōizumi-machi, a western suburb of Tōkyō and the development of the cultural landscape)," Chirigaku hyōron, v. 15 (1939), no. 7, 1-16.
The writer reports on his discovery of a groundwater fall-line during a hydrologic survey at Ōizumi, west of Tōkyō, and discusses the effects of that phenomenon upon the development of the settlement pattern in that neighborhood.

515. Yoshimura Shinkichi 吉村信吉 and Misawa Katsue 三澤勝衛, "Kami-Suwa onsen kenkyū 上諏訪溫泉研究 (A study of the Kami-Suwa hot springs)," Chirigaku hyōron, v. 7 (1931), no. 3, 12-21; no. 4, 1-24; no. 5, 70-90; v. 8 (1932), no. 4, 19-34; no. 5, 45-73; no. 6, 22-44.
A technical investigation of the physio-chemical properties of the natural hot springs at Kami-Suwa in Nagano prefecture including observations on underground temperature, the effect of the springs on air temperature, and an analysis of spring waters.

516. Yoshimura Shinkichi 吉村信吉, Sakamoto Keiichi 坂本圭一, and Tanaka Tōkichirō 田中藤吉郎, "Tōkyō-shi Yodobashi chikasui bakufusen (dainihō) 東京市淀橋地下水瀑布線（第2報）(The Yodobashi groundwater fall-line in Tōkyō city, part 2)," Chirigaku hyōron, v. 17 (1941), no. 6, 1-13.
Report on a detailed survey of groundwater conditions in Yodobashi ward, north-central Tōkyō. The location and significance of the fall-line is considered in regard to local fluctuations in groundwater levels, seasonal variations, PH factors, well depths, etc.

517. Yoshimura Shinkichi 吉村信吉, "Musashino daichi tōbu Ōizumi Hōya fukin daichi no asai kubochi chikei 武藏野臺地東部大泉保谷附近臺地の淺い窪地地形 (Shallow depressions on a plain near Ōizumi and Hōya in the eastern part of the Musashino upland)," Chirigaku hyōron, v. 19 (1943), no. 5, 15-32.
An investigation of local topography and the influence of groundwater in certain areas of the Musashino upland. The development of settlement at these places is also discussed.

E. Flora

518. Asai Jun 淺井淳, Nihon sekitan tokuhon 日本石炭讀本 (A reader on Japanese coal), Tōkyō, Kokon Shoin, 1941, 8+13+439 pp., 43 illus.
Essentially a history of coal mining in Japan. Activities during the Tokugawa and Meiji eras are recorded in considerable detail, whereas developments during the Taishō and Shōwa periods are summarized more simply. However, references to current mining problems in Japan occur frequently throughout the work.

519. Honda Seiroku 本多靜六, "Nihon shokubutsu chiri ni tsuite 日本植物地理に就て (On plant geography of Japan)," Chigaku zasshi, v. 14, no. 157 (1902), 8-18; no. 158, 51-76; no. 160, 223-237; no. 161, 297-324, 1 map.
Written by a dendrologist, this is a good article on the general geography of Japanese flora. Plants are listed on the basis of broadly delimited climatic zones and the principal species of each zone are described in some detail.

520. Kano Tadao 鹿野忠雄, "Kōtōsho seibutsu chirigaku ni kansuru shomondai 紅頭嶼生物地理學に關する諸問題 (Various problems concerning the biogeography of Kōtōsho)," Chirigaku hyōron, v. 11 (1935), no. 11, 40-49; no. 12, 47-75; v. 12 (1936), no. 1, 33-46; no. 2, 70-93; no. 10, 59-83; no. 11, 41-64; no. 12, 45-71.
Kōtōsho (Lan Yu or Botel Tobago) is a small island to the southeast of Formosa. Following an outline of the physical factors of the natural environment, Kano describes in detail the flora and fauna of the island, treating also distribution and classification. Comparing the biogeography of adjacent regions, he finds the closest similarity with the flora and fauna of the Batan islands, to the south.

521. Kishida Hisakichi 岸田久吉 and Yabe Yoshisada 矢部吉禎, Fuji no dōbutsu, Fuji no shokubutsu 富士の動物,富士の植物 (Animals of Mt. Fuji and plants of Mt. Fuji), Tōkyō, Kokon Shoin, 1928, 9+2+514 pp., and 4+2+200 pp., 107 illus.
The first section of this work, by Kishida, is a listing and description of the animal life found in the Mt. Fuji district. Nine of the ten chapters are devoted to classification of the fauna, beginning with mammals and end-

ing with spiders. Kishida's last chapter deals with the local distribution of these forms. Yabe's approach to the discussion of Fuji flora in the second portion of the book is essentially ecological. He gives a brief background of the history of botanical study in the area, sketches in the physical environment including geology and meteorology, and then lists and describes the flora on the basis of life zones. Notes on plants to be observed along the main climbing routes up Mount Fuji conclude the volume.

522. Kurihara Hisashi 栗原壽, "Izu Miyakejima no shokubutsu keikan 伊豆三宅島の植物景觀 (The vegetation of Miyakejima, Izu islands)," Chirigaku hyōron, v. 16 (1938), no. 3, 1-13, 11 photos.
An outline of the vegetation of Miyakejima, a volcanic island in the Izu island group, south of Tōkyō.

523. Makino Tomitarō 牧野富太郎, Makino Nihon shokubutsu zukan, kaiteiban 牧野日本植物圖鑑, 改訂版 (Makino's illustrated cyclopedia of Japan's flora, revised edition), Tōkyō, Hokuryūkan, 1949, 5+2+13+1070+29+72+35+11 pp.
A beautifully illustrated, scientific compilation. Approximately 3,200 species are described. In the index Japanese and Chinese names appear in Japanese syllabic order, Latin and English names in alphabetical order. The first edition of this work appeared in 1940.

524. Masamune Genkei 正宗嚴敬, "Taiwan no kōzan shokusō to sono kigen 臺灣の高山植相とその起源 (On the alpine vegetation of Formosa and its origin)," Chirigaku hyōron, v. 14 (1938), no. 7, 1-8.
A botanical survey of the vegetation at high altitude in the central mountains of Formosa. The author determined the type and distribution of sub-alpine and alpine-like plants. In this monograph he speculates on the origin of these plant forms in terms of past climate change. His conclusions are based upon observed botanical evidence.

525. Nō Toshio 能登志雄, Shūraku no chiri 聚落の地理 (Geography of rural settlements), Tōkyō, Kokon Shoin, 1952, 164 pp., 3 tables, 15 photos, and 26 illus.
A treatise on the methodology of settlement geography. On the basis of posit materials he has collected on Japanese rural settlement, the author attempts to develop further a landscape theory after the views of O. Schlüter and C. O. Sauer.

526. Shibata Keita 柴田桂太, ed., Shigen shokubutsu jiten 資源植物事典 (A cyclopedia of plants and plant products),* Tōkyō, Hokuryūkan, 1949, 24+876+35+64 pp., 445 illus. *English title.
Under Shibata's editorship, eleven specialists collaborated in the preparation of this encyclopedia which describes the distribution, botanical form, production, and the parts and methods of use of some 875 Japanese plants. Has an index of the Japanese names in the order of the syllabary, and an index of the Latin names in alphabetical order. Well illustrated.

527. Takeda Hisakichi 武田久吉, Nihon no shizembi 日本の自然美 (Natural beauty in Japan), Tōkyō, Fugaku Honsha, 1946, 2+126 pp., 21 illus.
A collection of the author's botanical papers, most of which were previously published in the Kagaku gahō (Scientific pictorial magazine) and other journals. The majority of the articles treat alpine plants and other mountain flora or are concerned with seasonal changes in the botanical landscape of Japan. The improvement and protection of scenic areas are also discussed.

528. Takeuchi Makoto 竹内亮, "Unzendake no shokubutsu keikan ni tsuite 雲仙岳の植物景觀に就て (On the vegetation of the Unzen volcanoes)," Chirigaku hyōron, v. 15 (1939), no. 4, 13-31.
A survey of the vegetation of Mt. Unzen, the Kyūshū volcano on the Shimabara peninsula east of Nagasaki. The survey is preceded by an introductory account of the mountain's geomorphology and climate. Takeuchi's description of the effect of volcanic activity upon the vegetation is of particular interest. With a bibliography.

529. Yasuda Atsushi 安田篤, Nihon sanchiirui zusetsu 日本產地衣類圖說 (Illustrations of the lichens of Japan), [=Saitō Hōonkai Gakujutsu Kenkyū Sōmubu 齋藤報恩會學術研究總務部 (Saitō Requital Foundation, General Affairs Section for Scientific Research), Gakujutsu kenkyū hōkoku 學術研究報告 (Scientific research reports), no. 2], Sendai, 1925, 6+9+118+8 pp., 25 plates.
This is a Japanese translation of Yasuda's work previously published in German. 197 species of lichen are described in relation to their distribution from northern Karafuto (Sakhalin) to southern Kyūshū.

F. Fauna

530. Hirase Shintarō 平瀬信太郎, Tennenshoku shashin Nihon kairui zufu 天然色寫眞日本貝類圖譜 (Japanese shells with illustrations in natural color), Tōkyō, Matsumura Sanshōdō, 1934, 14+2+1+129+218 pp., 129 color plates.
Scientific classification and description of the physical characteristics of Japan's Mollusca. Indices in Latin and in the Japanese syllabary, and a full English translation of text and plate captions.

531. Honda Masatsugu 本田正次, ed., Nihon-rettō-hen --Shizen kagaku kansatsu to kenkyū sōsho, dainishū 日本列島篇 自然科學觀察と研究叢書 第二輯 (The Japanese archipelago -- Observations and studies in the natural sciences, compilation no. 2), Tōkyō, Sangabō, 1944, 2+5+489 pp., maps, illus.
A general compendium of factual and descriptive information on the Japanese islands. Coverage includes aspects of both the natural and social sciences.

532. Kano Tadao 鹿野忠雄, "Kōtōsho no dōbutsu chirigakuteki kenkyū fu Wāresu-sen hokutan no mondai 紅頭嶼の動物地理學的研究 附ワーレス線北端の問題 (Zoogeographical study of Botel Tobago and the problem of the northern portion of Wallace's line)," Chirigaku hyōron, v. 9 (1933), no. 5, 23-41; no. 6, 21-37; no. 7, 39-61; no. 8, 27-53.

The nature of fauna on the island of Kōtōsho, southeast of Formosa, is closely analyzed, together with the historical geology of this region of the Pacific and the delimitation of life zones. Kano discusses the possibility of the one-time existence of a land-bridge between Formosa and Luzon.

533. Kano Tadao 鹿野忠雄, *Taiwan Tsugitaka san'i no dōbutsu chirigakuteki kenkyū* 臺灣次高山彙の動物地理學的研究 (Zoogeographical studies of the Tsugitaka mountains of Formosa), Tōkyō, Achikku Myūjiamu, 1940, 145 pp., 12 photos, and 27 illus.

This book is a major contribution to the knowledge of the biogeography of Southeast Asia. The ecology, taxonomy, and history of the fauna of Formosa are treated, with particular emphasis upon the vertical distribution of species in the Tsugitaka range. With a general summary of the geography and geology of the central Formosa mountains.

534. Okada Yaichirō 岡田彌一郎, "Fujisan no dōbutsukai 富士山の動物界 (The world of fauna of Mt. Fuji)," *Chigaku zasshi*, v. 52, no. 618 (1940), 16-41, 9 photos, 3 illus.

Most scientific aspects of Mt. Fuji have been exhaustively investigated with the exception of zoology, a field which has been relatively neglected. Okada presents an introductory survey of Fuji fauna, treated in terms of life zones and species, and based on altitude. The study includes the animal forms inhabiting the districts around the foot of the mountain.

535. Okazaki Tsunetarō 岡崎常太郎, *Tennenshoku shashin konchū nanahyakushu* 天然色寫眞昆蟲 700 種 (Seven hundred species of insects in natural color photographs), Tōkyō, Matsumura Sanshōdō, 1930, 8+4+144+16 pp., 62 photo plates.

Japan's most common insects, scientifically classified and briefly described. Index in the Japanese syllabary.

536. Ōshima Masamitsu 大島正滿, "*Yamame oyobi Amago no bumpu kyōkaisen ni tsuite* ヤマメ及びアマゴの分布境界線に就いて (On the distribution of *Oncorhynchus masou* and *Oncorhynchus formosanus* [river trout] in Japan proper)," *Chirigaku hyōron*, v. 6 (1930), no. 7, 634-656.

This is a short scientific discourse on the distribution and habitat of two species of fresh water trout found in Honshū and Kyūshū. The genus under study had been identified as also being indigenous to Formosa.

537. Uchida Seinosuke 内田清之助 and others, *Nihon konchū zukan, kaiteiban* 日本昆蟲圖鑑, 改訂版 (Iconography of Japanese insects, revised edition), Tōkyō, Hokuryūkan, 1950, 13+1738 pp., 15 colored plates.

An illustrated encyclopedia of Japanese entomology compiled by 25 specialists. Included are insects of Sakhalin, Korea, Formosa, and the former mandated islands. About 5,000 species are described. The first edition was printed in 1932.

538. Uchida Seinosuke 内田清之助 and others, *Nihon dōbutsu zukan* 日本動物圖鑑 (Illustrated encyclopedia of the fauna of Japan), Tōkyō, Hokuryūkan, rev. ed., 1947, 1898+108+89 pp., 12 plates.

A collaborative work by 20 zoologists and zoogeographers, classifying and describing over 5,000 species of fauna common in Japan, Korea, Manchuria, Formosa, and the former mandated islands. Explanations of the species, ecology, distribution, and utility from the human standpoint are given. With indices of Japanese and scientific names.

G. Soils

539. Harada Masao 原田正夫, *Kantō rōmu no seisei ni tsuite* 関東ロームの生成について (On the origin of Kantō loam), Tōkyō, Yōkendō, 1943, 140 pp.

This study was first published as Report No. 3 of Tōkyō Teikoku Daigaku Nōgakubu, Nōgei Kagakuka, Dojō Hiryō Kenkyūshitsu (The Institute of Soil and Fertilizer Science, Department of Agricultural Science, Faculty of Agriculture, Tōkyō Imperial University). It is a study of the origin and distribution of the brown volcanic soils termed Kantō loam, found especially in the Tōkyō region. This work resulted from a long period of observation, sampling, and testing by the author. A valuable report for the physical geographer or land use specialist concerned with soil science in Japan.

540. Hosoda Katsuki 細田克己, *Hompō ni okeru iwayuru kurotsuchi no kenkyū* 本邦に於ける所謂黒土の研究 (Investigation of the so-called black soils in Japan), Tottori, Tottori Kōtō Nōgyō Gakkō, 1928, 238 pp., 19 photos.

Summarizing ten years of laboratory and field research this is an excellent report on the unproductive black soils of Japan which occupy a total area estimated at over 200,000 square kilometers. Most of the black soil acreage remains uncultivated in spite of the fact that much of it occurs in lowland areas where other farm land is at a premium. In large part this study is a technical analysis of the physical, chemical, and micro-biological properties and characteristics of these soils. A general summary is included. Valuable to the geographer and others concerned with Japanese land use problems.

541. Ishidō Shin 石堂信, *Ishikawa-ken dojō gaiyō* 石川縣土壤概要 (Outline of soils in Ishikawa prefecture), Ishikawa, Ishikawa Kenritsu Nōji Shikenjō, 1935, 2+164 pp.

A description of the soils in Ishikawa prefecture and a listing of tests for the respective soil types to indicate fertilizer requirements.

542. Kamoshita Kan 鴨下寛, *Nihon dosei chōsaron* 日本土性調査論 (Soil surveys in Japan), Tōkyō, Kawade Shobō, 1948, 2+83 pp.

An introduction to Japanese soil science and soil survey methods, written by a leading specialist. With an outline history of the development of soil surveying in Japan.

543. Kanno Ichirō 菅野一郎, Dojō chōsahō 土壌調査法(Methods of soil surveying), Tōkyō, Kokon Shoin, 1953, 2+4+298 pp., 48 illus., 50 tables.

A general textbook of soil science covering both theoretical and practical field aspects of this subject. Chapters concern other countries and foreign techniques as well as Japan and Japanese methods.

544. Kiuchi Shirōbei 木内四郎兵衛, "Manshū ni okeru tōketsu genshō 満洲に於ける凍結現象 (Phenomena relating to the perpetually frozen subsoil in Manchuria)," Chirigaku hyōron, v. 19 (1943), no. 7, 18-31; no. 8, 40-45.

The distribution of permafrost is outlined, with assumptions based on field surveys made at several key points. The origin of the subsoil in Manchuria is also considered with frequent references to the theories of other scholars, especially to the studies of the Russians in adjacent regions of Siberia.

545. Nagasaki-ken Nōji Shikenjō 長崎縣農事試驗場 (Nagasaki Prefecture Agricultural Testing Station), Nagasaki-ken dosei yōran 長崎縣土性要覧 (Outline of the soils of Nagasaki prefecture), Isahaya, Nagasaki Kenritsu Nōji Shikenjō, 1930, 89 pp.

Prior to 1928 Nagasaki soils had been studied and mapped on a gun basis. This pamphlet presents geological and soil tables by machi and mura, and includes the results of soil productivity experiments carried out on more than 100 representative soil types within the prefecture. Its purpose is to assist farmers in determining fertilizer requirements.

546. Nakao Seizō 中尾清藏, "Kikaiteki sosei narabi ni kōbutsu sosei yori mitaru iwayuru Kantō rōmu 機械的組成並に鑛物組成より見たる所謂關東ローム (The so-called Kantō loam from the viewpoint of its mechanical constitution and mineral composition)," Chishitsugaku zasshi, v. 36, no. 426 (1929), 91-102; v. 38, no. 450 (1931), 97-121; v. 39, no. 468 (1932), 580-586; no. 471, 747-758.

A technical soil study of the physical and chemical characteristics of the Kantō loam. Samples were taken for analysis from a variety of sites both east and west of Tōkyō.

547. Nōshōmushō Chishitsu Chōsajo Doseika 農商務省地質調査所土性課 (Soil Section, Bureau of Research on the Nature of Soils, Agriculture and Commerce Ministry), Dosei shiken seiseki 土性試驗成蹟 (Report of the soil survey), Tōkyō, Nōshōmushō Chishitsu Chōsajo, 1902, 6+200 pp.

An early soil study. Two hundred samples taken at points throughout Japan are analyzed and their characteristics described.

548. Shioiri Matsusaburō 鹽入松三郎 and Aomine Shigenori 青峰重範, Kyūkanki ni okeru suiden dojō kansō no kōka ni tsuite 休閑期に於ける水田土壌乾燥の効果に就て (On the effect of drying paddy soils in the season when the fields are empty), Tōkyō, Nōrinshō Nōji Shikenjō, 1940, 30 pp.

Rice-paddy soils often contain insufficient nitrogen with resultant low productivity causing serious economic hardship. The authors report in this pamphlet on their own investigations of this problem while conducting studies for the Nōji Shikenjō or Agricultural Experiment Station in Tōkyō. Thorough drying of the paddy soils, periodically, appears to be a key to improving nitrogen content. The report cites specific methods and effects and should be of definite interest to the land use specialist.

549. Watanabe Shigeo 渡邊重雄, Tochigi-ken no dojō 栃木縣の土壌 (Soils of Tochigi prefecture), Utsunomiya, Shūeidō, 1926, 2+2+3+4+126 pp., 1 map, 3 photos.

A classification and description of the soils of Tochigi prefecture followed by a discussion of fertilizer requirements and testing methods. Economy measures in the application of fertilizers to acidic soils are pointed out in the closing section. The accompanying soil map is clearly explained.

H. Climate and weather

550. Arakawa Hidetoshi 荒川秀俊, Nihon no kikō 日本の氣候 (The climate of Japan), Tōkyō, Heibonsha, 1948, 2+157+21 pp., 10 tables, 30 maps.

A general outline of the climate of Japan with a meteorological analysis of its regionality. Also, several minor topics are treated in some detail; e. g., the climatic causes of crop failure in Tōhoku, and the micro-climate of urban areas.

551. Chūō Kishōdai 中央氣象臺 (Central Meteorological Observatory), Shin-Nihon kikō zuchō 新日本氣候圖帖 (A new climatological atlas of Japan), Tōkyō, Sangyō Kishōkai, 1948, 42 maps+12 pp.

This general climatological atlas is a reliable and handy reference work for the geographer. Most of the maps are drawn on a nationwide base and show meteorological elements and their distribution pattern or mean condition by month. The second portion contains supplementary climatic statistics for principal weather stations. Headings, place names, and explanatory text are in both Japanese and English.

552. Chūō Kishōdai 中央氣象臺 (Central Meteorological Observatory), Yuki no kikōzu 雪の氣候圖 (Climatological maps of snow), Niigata, Nihon Sekisetsu Rengō, 1949, 32+20 pp., 30 maps.

A reliable climatic atlas of snow in Japan. Thirty small-scale colored maps show by isopleth methods: the first and last date of snowfall, duration of continuous snow cover, maximum and mean depths, and other factors related to the nature and distribution of snowfall, snow accumulation, and melting. Supplementary graphs give mean daily snowfall at 37 observatories during a 15 year period. The statistics upon which the maps are based are included in table form together with more detailed data. Map titles and other headings are in English and Japanese.

553. Daigo Yoshiyasu 大後美保, Nihon nōgyō kishōzu binran 日本農業氣象圖便覽 (Handy atlas of Japanese agricultural meteorology), Tōkyō, Kyōritsu Shuppan Kabushiki Kaisha, 1947, 9+234 pp., 420 maps.
An interesting and useful atlas of Japanese agricultural phenology. All 420 maps on climate, seasonal phenomena in farming, flora and fauna, etc., are drawn on the same base at a scale of 1:25,000,000 by the use of isopleth techniques. Map headings are in English. With 234 pages of explanatory material.

554. Daigo Yoshiyasu 大後美保, Nihon sakumotsu-kishō no kenkyū 日本作物氣象の研究 (Studies in the meteorology of Japanese farm products), Tōkyō, Asakura Shoten, 1945, 11+655 pp., 251 maps.
The writer's many studies on agricultural phenology are summarized in this book which discusses the relationship between crop yields and climatic conditions. Twenty-five of Japan's major crops are considered. With abundant statistical data, bibliography, and index.

555. Fujiwara Sakuhei 藤原咲平 and Kawaguchi Sanjūrō 川口三重郎, "Nihon ni okeru shitsudo no bumpu ni tsuite 日本に於ける濕度の分布について (On the distribution of humidity in Japan)," Kishō shūshi, v. 5 (1927), no. 7, 161-166.
Humidity conditions for Japan as a whole are shown in terms of annual, monthly, and daily data. For selected locations, the patterns of annual variation in humidity are indicated.

556. Fukui Eiichirō 福井英一郎, "Atami-shi fukin ni okeru kion bumpu shōkikō kenkyū no gimmi 熱海市附近に於ける氣温分布(小氣候研究の吟味) (Temperature distribution in the vicinity of Atami city: A study in the investigation of micro-climate)," Chirigaku hyōron, v. 16 (1940), no. 3, 26-37.
Observation and interpretation of the temperature regimen in the area of the sea-side resort city of Atami, Shizuoka prefecture. Air temperature ranges are shown by isotherm techniques. Fukui explains some influences of the local landforms upon temperature readings.

557. Fukui Eiichirō 福井英一郎, "Hokkaidō no kikōgakuteki kenkyū 北海道の氣候學的研究 (A climatological study of Hokkaidō)," Chirigaku hyōron, v. 5 (1929), no. 9, 1-22.
Data from 80 weather stations and meteorological observatories on Hokkaidō and adjacent small islands were used, and a part is presented in this climatic analysis. Presentation is rather static in form, although the coverage is relatively comprehensive. In conclusion the climatic divisions delimited by the author are shown.

558. Fukui Eiichirō 福井英一郎, "Hompō daitoshi ni okeru shigaisen no bumpu 本邦大都市に於ける紫外線の分布 (Distribution of ultra violet radiation around the great cities of Japan)," Chirigaku hyōron, v. 15 (1939), no. 5, 55-76.
An investigation of the intensity of ultra-violet insolation received in the downtown areas of three large cities (Tōkyō, Ōsaka, and Nagoya) in comparison with the amount received in the suburbs. Fukui used L. Hill's methylene blue-aceton method of measurement.

559. Fukui Eiichirō 福井英一郎, "Hompō ni okeru kōsui henkaryō no kenkyū 本邦に於ける降水變化量の研究 (Studies of rainfall variability in Japan)," Chirigaku hyōron, v. 10 (1934), no. 9, 1-17, 1 map.
Rainfall variability at 96 stations in Japan and adjacent areas was calculated by means of Maurer's method, and the variability factor was mapped. This is an analysis of the map and the research process behind it.

560. Fukui Eiichirō 福井英一郎, "Hompō tōzai ryōkaigan ni okeru kion no hikaku 本邦東西兩海岸に於ける氣溫の比較 (Comparison of air temperature on the eastern and western coasts of Japan)," Chirigaku hyōron, v. 6 (1930), no. 7, 266-276.
A comparative climatic study of the Pacific and Japan Sea coasts. Annual temperature variations, seasonal extremes, latitudinal and oceanic influences, etc., are dealt with, and Fukui compares his conclusions with Koeppen's classification for the same regions.

561. Fukui Eiichirō 福井英一郎, "Izu Shimoda fukin no uryō bumpu 伊豆下田附近の雨量分布 (Distribution of rainfall around Shimoda, Izu peninsula)," Chirigaku hyōron, v. 12 (1936), no. 6, 1-27.
A report on the effects of topography upon local rainfall. This study is based on meteorological observations made by personnel of the Tōkyō Bunrika Daigaku Chirigaku Kenkyūshitsu (the Geographical Laboratory of Tōkyō Bunrika University at Izu Kaiyō Seibutsu Shikenjo (the Izu Marine Biological Station). Twenty maps show the distribution of precipitation for each rain of significance occurring during the study period. The results of ascending and descending air currents and the role of adiabatic cooling are carefully studied and conclusions of interest to micro-climatologists, climatologists and meteorologists are set forth.

562. Fukui Eiichirō 福井英一郎, Nampōken no kikō 南方圏の氣候 (The climate of the South Seas region), Tōkyō, Tōkyōdō, 1942, 3+4+279+14+16 pp., 4 maps and numerous illus.
This book was written for non-professional readers with the aim of improving the understanding of the South Seas region which was under Japanese occupation in 1942. An index map of place names and climatic table of 100 places within the region are added.

563. Fukui Eiichirō 福井英一郎, "Nemuro chihō no kikō to nōgyō 根室地方の氣候と農業 (Climate and agriculture in the Nemuro district)," Chirigaku hyōron, v. 5 (1929), no. 12, 1-9.
Correlates the climatic conditions and agricultural practices characteristic of this coastal region of eastern Hokkaidō.

564. Fukui Eiichirō 福井英一郎, "Nihon no kikōku (dainihō) 日本の氣候區（第2報）(Climatic divisions of Japan: second report)," Chirigaku hyōron, v. 9 (1933), no. 1, 1-19; no. 2, 19-37; no. 3, 11-35; no. 4, 1-30.

A general climatic survey of Japan including the Kuriles, Karafuto, Korea, and Formosa. Such factors as the effect of latitude, topography, and the role of monsoonal circulation are discussed. Fukui delimits three major climatic zones and a number of sub-divisions.

565. Fukui Eiichirō 福井英一郎, "Sakyū chitai ni okeru sekisetsu bumpu ni tsuite 砂丘地帯に於ける積雪分布に就いて (On snow distribution on the surface of sand dunes)," Chirigaku hyōron, v. 12 (1936), no. 5, 1-23.
The peculiar characteristics of snowfall distribution and accumulation in the Hokuriku region are examined in this article. The author establishes the relationship between local variations in accumulation and the effect of topographic and other physiographic features upon prevailing winds. Data obtained in the field at observation stations near Kashiwazaki and Niigata are presented to support his conclusions.

566. Fukui Eiichirō 福井英一郎, "Sekisetsu no keinen henka 積雪の經年變化 (Secular variation of snow accumulation)," Chirigaku hyōron, v. 18 (1942), no. 5, 1-17.
A regional study of fluctuations in yearly snow accumulation. With a map indicating the variability factor and the distribution of heavy snowfall for all of Japan.

567. Fukui Eiichirō 福井英一郎, "Waga kuni ni okeru kikō bunrui ni tsukite 我邦に於ける氣候分類に就きて (On a classification of Japanese climates)," Chirigaku hyōron, v. 4 (1928), no. 9, 1-13.
A semi-technical climatological monograph. Fukui used a formula based on temperature and precipitation variables to derive nine regional climatic categories for Japan proper. Rainfall divisions, calculated from quantitative and monthly variability data and numbering seven classes and four regional types, are then identified and explained. A general summary follows.

568. Fukui Eiichirō 福井英一郎 and Imamura Gakurō 今村學郎, "Yakushima no kenkyū sono ni: Satsunan shotō no kikō ni kansuru nisan no hōkoku 屋久島の研究その2. 薩南諸島の氣候に關する二・三の報告 (Study of Yakushima, part II: Some notes on the climate of the Satsunan islands)," Chirigaku hyōron, v. 8 (1932), no. 4, 35-48.
The writers point out the existence of a sharp climatic difference between southern Kyūshū and the neighboring Satsunan islands (Ōsumi-guntō); a reacknowledgment and further refinement of a climatic boundary between Kyūshū and the Ryūkyūs which had been earlier identified by these researchers.

569. Fukui Eiichirō 福井英一郎 and Koizumi Gihei 小泉義兵, "Dai-Tōkyō ni okeru uryō no bumpu 大東京に於ける雨量の分布 (The distribution of precipitation in Tōkyō)," Chirigaku hyōron, v. 14 (1938), no. 6, 1-25.
Fifty observation stations were established within the 35 wards of metropolitan Tōkyō and the distribution, amount, and frequency of precipitation from July to November, 1952, were mapped and recorded. Observations were made for too short a period to be of much climatological significance. The writers' conclusions concern long-range rainfall trends.

570. Fukui Eiichirō 福井英一郎 and Wada Norio 和田憲夫, "Hompō no daitoshi ni okeru kion bumpu 本邦の大都市に於ける氣温分布 (Horizontal distribution of the air temperature in the greater cities of Japan)," Chirigaku hyōron, v. 17 (1941), no. 5, 22-40.
A report of carefully-conducted observations on air temperature made in and around three of Japan's largest cities, Tōkyō, Ōsaka, and Nagoya. The writers present their conclusions on the reasons for increased temperatures in urban areas.

571. Hoyanagi Mutsumi 保柳睦美, "Hokushi Mōkyō no kikō henka to nenrin bunseki 北支蒙疆の氣候變化と年輪分析 (Climatic changes and tree-ring analysis in Northern China and Inner Mongolia)," Chirigaku hyōron, v. 16 (1940), no. 12, 1-17.
During field investigations in Inner Mongolia, Hoyanagi found evidence of more abundant rainfall in the past by analyzing the annual growth rings of elm trees in that region. This article summarizes his findings and conclusions.

572. Ichihara Tadashi 市原正, "Nihon no kōi kisetsu ni tsuite 日本の更衣季節に就いて (On seasons for changing clothes in Japan)," Chigaku zasshi, v. 47, no. 555 (1935), 6-12.
A rather unique socio-climatic study of seasonal change in Japan founded on a correlation of air temperature data and observations on personal habits in dress. Investigation was made in terms of both Japanese and Western-style clothing, transition periods when changes in seasonal types of clothing occurred, and the duration of the period during which a particular type of seasonal garb is predominantly worn. The regional patterns which emerge from this social approach to climatic interpretation are presented in graphic form.

573. Iwasaki Kenkichi 岩崎健吉, "Tosa Murotozaki fukin kaigan ni okeru bōfūrin no bumpu ni tsuite 土佐室戸岬附近海岸に於ける防風林の分布に就いて (On the distribution of wind breaks on the coast near Cape Muroto)," Chirigaku hyōron, v. 15 (1939), no. 2, 28-51.
Concerning the nature and utilization of wind breaks in the region of Cape Muroto in southeastern Shikoku.

574. Kanao Sōhei 金尾宗平, "Hompō ni okeru nōji tōchakushusenzu ni tsuite 本邦に於ける農事等着手線圖に就て (On the isochrono-curvi of the commencement of agricultural work)," Chirigaku hyoron, v. 4 (1928), no. 12, 48-54.
Local planting and harvesting times for selected crops and for the hatching of silkworms are plotted and the resulting isochronal map analyzed in relation to regional climatic patterns and conditions.

575. Nagai Masatarō 長井政太郎, "Sekisetsu ni kansuru bumpu zurui ni tsuite 積雪に關する分布圖類について (On the geographical distribution of snowfall and related studies)," Chirigaku hyōron, v. 10 (1934), no. 6, 1-24.
Winter-long observations on snow conditions were made by school children throughout Yamagata prefecture, and Nagai utilized that source of data to make this detailed study of snowfall. After discussing snowfall frequency, distribution, accumulation, thawing, etc., he divides Yamagata into three generalized snowfall regions.

576. Nagai Masatarō 長井政太郎, "Yamagata-ken-nai sekisetsu no bumpu to sono chirigakuteki igi 山形縣内積雪の分布とその地理學的意義 (The distribution of accumulated snow in Yamagata prefecture and its geographical meaning)," Chirigaku hyōron, v. 7 (1931), no. 1, 42-54; v. 10 (1934), no. 6, 1-24.

Nagai divides Yamagata prefecture into three climatic regions based in part upon an analysis of his maps on the depth of snowfall and his iso-chronologic charts on thawing, cherry blossom bloom, etc.

577. Nishioka Hideo 西岡秀雄, Kandan no rekishi 寒暖の歴史 (History of warmth and coldness), Tōkyō, Kōgakusha, 1949, 4+5+205+8 pp., 14 maps, 8 illus., and 24 tables.

An explanation of the author's theory on the climatic changes which have affected Japan since prehistorical times. Citing evidence ranging back three centuries, Nishioka recognizes cyclical temperature variations which have occurred about every 700 years. He founds his arguments on sustentative fact and hypothesis gleaned from his study of archeological materials, old documents, growth rings of trees, geology and other physical sciences, social anthropology, history, etc.

578. Obara Nobuhiko 尾原信彦, "Shimoda-machi Nabetagawadani ni okeru fūkō to uryō no kankei 下田町鍋田川谷に於ける風向と雨量の關係 (The relationship between wind direction and rainfall along the valley of Nabeta river near Shimoda)," Chirigaku hyōron, v. 13 (1937), no. 1, 38-62.

A study of local climate and weather characteristics along a small stream valley inland from Shimoda on the southeastern coast of the Izu peninsula. Positive opinions on the influence of topographic features are presented and supported by observational data.

579. Okada Takematsu 岡田武松, "Hompō kion no chirigakuteki bunseki 本邦氣温の地理學的分析 (Geographical analysis of air temperature in Japan)," Chirigaku hyōron, v. 7 (1931), no. 4, 68-80.

Experimental calculation of air temperature using a theoretical formula based upon geographical factors.

580. Okada Takematsu 岡田武松, "Kion no nichikōsa ni oyobosu kairiku no eikyō ni tsuite 氣温の日較差に及ぼす海陸の影響に就いて (On the effect of the areal ratio of land and sea on the daily range of air temperature)," Chirigaku hyōron, v. 7 (1931), no. 3, 1-11.

After comparing long-term statistics on the daily variation of air temperature with experimental computations, Okada concluded that the effect of areal differences in the distribution of land and sea on diurnal temperature change is restricted to a surprisingly limited area.

581. Ono Suminosuke 小野澄之助 and Nishimura Denzō 西村傳三, "Nihontō fukin no heikin kion 日本島附近の平均氣温 (Average air temperature in the neighborhood of Japan)," Kishō shūshi, v. 6 (1928), no. 9, 317-325.

Based on abundant observational data, the writers advance a formula for determining average air temperature at any position in Japan as a function of such factors as latitude, longitude, altitude, etc.

582. Sapporo Kanku Kishōdai 札幌管區氣象臺 (Sapporo District Meteorological Observatory), Hokkaidō no kikō 北海道の氣候 (The climate of Hokkaidō), Sapporo, 1953, 12+76+24+127+48+108+24 pp.

Designed for the layman, this is a detailed and useful compilation of climatic data and information published by the Sapporo Meteorological Observatory. With 33 climatic maps and comprehensive statistical tables.

583. Sasakura Kōzō 佐々倉航三, "Kikōjō yori mitaru Shōwa rokunen no kome kyōsaku ni tsuite 氣候上より見たる昭和六年の米凶作に就て (On the poor rice crop of 1931 considered climatically)," Chirigaku hyōron, v. 8 (1932), no. 3, 33-43.

A climatic study of the causes of crop failure. The influence of air temperature and insolation upon rice growth is stressed.

584. Sasakura Kōzō 佐々倉航三, "Nihon ni okeru kayatsuri jōkyō to kikō yōso to no kankei ni tsuite 日本に於ける蚊帳吊状況と氣候要素との關係に就いて (On the relationship between the use of mosquito nets and climatic elements in Japan)," Chirigaku, v. 7 (1939), no. 9, 59-67.

An interesting study of the relationship between the use of mosquito nets and certain climatic elements, principally air temperature. Observations were made at Tōkyō and at other selected points throughout Japan.

585. Sasakura Kōzō 佐々倉航三, "Shinano Ōmachi fukin no kion bumpu no nitchū henka ni tsuite 信濃大町附近の氣温分布の日中變化に就いて (On the diurnal variation of the distribution of air temperature in the town of Ōmachi in Shinano and its vicinity)," Chirigaku hyōron, v. 8 (1932), no. 1, 1-11.

A micro-climatic study. Ōmachi is located at the eastern foot of the northern Japanese Alps in Nagano prefecture. The influences of both cultural and natural elements of the landscape upon temperature variation are pointed out.

586. Sasakura Kōzō 佐々倉航三, Shōkikōgaku 小氣候學 (Kleinklima), Tōkyō, Kokon Shoin, 1950, 3+3+3+167 pp., 93 illus.

This well-illustrated volume presents Sasakura's concept of regional climatology. Detailed climatic observations confined to local areas constitute the principal basis for his theoretical conclusions. He correlates climatic factors with terrain elements and with various aspects of human geography, citing many specific examples concerning Japan. In the concluding chapter his observations on phenology are particularly noteworthy.

587. Sasakura Kōzō 佐々倉航三 and Awaji Shōzō 淡路正三, "Tōkyō ni okeru keidofūkō to chijōfūkō to no kankei ni tsuite 東京に於ける傾度風向と地上風向との關係に就て (On deflection angles and gradient winds at Tōkyō)," Chirigaku hyōron, v. 9 (1933), no. 9, 29-35.

An analysis of gradient wind data derived from synoptic charts of the Tōkyō Meteorological Observatory for the period 1929-31. The Guldberg-Mohn coefficient of resistance is calculated for the Tōkyō area and the prevailing wind directions and variability are noted.

588. Sasakura Kōzō 佐々倉航三, "*Tōkyō-shi oyobi sono fukin ni okeru shosō kijitsu no bumpu no ichirei* 東京市及其附近に於ける初霜起日の分布の一例 (On the distribution of the dates of the first hoarfrost in Tōkyō and vicinity)," *Chirigaku hyōron*, v. 15 (1939), no. 8, 47-53.

Based on observations made at a dozen stations in the neighborhood of Tōkyō, a map was prepared showing the location, date, and time of the first hoarfrost. This was correlated with data on minimum temperatures and weather conditions. Conclusions indicate that frost more often occurred on clear nights and that places with dense population recorded higher air temperatures and later and less frequent frost.

589. Sasakura Kōzō 佐々倉航三 and Misawa Katsue 三澤勝衛, "*Shōwa rokunen hachiku ryōgetsu no Nagano-ken Suwa-gun oyobi sono fukin ni okeru kōsuiryō no bumpu ni tsuite* 昭和六年八.九兩月の長野縣諏訪郡及其附近に於ける降水量の分布に就いて (On the distribution of precipitation during August and September, 1931, in Suwa district, Nagano prefecture, and its neighborhood)," *Chirigaku hyōron*, v. 8 (1932), no. 10, 1-20.

Observations on the local distribution of rainfall in a section of Nagano prefecture to determine the influence of topography upon precipitation. Data was collected at 27 rain-gauge stations for a two-month period.

590. Sasakura Kōzō 佐々倉航三, "*Shōwa shichinen hachikugatsu Nagano-ken Suwa-gun oyobi sono fukin ni okeru uryō kansoku hōkoku* 昭和七年八.九月長野縣諏訪郡及其附近に於ける雨量觀測報告 (Report on the distribution of precipitation during August and September, 1932, in Suwa district, Nagano prefecture, and its neighborhood)," *Chirigaku hyōron*, v. 10 (1934), no. 7, 15-34.

Using observational data recorded at 46 stations in Suwa-gun and vicinity, Sasakura drew maps showing for a two-month period the distribution of precipitation, wind direction, and the frequency of thunder storms. This weather information he then correlated with the topographic situation and drew conclusions regarding the interrelationship of these two geographic factors.

591. Sasakura Kōzō 佐々倉航三, "*Tōkyōshi naigai no kionsa ni tsuite* 東京市内外の氣温差に就て (On differences of air temperature within, and in the vicinity of, Tōkyō)," *Chirigaku hyōron*, v. 7 (1931), no. 7, 29-37.

A micro-climatic study of the effect of a great metropolitan center upon air temperature.

592. Sekiguchi Takeshi 関口武, "*Hompō fukin no kōsui nissū no bumpu narabi ni sore ni yoru kikō kubun* 本邦附近の降水日数の分布並にそれによる氣候區分 (Distribution of rainy days in Japan and the climatic divisions dependent on them)," *Chirigaku hyōron*, v. 21 (1949), no. 12, 6-12.

An analysis of the frequency and distribution of rainfall in Japan and its annual variability led the author to propose a breakdown of climatic regions determined by the number of rainy days.

593. Sekiguchi Takeshi 関口武, "*Iwayuru 'Ohōtsuku' kai kikai ni tsuite* 所謂「オホーツク」海氣塊に就て (On the so-called 'Okhotsk' air mass)," *Chirigaku hyōron*, v. 19 (1943), no. 2, 22-32.

From an analysis of synoptic weather charts, Sekiguchi denies the existence of an independent "Okhotsk" air mass. He identifies the usual air conditions of that region as representing an air-mass complex, resulting from the interaction of three other air masses, which he identifies and describes.

594. Sekiguchi Takeshi 関口武, "*Kaze no chihōmei no kenkyū, sono ni: yamaji (yamaze) to yamase* 風の地方名の研究その2 ヤマジ(ヤマゼ)とヤマセ (On the local names of winds, part 2: yamaji (yamaze) and yamase)," *Chirigaku hyōron*, v. 17 (1941), no. 10, 50-68.

An ethnographical and meteorological discussion of regional terminology for the wind; particularly an explanation of the nature of the "*yamaji*" winds of the Inland Sea region and "*yamase*" winds of the Japan Sea coast.

595. Sekiguchi Takeshi 関口武, "*Kikō hyōgen ni okeru suibun heikō no mondai* 氣候表現に於ける水分平衡の問題 (The problem of the water balance as a method of climatic representation)," *Chirigaku hyōron*, v. 22 (1949), no. 9, 16-22.

A discussion of theories on the equilibrium of water as a synthetic expression of climate. Isopleth maps of Japan show Lang's precipitation factor, E. de Martonne's index of aridity, and Angström's coefficient of humidity.

596. Sekiguchi Takeshi 関口武, "*Kikōteki ni mita Nihon kakuchi no mizu no kafusoku* 氣候的に見た日本各地の水の過不足 (Water balance in various areas of Japan viewed climatologically)," *Chirigaku hyōron*, v. 23 (1950), no. 7, 1-12.

An investigation of the evaporation-precipitation ratio and a quantitative interpretation of humidity and aridity factors in Japan, as an aid in determining the hydrologic resource.

597. Sekiguchi Takeshi 関口武, "*Nihon no nenkō ni tsuite* 日本の年候に就いて (On the annual climate in Japan)," *Chirigaku hyōron*, v. 24 (1951), no. 6, 1-11.

Köppen's climatic formula is applied to Japan to indicate annual climatic types, and the resulting geographical patterns are studied. Variations in the annual climate at selected stations are analyzed.

598. Suda Kanji 須田皖次, "*Tōhoku no reigai* 東北の冷害 (On the poor crops due to cool summers in northeastern Japan)," *Chigaku zasshi*, v. 59 (1950), no. 3, 20-28.

Rice yields in Tōhoku are frequently much reduced by low summer temperatures. During the past forty years a number of meteorologists, agriculturists, and oceanographers have studied the causes of unusually cool summers in that region and several hypotheses were published concerning such diverse climatic influences as cold currents (the *oyashio*), sea ice, sun spots, volcanic activity, typhoons, and atmospheric and oceanic circulation. The author here briefly explains these theories and proposes new research approaches to this problem and some methods of forecasting.

599. Taguchi Tatsuo 田口龍雄, Yuki 雪 (Snow), Tōkyō, Kokon Shoin, 1940, 9+201 pp., 87 illus.
An account of both scientific and popular knowledge concerning snow and human adaptation to it, particularly as regards Japan. The first of three sections treats snowfall as a meteorological phenomenon; the second pertains to snow accumulation and related effects such as drifting, melting, etc.; and the last portion is a description of man's adjustment to life in snow country.

600. Tsukuda Jūkichi 佃 十吉 and Yoshikawa Taizō 吉川泰三, "Nihon kinkai ni okeru unryō no bumpu ni tsuite 日本近海に於ける雲量の分布に就いて (On cloudiness over the adjacent seas of Japan)," Umi to sora, v. 10 (1930), no. 9, 329-343.
Data on the cloudiness over Japan is abundant but this is one of the few studies of offshore cloudiness. Observations on cloud cover are recorded from a variety of sources and the monthly variation in cloudiness over Japanese waters is indicated by means of isopleth maps.

601. Yaguwa Risuke 八鍬利助, Machida Minoru 町田實, and Ōkita Jūzō 大喜多重三, "Hyōgo-ken-nai no uryō chōsa 兵庫縣内の雨量調査 (A survey of the rainfall in Hyōgo prefecture)," Umi to sora, v. 6 (1926), no. 4, 72-77; no. 7, 116-123; no. 12, 228-243.
One of the first detailed studies of rainfall on a prefectural base. The contrast in the precipitation of the coast of the Japan Sea and of the Inland Sea is impressively shown.

602. Yamaguchi Sadao 山口貞夫 and Tsujimura Tarō 辻村太郎, "Nihon-guntō fukin ni okeru tōsho no bunrui oyobi bumpu 日本群島附近に於ける島嶼の分類及分布 (Classification and distribution of the islands in the vicinity of the Japanese archipelago)," Chirigaku hyōron, v. 11 (1935), no. 8, 25-50.
Only islands smaller than Etorofu (approx. 3,000 sq. km.) in the Kuriles are considered. The authors attempt a classification of islands based on a geomorphological analysis, following the general delimitation of oceanic and continental categories.

603. Yazawa Taiji 矢澤大二, "Hompō kakuchi no kikōteki shihyōsei ni tsuite 本邦各地の氣候的示表性について (On climatic representativeness of places in Japan)," Chirigaku hyōron, v. 21 (1948), nos. 7 and 8, 25-32.
An analysis of local climatic data in an effort to select the minimum number of weather stations necessary to reflect statistically the representative regional climates characteristic of Japan. Research was based upon the average monthly temperature and precipitation records available for 33 observation stations for a 50-year period.

604. Yazawa Taiji 矢澤大二, "Keikan ni arawareta kikōteki tokusei 景觀に現われた氣候的特性 (Climatic characteristics reflected by the landscape)," Chirigaku hyōron, v. 23 (1950), no. 1, 1-9.
A consideration of regional climatic characteristics in Japan as reflected in the natural and cultural landscape. Most of the article is concerned with the distribution of artificial wind breaks, particularly pine plantings.

605. Yazawa Taiji 矢澤大二, Kikō keikan 氣候景觀 (Climatic landscape), Tōkyō, Kokon Shoin, 1953, 227 pp., 21 photos, 17 tables, and 58 illus.
Yazawa introduces his concept of "climatic landscape" founded on a correlation of natural and cultural features with the occurrence of climatic phenomena. Largely based on observational and statistical data collected by the author in Nagano prefecture and on the island of Ōshima.

606. Yazawa Taiji 矢澤大二, "Kikō keikan no kenkyū 氣候景觀の研究 (A study of the climatic landscape)," Chirigaku hyōron, v. 23 (1950), no. 12, 32-38.
A study of windbreaks as an indicator of prevailing winds. Perhaps of more interest to the cultural geographer than to the climatologist.

607. Yazawa Taiji 矢澤大二, "Sōkan kikōgaku no kenkyū 綜觀氣候學の研究 (Introduction to synoptic climatology)," Chirigaku hyōron, v. 22(1949), no. 2, 1-10.
Yazawa proposes an approach to climatic interpretation based on the graphic presentation of synoptic phenomena as opposed to the usual analysis of purely statistical data.

608. Yazawa Taiji 矢澤大二, "Tōkyō kinkō ni okeru bōfūrin no bumpu ni kansuru kenkyū 東京近郊に於ける防風林の分布に關する研究 (Studies concerning the distribution of windbreaks in the vicinity of Tōkyō)," Chirigaku hyōron, v. 12 (1936), no. 1, 47-66; no. 3, 56-76.
Windbreaks around suburban and rural residences are a conspicuous feature of the Kantō landscape in the Tōkyō area. Yazawa here describes the variety and utility of these "wind mantles" and presents maps of their distribution and alignment compiled from air-photo analysis.

609. Yoshimura Shinkichi 吉村信吉, Iwasaki Kenkichi 岩崎健吉, and Itō Gōhei 伊藤鄕平, "Shizuoka-ken Kunō nanroku ichigobatake ni okeru bikikō no kansoku 靜岡縣久能南麓苺畑に於ける微氣候の觀測 (Micro-climatic observations of strawberry fields at the southern foot of Mt. Kunō, Shizuoka prefecture)," Chigaku zasshi, v. 47, no. 554 (1935), 6-16, 12 graphs.
The Mt. Kunō area in Shizuoka prefecture is known for strawberry raising and the special cultivation techniques employed. The berries are grown on stone and concrete terraces oriented to receive the maximum insolation and afford the best protection from adverse wind and temperature conditions. This is a micro-climatic study of the effect of this artificial planting method upon wind and air temperature conditions and the influence of these factors upon plant growth in the strawberry beds. Contains 12 graphs on wind and temperature measurements.

I. Destruction by physical forces

610. Fukui Eiichirō 福井英一郎, "Nihon no kikō saigai 日本の氣候災害 (Climatic disasters in Japan)," Chigaku zasshi, v. 61, no. 4 (1952), 6-9.

Natural disasters, including crop failure, which stem from meteorological causes are discussed in terms of their seasonality, range of severity, frequency of occurrence, and distribution in Japan as a whole. Damage from wind, flood, snow, drought, and unseasonal cold is considered briefly.

611. Miyazaki Kenzō 宮崎健三, "Nadare no chirigakuteki kōsatsu 雪崩の地理學的考察 (Geographical observations of avalanches)," Chirigaku hyōron, v. 14 (1938), no. 9, 1-14.

A study of avalanches and their effect on human activity, particularly in regard to railroad operations. Although beginning with a general discussion of the nature of avalanches, the article is mainly an analysis of snowslides occurring along the Takayama railroad line in Toyama and Gifu prefectures. Observations were made on the distribution and frequency of avalanches and the author determined the periods and times of day when the slides are most likely to take place.

612. Nakamura Keisaburō 中村慶三郎, Yamakuzure 山崩 (Landslides), Tōkyō, Iwanami Shoten, 1934, 3+4+254 pp., 2 maps, and 136 illus.

An able study of the causes and effects of landslides. The first portion is devoted to a general discussion of landslide phenomena, while the second part of the book deals with the investigation of seven specific slide areas in the Hokuriku district where the author lived and studied this subject for many years. With two separate map sheets.

613. Nakamura Keisaburō 中村慶三郎, "Yamakuzure no kenkyū 山崩の研究 (A study of landslides)," Chirigaku hyōron, v. 13 (1937), no. 4, 1-20.

Concerns landslides, their causes, distribution, and frequency in Japan with emphasis on the nature of spring landslips in the Hokuriku district. Describes six examples, one in Toyama and five in Niigata.

614. Sumitani Keisuke 炭谷恵副, "Kagawa-ken no kambatsu, kangai kenkyū (1) 香川縣の旱魃, 旱害研究(1) (Studies of dry-weather crop damages in Kagawa prefecture, part I)," Chirigaku hyōron, v. 13 (1937), no. 11, 1-37; no. 12, 25-63.

Stimulated by the record-making drought of 1934, Sumitani initiated this study of irrigation problems and crop damage resulting from periodic and severe dry weather in Kagawa prefecture, Shikoku. The frequency and length of droughts are shown in a chronological list and the geographical distribution of districts with a history of relatively good and poor crop yields is indicated. Various aspects of the problem are considered and in conclusion the writer suggests some improved methods of water conservation.

615. Tanaka Kaoru 田中薫, "Nadare no bunrui to hompō chūbu kōsanchi ni okeru nadare no keitai ni tsuite 雪崩の分類と本邦中部高山地に於ける雪崩の形態に就いて (On the classification of avalanches and their forms in the mountain districts of central Japan)," Chirigaku hyōron, v. 4 (1928), no. 5, 1-22.

Studies of avalanches and various systems of their classification, including Tanaka's observations in the mountains of central Japan where he related the avalanches to meteorological changes.

616. Yamaguchi Yaichirō 山口彌一郎, Tsunami to mura 津波と村 (Tidal waves and settlement), Tōkyō, Kōshunkaku Shobō, 1943, 5+3+221 pp., 7 maps, 4 photos.

The author reports on ten years of research on the effects of tsunami or seismic tidal waves upon coastal settlement, especially along the Sanriku coast of Tōhoku. Emphasis is upon village reconstruction and family rehabilitation following tsunami disasters.

J. Oceanography

617. Konishi Kanō 小西和, Setonaikairon 瀬戸内海論 (A study of the Inland Sea), Tōkyō, Bunkaidō, 1912, 16+986+14 pp.

An elaborate compilation of general materials and scientific studies on the Inland Sea region derived from both published literature and field observation. It provides a valuable reference book on the physical and human geography, the geology, oceanography, climate, and flora and fauna of the region. Ten of 15 chapters treat physical rather than cultural topics.

618. Kōno Kazuo 河野一夫 and Mori Michiyasu 森通保, "Marugame heiya no chikasui narabi ni tameike no rikusuigakuteki kenkyū 丸亀平野の地下水並に溜池の陸水學的研究 (A limnological study of the underground water and irrigation ponds of the Marugame plain)," Chirigaku hyōron, v. 19 (1943), no. 6, 39-50.

The Marugame plain in northern Shikoku is noted for its numerous artificial storage ponds used for irrigation. The first section of this article deals with the hydrology of that plain, and the latter part pertains to the pond plankton.

619. Niino Hiroshi 新野弘, "Murotozaki oki Tosabae no teishitsu ni tsuite 室戸岬沖土佐碆の底質に就いて (On soundings at Tosabae off Murotozaki)," Chirigaku hyōron, v. 11 (1935), no. 8, 1-9.

Geological investigation of the nature and structural significance of Tosabae bank and the adjacent marine trench off Cape Muroto, southern Shikoku.

620. Niino Hiroshi 新野弘, "Shikoku Taiheiyō-gan okiai oyobi Ōsumi kaikyō kaitei ni okeru suiseigan no rotō 四國太平洋岸沖合及び大隅海峽海底に於ける水成岩の露頭 (Outcrops of sedimentary rocks on the sea bottom off Shikoku and Ōsumi strait along the Pacific Ocean coast of Japan)," Chirigaku hyōron, v. 12 (1936), no. 9, 44-57.

A technical study of neo-Tertiary submarine geology and associated sedimentation.

621. Obara Nobuhiko 尾原信彦, "Habu-wan no kaiyōgakuteki kenkyū 波浮灣の海洋學的研究 (Oceanographical researches of the landlocked water of Habu)," Chirigaku hyōron, v. 17 (1941), no. 3, 29-46.

A study of water temperature, chlorine content, and related phenomena at Habu bay on Oshima island south of Tōkyō. This bay, an unusual example of a one-time low-level caldera lake breached by the sea, is of interest to both geomorphologist and oceanographer.

622. Obara Nobuhiko 尾原信彦, "Mikomotojima ni okeru maijitsu kansoku ni yorite shiraretaru hyōsōsui no kando no zōgen no gen'in ni tsuite 神子元島に於ける毎日觀測によりて知られたる表層水の鹹度の増減の原因に就いて (On the conditions governing the salinity fluctuation of the upper layer of sea water found through daily observations at Mikomoto island)," Chirigaku hyōron, v. 16 (1940), no. 2, 34-51.

A technical oceanographic investigation into the periodicity of fluctuations in the chlorine content of surface layers of sea water. The paper is based on two years of observational research.

623. Obara Nobuhiko 尾原信彦, "Shimoda-wan no suishitsu ni kansuru kenkyū 下田灣の水質に關する研究 (An oceanographical study of the sea water of Shimoda bay)," Chirigaku hyōron, v. 16 (1940), no. 9, 10-37.

This is a part of a technical oceanographic report published earlier in Tōkyō Bunrika Daigaku Rinkai Jikkenjo hōkoku (Proceedings of the Coastal Experiment Station of Tōkyō Bunrika University). It describes tests and observations of the waters of Shimoda bay conducted at the southern tip of the Izu peninsula in an effort to determine the differences in chemical and biological content between the bay water and that of the open ocean.

624. Ogura Shinkichi 小倉伸吉, "Izu shotō oyobi Mariana shotō fukin no chōryū ni tsuite 伊豆諸島及マリアナ諸島附近の潮流に就いて (On the tidal currents in the vicinity of the Izu and Mariana islands)," Chirigaku hyōron, v. 6 (1930), no. 7, 564-576.

A theoretical computation of the speed of the Izu-Mariana tidal current.

625. Suirobu 水路部 (Hydrographic Office), "Nihon kinkai no chōseki 日本近海の潮汐 (Tides in the seas adjacent to Japan)," Tōkyō, Suirobu, 1915, 216 pp., 30 illus.

A classic oceanographic work. This publication resulted from Dr. S. Ogura's research with the Kaigun Suirobu 海軍水路部 (Coastal Survey of the Imperial Navy) and is based upon abundant and accurate data.

626. Wada Yūji 和田雄次, "Nihon-kai no kanryū 日本海の環流 (Circular currents in the Sea of Japan)," Chigaku zasshi, v. 25, no. 292 (1913), 1-13; no. 293, 6-16; no. 294, 20-26.

Summarizing his research and thinking on the subject of currents in the Sea of Japan, the author has here collected notes from his earlier lectures. For many years Wada released bottles to trace the currents and studied drifting wrecks and debris. He also observed mines which drifted to Japan from Vladivostok during the Russo-Japanese War. Of value to those interested in this phase of oceanography.

K. Limnology

627. Tanaka Akamaro 田中阿歌麿, Koshōgakujō yori mitaru Suwa-ko no kenkyū 湖沼學上より見たる諏訪湖の研究 (Limnological study of Lake Suwa), Tōkyō, Iwanami Shoten, 1918, 2 v.

References to topography and to various aspects of human geography in the Suwa basin broaden the scope of this study beyond purely technical limnology. A monumental work by a pioneer Japanese limnologist.

628. Tanaka Akamaro 田中阿歌麿, Nihon Kita-Arupusu koshō no kenkyū 日本北アルプス湖沼の研究 (Study of the lakes in the North Japanese Alps), Tōkyō, Kokon Shoin, 1930, 2+2+23+1036 pp., illus.

The author is a pioneer limnologist in Japan and wrote an earlier volume on Lake Suwa and Lake Nojiri. In this book are collected a great amount of observational data together with Tanaka's descriptions and thoughts on the lakes in the Hida mountains or Japanese Alps. The Three Lakes of Nishina, and the lakes Aoki, Nakazuna, and Kizaki are among those most intensively studied, although other lakes in the upstream regions of the Azusa, Kiso, Hime, and Jinzū rivers are treated in considerable detail. Many graphs and photographs are included.

629. Tanakadate Hidezō 田中館秀三, "Ezo-Fuji Hangetsu-ko fu kosui no shimbunrui 蝦夷富士半月湖附湖水の新分類 (Hangetsu-ko lake north of Ezo-Fuji volcano and a new classification of lake water)," Chirigaku hyōron, v. 3 (1927), no. 9, 14-25.

A limnological study of a stagnant crater lake in a parasitic cone of Ezo-Fuji volcano, Hokkaidō.

630. Tanakadate Hidezō 田中館秀三, "Hokkaidō hontō no kazanko 北海道本島の火山湖 (On the volcanic lakes in Hokkaidō)," Chigaku zasshi, v. 34, no. 398 (1922), 1-16; no. 400, 17-20, 2 photos, 3 illus.

A geomorphological and limnological study of the following volcanic lakes in Hokkaidō: Ōnuma, Konuma, Junsainuma, Kokurinuma, Iwaonupuri-Naganuma and Iwaonupuri-Nakanuma.

631. Tanakadate Hidezō 田中館秀三, "Hokkaidō no kazanko 北海道の火山湖 (Volcanic lakes in Hokkaidō)," Chigaku zasshi, v. 36, no. 419 (1924), 22-36; no. 420, 13-22.

In his earlier article, "Hokkaidō hontō no kazanko" (see preceding entry), Tanakadate considered the volcanic lakes individually. This is a generalized treatment of the same topic.

632. Watanabe Muneshige 渡邊宗重, "Ezo-Fuji Hangetsu-ko kosui no sanso gan'yūryō 蝦夷富士半月湖湖水の酸素含有量 (Oxygen content of the lake water of Hangetsu-ko at Ezo Fuji volcano)," Chirigaku hyōron, v. 7 (1931), no. 5, 27-48.

A technical limnological investigation of lake Hangetsu, a volcanic lake in Hokkaidō. Annual and vertical changes in the oxygen content of the lake water and their effect upon plant and animal life are primarily considered.

633. Watanabe Muneshige 渡邊宗重 ,"Ezo-Fuji Hangetsu-ko kosui no tansangasu gan'yūryō 蝦夷富士半月湖湖水の炭酸瓦斯含有量 (Carbon dioxide content of the water of Hangetsu lake in the volcano Ezo-Fuji)," Chirigaku hyōron, v. 7 (1931), no. 8, 1-15.

Seasonal and depth variations in the carbon dioxide content of this volcanic lake in Hokkaidō are observed and correlated with Watanabe's analysis of the oxygen content of Hangetsu-ko, as reported in a previous number of this journal (see preceding entry).

634. Yoshimura Shinkichi 吉村信吉, "Akita-ken Okatsu-gun Ketakura-numa, Kai-numa kogun no koshōgakuteki yosatsu kenkyū 秋田縣雄勝郡桁倉沼, 貝沼湖群の湖沼學的豫察研究(Limnological reconnaissance of the Ketakura and the Kai lake groups in Okatsu district, Akita prefecture)," Chirigaku hyōron, v. 10 (1934), no. 8, 53-73; no. 9, 33-44.

A limnological survey of six lakes in the middle portion of the Minasegawa drainage basin. In addition to water characteristics, biological factors are considered.

635. Yoshimura Shinkichi 吉村信吉, "Bandai kazan shishū no kazanko no chihō koshōgakuteki yosatsu kenkyū 磐梯火山四周の火山湖の地方湖沼學的豫察研究 (Reconnaissance of the regional limnology of the volcanic lakes surrounding the volcano Bandai)," Chirigaku hyōron, v. 8 (1932), no. 10, 56-76; no. 11, 46-66; no. 12, 29-72.

A classification of the lake types in the Bandai district of Fukushima prefecture based on an analysis of detailed, technical field data.

636. Yoshimura Shinkichi 吉村信吉, "Kitakami-gawa karyū Toyoma-gun Hatazaki no nioppori no koshōgakuteki kenkyū 北上川下流登米郡畑崎の二落堀の湖沼學的研究 (Limnological investigations of two ponds on Hatazaki, Toyoma district, Miyagi prefecture)," Chirigaku hyōron, v. 15 (1939), no. 10, 1-19; no. 12, 41-59.

An investigation of the chemical constitution and pH factor of the water of two lakes formed by past flooding of the Kitakami river. Yoshimura discovered water under ice which contained no apparent oxygen.

637. Yoshimura Shinkichi 吉村信吉,"Koshō no sanso ganryō no nenchū henka 湖沼の酸素含量の年中變化 (Annual changes of oxygen content in lake water)," Chirigaku hyōron, v. 3 (1927), no. 6, 24-49.

A short limnological discourse on the reasons for seasonal variation in oxygen content and the effect of such fluctuations upon phyto-plankton. Case examples are cited for lakes both in Japan and abroad.

638. Yoshimura Shinkichi 吉村信吉, "Kosui no sanso no kenkyū hōhō. Sanso ganryō ni yoru bunrui. Shōseisō 湖水の酸素の研究方法・酸素含量による分類・小成層 (A method for the study of oxygen in lake water. A classification by the amount of oxygen. Micro-stratification)," Chirigaku hyōron, v. 3 (1927), no. 12, 11-46.

Yoshimura summarizes past approaches to this problem and introduces his own system of lake-water classification on the basis of oxgen content. He also points out new observations on "micro-stratification of oxygen content" at low levels.

639. Yoshimura Shinkichi 吉村信吉, "Kyūshū nambu kazanko no rikagakuteki oyobi seibutsugakuteki yosatsu kenkyū 九州南部火山湖の理化學的及び生物學的豫察研究 (Physical, chemical, and biological reconnaissance of volcanic lakes in southern Kyūshū)," Chigaku zasshi, v. 42, no. 497 (1930), 15-21; no. 498, 20-31; no. 501, 30-39, 7 illus.

Yoshimura discusses the limnology of several volcanic lakes in southern Kyūshū, with emphasis upon the chemical constitution of the lake water and related technical aspects, such as water color, transparency, temperature, and the nature of plankton. Those lakes studied most closely are Ikedako, Unagiike, Kagamiike, and Miike. With seven tables of data and list of references.

640. Yoshimura Shinkichi 吉村信吉, "Nihon no koshō no chihō koshōgakuteki tembō dainihō 日本の湖沼の地方湖沼學的展望第2報 (A view of the regional limnology of Japanese lakes -- second report)," Chirigaku hyōron, v. 13 (1937), no. 4, 21-45; no. 6, 28-48; no. 7, 49-62.

A classification of Japanese lakes using the limnological criteria and methods of E. Naumann (Regionalen Limnologie). Yoshimura presents, in addition, his own variation of that classification system.

641. Yoshimura Shinkichi 吉村信吉, "Nihon no kosui no suiso ion nōdo 日本の湖水の水素イオン濃度 (The hydrogen ion concentration in the lake waters of Japan)," Chirigaku hyōron, v. 7 (1931), no. 10, 50-78; no. 11, 43-69.

A classification of the lakes in Japan on the basis of such technical limnological criteria as the pH factor and relationships between the chemical constitution of the water and the plankton population.

642. Yoshimura Shinkichi 吉村信吉, "Oga hantō sankazanko no koshōgakuteki yosatsu kenkyū 男鹿半島三火山湖の湖沼學的豫察研究 (Limnological reconnaissance of three crater lakes of the Oga peninsula)," Chirigaku hyōron, v. 11 (1935), no. 10, 34-54.

A survey of the water characteristics and a limnological classification of three volcanic lakes on the Oga peninsula in Akita prefecture.

643. Yoshimura Shinkichi 吉村信吉, "Saitama-ken Takasukanuma ni okeru suichū ni yōzon-suru chisso ganryō no nenchū henka 埼玉縣高須賀沼に於ける水中に溶存する窒素含量の年中變化 (Annual variation of dissolved nitrogen compounds in Takasukanuma, Saitama prefecture)," Chirigaku hyōron, v. 6 (1930), no. 7, 145-174.

Another in the series of technical limnological papers published by this authority.

644. Yoshimura Shinkichi 吉村信吉 ,"Tottori-ken Tanegaike no koshōgakuteki yosatsu kenkyū 鳥取縣多鯰ヶ池の湖沼學的豫察研究 (A limnological study of Tanegaike in Tottori prefecture)," Chirigaku hyōron, v. 5 (1929), no. 11, 13-36; v. 6 (1930), no. 3, 42-54.

A map and diagrammatic summary of pH conditions, oxygen and plankton content, and other distributional aspects of the lake water and of the lake-basin morphology.

645. Yoshimura Shinkichi 吉村信吉, "Tsugaru jūniko kosui no suiso ion nōdo, yōkaisei sanso, tanso 津輕十二湖湖水の水素イオン濃度,溶解性酸素,炭素 (Hydrogen ion concentration, dissolved oxygen, and carbon dioxide content of the waters of the twelve Tsugaru lakes)," Chirigaku hyōron, v. 13 (1937), no. 10, 40-52; no. 11, 63-82.
Describes the characteristics of the Tsugaru lakes in Iwate prefecture by chemical content and physical form.

646. Yoshimura Shinkichi 吉村信吉, "Tsugaru jūniko no suion, tōmei-do 津輕十二湖の水温,透明度 (Water temperature and transparency of the Tsugaru twelve lakes)," Chirigaku hyōron, v. 11 (1935), no. 4, 26-50; no. 5,31-48.
Another of the writer's several limnological studies, in this case concerning differences in water temperature between a group of twelve otherwise similar lakes in Iwate prefecture. Yoshimura types the lakes according to his system of classification.

647. Yoshimura Shinkichi 吉村信吉 and Kawada Saburō 川田三郎, "Fuji goko no koshōgakuteki kenkyū 富士五湖の湖沼學的研究 (Limnology of the five lakes at the foot of Mt. Fuji)," Chirigaku hyōron, v. 18 (1942), no. 6, 1-15.
A report of field investigations. The authors made 204 soundings at Lake Kawaguchi and 98 at Lake Shōji and on the basis of this data completed isodepth maps of these lakes. The article describes the general nature of the lake basins and their waters.

648. Yoshimura Shinkichi 吉村信吉 and Koba Kazuo 木場一夫, "Aomori-ken Iwasaki-mura Matsukami jūniko no koshōgakuteki yosatsu kenkyū 青森縣岩崎村 松神十二湖の湖沼學的豫察研究 (Preliminary report of the limnological investigation of the lakes at Matsukami, Iwasaki village, Aomori prefecture)," Chirigaku hyōron, v. 9 (1933), no. 12, 50-72.
The chemical constitution, temperature range, depths, and related characteristics of the lakes at Iwasaki village are compared with those of other lakes in Japan.

649. Yoshimura Shinkichi 吉村信吉, Koba Kazuo 木場一夫, Obara Nobuhiko 尾原信彦, and Nagatsu Ichirō 長津一郎, "Tsugaru jūniko no kobon keitai 津輕十二湖の湖盆形態 (Morphology of the Tsugaru jūniko lake group)," Chirigaku hyōron, v. 10 (1934), no. 11, 30-51; no. 12, 39-63.
On the form and hydrology of some 33 small lakes scattered in the valleys and on the coastal terraces of southwestern Aomori prefecture.

650. Yoshimura Shinkichi 吉村信吉, Yoshii Narao 吉井楢雄, and Kojima Shigeichi 小島茂一, "Aburatsubo-wan no suion, embun oyobi yōkaisei sanso no nenchū henka 油壺灣の水温, 鹽分及溶解性酸素の年中變化 (Annual variation of water temperature, salinity, and dissolved oxygen in Aburatsubo bay)," Chirigaku hyōron, v. 6 (1930), no. 8, 1-26.
The small bay of Aburatsubo, near the south end of the Miura peninsula in Kanagawa prefecture, is studied in regard to the effect of precipitation upon salinity, the stratification of water temperature, the relationship between oxygen content and phyto-plankton, etc.

Chapter IX

Historical and Cultural Geography

Modern geography in Japan, as elsewhere mentioned, derives from two initial sources - geology and history. The inheritance from geology has been the stronger force but the historical approach has also been highly significant. Published studies in historical geography are relatively quite numerous. Many of them are of very high quality. Others are but poor attempts to give a geographical slant to already published historical studies. In this bibliography, a rather large proportion of the best studies in historical geography has been included, not only because of the high quality of the studies but also because of the poor availability of the original documents and the difficulty of handling them.

Settlement studies make up another important development in Japanese geography and one which has attracted a persistent interest since the middle 1920s. Both in morphology and genesis, rural settlement in Japan has been worked out in considerable detail. Most authors have considered both phases in their publications. Urban studies are much less well developed.

In general, other aspects of cultural geography have been relatively neglected. In political geography, much that has been published is geopolitics in an extreme form. The work in linguistic geography has been done mostly by linguists and lacks a true geographical flavor. Much the same can be said of other cultural fields - the main contributions have been made by other than geographers and so are not entirely satisfying to the geographer. The boundary between cultural and economic geography must be drawn arbitrarily. The problem has been partially overcome here by cross indexing. The basic problem, however, is that an inordinately large proportion of Japanese geographical publication is made up of general, intensive studies of small areas in which all geographical aspects are considered and integrated. The number of studies covering all Japan is extremely small. Fewer still are those studies pertaining to areas outside of Japan. The reason often given for this dominance of small area studies is that Japanese culture is very complex. This, however, does nothing to explain the situation.

A. Historical geology

651. Akabori Eizō 赤堀英三,"Nihon ni okeru sekki no bumpu ni tsuite 日本に於ける石器の分布に就て (On the distribution of stone implements in Japan)," Chirigaku hyoron, v. 4 (1928), no. 6, 36-51.
An ethno-geographical study based upon specimens collected over a period of 40 years by the Tōkyō Daigaku Jinruigaku Kyōshitsu 東京大學人類學教室(Institute of Anthropology, Tōkyō University). It presents the classification and distribution of arrow points, axes, spoons, and other stone tools, utensils, and weapons.

652. Aritaka Iwao 有高巖, "Kodai ni okeru Nihon to Semman to no kōtsū 古代に於ける日本と鮮満との交通 (Communication between Japan and Korea and Manchuria in ancient times)," Chiri kyōiku, v. 26 (1937), no. 6, 13-17.
An outline of the history of communication between Japan and the Asian continent from antiquity to the end of the Han dynasty with comments on folklore, archaeology, ethnography, and linguistics. Based mainly on the study of old documents.

653. Asaka Yukio 浅香幸雄, "Kinsei sonraku kōzō no tenkai 近世村落構造の展開 (Development of village structure in modern times)," Shin-chiri, v. 4 (1950), nos. 3, 4, and 5, 30-48.
An investigation of the internal structure of the small rural community in the modern period as contrasted with village make-up of the past. Asaka considers both the physical and social aspects of rural settlement and believes that basic changes in the constitution of rural life began to come about during the Genroku period or thereabouts. He cites specific changes in land use, political division, etc., which resulted in part from a general increase in population at that time. For detailed study he selected Kikoba, a buraku on the Miura peninsula in Kanagawa prefecture.

654. Doi Senkichi 土井仙吉, "Chihō shōtoshi no seichō katei 地方小都市の成長過程 (The development of a small local town)," Jimbun chiri, v. 4 (1952), no. 2, 24-37.
Ikeda in Tokushima prefecture is discussed in this paper as an example of a local commercial and political center. The author examines the favorable geographical factors promoting the town's development in an area of very early settlement and its rise as a strategic castle town under the Hachisuga clan. Tobacco production and processing became the town's principal industry following the close of the feudal era, and this activity, together with the influence of the railroad upon marketing in the Ikeda district, is described in some detail.

655. Fujioka Kenjirō 藤岡謙二郎, Chiri to kodai bunka 地理と古代文化 (Geography and ancient culture), Kyōto, Ōyashima Shuppan Kabushiki Kaisha, 1951, 2+5+196 pp., 41 illus.
Fujioka draws on a great deal of fragmental research material in an attempt to interpret prehistoric settlement and culture in terms of the physical environment, especially in regard to the influence of local landforms and other site factors upon living conditions and social development.

656. Fujioka Kenjirō 藤岡謙二郎, "Keikan hensenshiteki ni mita Chūbu chihō no senshi chiri -- senshi chishi jojutsu no ichihōhō 景観變遷史的に見た中部地方の先史地理 — 先史地誌叙述の一方法 (Prehistoric geography of the Chūbu district from the point of view of landscape changes -- a method of describing prehistoric topography)," Jimbun chiri, v. 5 (1953), no. 1, 1-25.
A summary of what is known about the cultural differences and the prehistoric contact between the Jōmon and Yayoi peoples in central Honshū. A map shows the location of archaeological finds.

657. Fujioka Kenjirō 藤岡謙二郎, "Kodai Kinai no bunka chiiki 古代畿内の文化地域 (The ancient cultural region of Kinai)," Shin-chiri, v. 2 (1948), no. 4, 1-8.
Fujioka delimits and discusses generally some of the ancient cultural centers and areas of the Kinai district such as the Kyōto and Nara basins, the Ōsaka plain, and the Yamato and Kizu valleys.

658. Fujita Motoharu 藤田元春, "Kodai Nihonkai-jō no sempaku 古代日本海上の船舶 (Ancient ships and vessels of the Japanese seas)," Chikyū, v. 22 (1934), no. 3, 10-23.
A brief history of shipping and sea communication from prehistoric times into the Tokugawa period.

659. Fujita Motoharu 藤田元春, "Mindai no Nihon kōtsūro 明代の日本交通路 (Japanese sea routes at the time of the Ming dynasty)," Chiri kyōiku, v. 22 (1935), no. 5, 64-79.
Information on the ancient sea routes between Japan and China and on those in Japanese waters is best obtained from old Chinese documents rather than from Japanese records. Fujita reports on this subject following a study of Chinese historical sources.

660. Furuta Ryōichi 古田良一, Edo-jidai no kōtsū 江戸時代の交通 (Communication in the Edo era) [=Iwanami kōza: Nihon rekishi, no. 6], Tōkyō, Iwanami Shoten, 1933, 47 pp.
A compact, academic survey of the routes and methods of transportation and communication in the Tokugawa period. No sources are cited, however.

661. Furuta Ryōichi 古田良一, "Nihonkai no engan kōro 日本海の沿岸航路(Coastal shipping in the Japan Sea)," Shakai chiri, no. 18 (1949), 2-6.
The history of coastal shipping in the Sea of Japan from the Nara to the Edo period, with conclusions on its economic significance.

662. Haga Zenjirō 芳賀善次郎, "Nagai-bonchi kaitaku no rekishi-chiriteki kōsatsu to kongo no mondai 長井盆地開拓の歴史地理的考察と今後の問題 (Some considerations on the historical geography of the Nagai basin and future problems of that area)," Shin-chiri, v. 2 (1948), no. 5, 34-44.
An historical account of the colonization of the Nagai basin. Ayugai-mura is selected as being a representative community in that area, and its establishment and growth is considered in some detail. The author mentions future problems recognizable in the present settlement picture.

663. Ishida Hiroshi 石田寛, "Yoshiigawa karyū ni okeru suiden kaihatsu katei no kōsatsu 吉井川下流に於ける水田開發過程の考察 (The process of development of rice fields in the plain of the lower Yoshii river)," Jimbun chiri, v. 3 (1951), no. 4, 31-41.

An historical outline of the development of rice farming in the wet, low-lying areas near the mouth of the Yoshii river in Okayama prefecture. Initial and later problems of land reclamation, land use and maintenance, flood protection, etc., are treated.

664. Ishikawa Seishō 石川成章, "Nihon kōgyō no konjaku 日本鑛業の今昔 (History of the mining industry in Japan)," Chikyū, v. 10 (1928), no. 3, 61-77.

An outline of the early history of mining and metallurgy in Japan from prehistoric times through the Nara and Heian periods.

665. Kagose Yoshiaki 籠瀬良明, "Hokuragawa kyūkadō no chūsei kaihatsu suiden 保倉川舊河道の中世開發水田 (Paddies brought under cultivation in the middle ages in the region near the old watercourse of the river Hokura)," Jimbun chiri, v. 2 (1950), no. 3, 36-47.

A study of topographic control over the form of old and new settlement, reclamation, and land use in the plain of Takada, Niigata prefecture. Kagose is particularly interested in settlement along the Hokura river where the "mother" villages are situated on natural levees or in old river meanders. The difficulties of reclamation and farming in this region are noted.

666. Kawahara Tsuneo 河原常雄, "Kōtsūro no hensen to shūraku no seisui ni kansuru ichikōsatsu 交通路の變遷と集落の盛衰に關する一考察 (A consideration of the transformation of communication routes and the growth and decline of villages)," Jimbun chiri, v. 2 (1950), no. 4, 39-49.

Traces the historical rise and decline of the town of Maibara in Shiga prefecture, primarily in terms of its functional development. Kawahara describes the factors in Maibara's growth from a small village in the seventeenth century into an important Biwa lake port and way station on the Nakasendō highway in the Tokugawa period, and its later shrinkage, in modern times, to a minor rail-junction town.

667. Kikuchi Toshio 菊池利夫, "Inton bushi ukeoi shinden to shite no Shinshima ryō Jūrokushima shinden ni tsuite 隱遁武士請負新田としての新嶋料十六嶋新田について (On the Jūrokushima shinden contracted for by the retired samurai)" Jimbun chiri, v. 4, no. 3 (1952), 10-23.

A study in historical geography. Jūrokushima is a farming community located on the lower reaches of the river Tone, the land of which was reclaimed some 350 years ago by war refugees. The political and geographical factors involved in the establishment of this Tokugawa settlement are considered.

668. Kitamura Toshio 喜多村俊夫, "Chihō ichiba no hatten, zanzon to sono yōin ni kansuru rekishi-chiriteki kenkyū 地方市場の發展, 殘存とその要因に關する歷史地理的研究 (A study in the historical geography of a local market town: the reasons for its development and continuance)," Jimbun chiri, v. 2, no. 4 (1950), 1-13.

An account of the commercial development of Yōkaichi, Shiga prefecture, as a local marketing center. Geographical, political, economic, and social factors affecting the town's functional role from pre-Tokugawa times are discussed.

669. Kitamura Toshio 喜多村俊夫, "Honjin o chūshin to suru kinsei shukueki no hatten ni kansuru rekishi-chiriteki kōsatsu 本陣を中心とする近世宿驛の發展に關する歷史地理的考察 (A historical geographic study on the development of the post town in modern times with inns as the chief object of consideration)," Chiri, v. 1 (1938), no. 2, 1-19.

From an analysis of the old registration records of honjin, or inns, at former stage towns along the Tōkaidō and Nakasendō within Ōmi province, Kitamura speculates on the local spheres of influence of such towns, the nature of traffic along the highways, economic changes, etc. The towns investigated are Ishibe, Tsuchiyama, Kusatsu, and Torihama.

670. Kitamura Toshio 喜多村俊夫, "Kinsei shukueki no hatten jōken ni kansuru rekishi-chiriteki kōsatsu 近世宿驛の發展條件に關する歷史地理的考察 (Historical geographic studies of the conditions of stage towns in the modern period)," Chiri kyōiku, v. 28 (1938), no. 6, 38-47.

The conditions determining the prosperity and decline of stage towns along main highways are analyzed through a study of the records preserved at honjin, the official inns at each way station.

671. Kitamura Toshio 喜多村俊夫, Ōmi keizaishi ronkō 近江經濟史論攷 (Papers on the economic history of Ōmi province), Kyōto, Taigadō, 1946, 2+8+457 pp.

Twelve short historical studies of interest to the economic geographer concerning various phases of land use, commerce, and communications in the Kyōto-Lake Biwa area. Half of the articles deal with economic activity during the years 1935 to 1944. The others are devoted to earlier historical periods, i.e., Tokugawa, etc.

672. Kitamura Toshio 喜多村俊夫, "Shukueki no baiyōken to shite no sukegō no chiriteki kenkyū 宿驛の培養圈としての助鄉の地理的研究 (Geographical study of the sukegō system as nurtured by a post town)," Chiri to keizai, v. 2 (1936), no. 1, 76-87; no. 2, 98-108; no. 3, 75-90.

An historical study of the sukegō system which regulated the local supply of porters and pack animals at the feudal stage towns. The writer speculates on the past geographic, economic, and social significance of that system in the old province of Ōmi, now part of Shiga prefecture.

673. Komaki Saneshige 小牧實繁, "Biwako kojō kōtsū no hensen 琵琶湖湖上交通の變遷 (Historical changes in communications on Lake Biwa)," Chiri kyōiku, v. 22 (1935), no. 3, 1-6; no. 4, 45-50; no. 5, 275-281.

Descriptions of travel and commerce on Biwako from ancient times into the early Meiji period. The nature of the ports and traffic on that lake is considered in relation to historical change in the social and economic scene of Japan as a whole.

674. Komaki Saneshige 小牧實繁, "Kodai Shikoku no shūraku ni tsuite 古代四國の聚落に就いて (On villages in ancient Shikoku)," Chikyū, v. 4 (1925), no. 3, 28-37.
Komaki identifies some of the gō place names in the Wamyōshō with present-day communities and discusses the geographic factors influencing the development of the old settlement systems in Shikoku.

675. Komaki Saneshige 小牧實繁, "Senshi shūraku chiri 先史聚落地理 (Prehistoric settlement geography)," Chikyū, v. 5 (1926), no. 4, 34-54.
A paper on methodology in the study of prehistoric settlement geography. Komaki adapts a number of Western concepts and ideas to the Japanese approach in this field.

676. Marumo Takeshige 丸茂武重, "Shōen shūraku ni kansuru nisan no kōsatsu 莊園聚落に關する二三の考察 (Some considerations on shōen villages)," Chiri kyōiku, v. 22 (1935), no. 5, 282-288.
The basic types of shōen villages are first described, followed by an explanation of the distribution of early shōen settlements. Various political and social aspects of the internal structure of shōen are touched on and the author cites a number of specific examples from his study of old documents.

677. Maruyama Kunio 丸山國雄, Nihon hoppō hattenshi 日本北方發展史 (History of Japan's northward development), Tōkyō, Suisansha, 1942, 5+8+327 pp.
Actually, a historical survey of Russo-Japanese diplomacy and competition prior to the Meiji restoration. Slightly nationalistic in slant but of possible value to the political geographer concerned with the Far East as a whole and in particular with historical background material on Sakhalin and the Kuriles.

678. Matsuo Toshio 松尾俊郎, "Ariakekai hokugan teichi ni okeru tochi to seikatsu ni kansuru jakkan no kiroku 有明海北岸低地に於ける土地と生活に關する若干の記錄 (Some notes concerning the land and life in the lowlands of the north coast of the Ariake sea)," Chirigaku hyōron, v. 7 (1931), no. 10, 31-49.
This is a brief introduction to the historical changes which have occurred in canal transportation on the delta at the head of Ariake bay. The rise and decline of port towns along that coast is also discussed.

679. Matsuo Toshio 松尾俊郎, "Jōkamachi to shite no Kawagoe ni tsuite 城下町としての川越について (On Kawagoe as an example of a castle town)," Chiri kyōiku, v. 31 (1939), no. 3, 1-15.
An interesting study of functional and structural change in the historical development of a former Kantō castle town. Kawagoe, located northwest of Tōkyō in Saitama prefecture, was a local political center in feudal times. It is now a small commercial and industrial satellite of Tōkyō.

680. Matsuo Toshio 松尾俊郎, "Kakō oyobi kasen kōtsū no hensen ni tsuite 河港及び河川交通の變遷に就いて (On historical changes in river ports and river communication)," Chiri kyōiku, v. 23 (1936), no. 4.
Primarily an investigation of the functional and physical changes undergone by the town of Kurihashi, located at the head of the Edogawa distributary of the Tone river in Saitama prefecture. Historical changes in the nature of river traffic are also discussed.

681. Mitomo Kunigorō 三友國五郎, "Senshi shūraku ni kansuru kōsatsu 先史聚落に關する考察 (A study of prehistoric settlements)," Chirigaku hyōron, v. 14 (1938), no. 5, 1-33; no. 6, 43-50.
Conjecture on prehistoric settlement based on Mitomo's study of prehistoric ruins and cultural remains at numerous sites throughout Japan.

682. Nihon Rekishichiri Gakkai 日本歷史地理學會 (Japanese Historical Geography Association), Nihon kōtsūshiron 日本交通史論 (Studies on the history of communication in Japan), Tōkyō, Nihon Gakujutsu Fukyūkai, 1925, 5+2+688 pp.
Thirty-six monographs on the historical geography and political and social history of pre-Meiji land and sea communications. The first fifteen articles by A. Fujita largely concern various aspects of the Tōkaidō road, from ancient times through the Edo period, e. g.; post and stage systems, engineering projects, key ferries and barriers, ports and related shipping, pilgrimages and military movements, etc. The remaining papers, by several different scholars, treat similar subjects in regard to other important routes. In some instances, the contents overlap.

683. Nishimura Shinji 西村眞次, Nihon kaigai hattenshi 日本海外發展史 (History of Japanese overseas development), Tōkyō, Tōkyōdō, 1942, 6+20+507 pp.
Based largely upon the contents of Sekai no Nihonshi (Japan's history in relation to the world) edited by Nishimura and published by the same firm twenty years earlier, this is a survey history of Japan's foreign relations and overseas contacts from ancient times until the close of the Shogunate era.

684. Nishimura Tomoichi 西邨知一, "Waga kuni ni okeru kōtsū hattatsu no gairyaku 我が國に於ける交通發達の概略 (Outline of the development of communication in Japan)," Chiri kyōiku, v. 22 (1935), no. 5, 191-195.
Brief historical description, restricted to the development of correspondence in Japan up to the Tokugawa period. Other aspects of communications are not included.

685. Noguchi Yasuichirō 野口保市郎, "Ine no kigen ni kansuru chirigakuteki kōsatsu 稻の起源に關する地理學的考察 (Geographical considerations on the origin of rice)," Chiri kyōiku, v. 19 (1933), no. 3, 4-10; no. 4 (1934), 36-42.

The writer discusses the origin of rice in Japan on the basis of folklore and legend and, more specifically, in terms of references to that subject found in the Kojiki and Nihon shoki. He also refers to the history of rice culture in India and China.

686. Ono Kōji 小野晃嗣, Kinsei toshi no hattatsu 近世都市の發達 (Development of cities during recent times) in Iwanami kōza, Nihon rekishi, v. 6, Tōkyō, Iwanami Shoten, 1933, 98 pp.

This is a historical description of urban conditions during the Tokugawa period. The author makes no attempt to carry his study into modern times as the title would suggest. As an introduction to his discussion of the character of cities and urban life under the Shogunate, Ono outlines the nature of town development during the medieval age. Contains valuable background material for the urban or economic geographer.

687. Ono Takeo 小野武夫, "Buraku to bōei 部落と防衛 (The community and self-defence)," Chiri kyōiku, v. 23 (1935), no. 3, 1-5.

Ono describes briefly several cooperative engineering projects and planting methods used by rural communities since ancient times as protection against natural or man-caused disasters such as field-fire, flood, windstorm, crop destruction by wild boar, etc.

688. Tamura Eitarō 田村榮太郎, "Kōgyoku Saimeiki no chimei to jimmei 皇極齊明紀の地名と人名 (Place names and personal names in the reigns of Empresses Kōgyoku and Saimei)," Shin-chiri, v. 3 (1949), no. 1, 23-29.

The identification and interpretation of place names which appear in the Nihon shoki.

689. Torii Tatsuzō 鳥居龍藏, Musashino oyobi sono shūi 武藏野及びその周圍 (Musashino and its adjacent areas), Tōkyō, Isobe Kōyōdō, 1924, 2+3+313 pp., 38 photos, 17 illus.

Twenty separate essays on the Musashino area as a whole or places therein. The essays principally concern ethnological history or archaeology. These writings by a famous Japanese archaeologist may be of interest to the anthropo-geographer.

690. Toyama Usaburō 外山卯三郎, Nippo bōeki shōshi 日葡貿易小史 (A short history of trade between Japan and Portugal), Tōkyō, Wakai Hitosha, 1942, 1+2+3+343 pp., 6 illus.

The first two chapters are Toyama's account of the arrival of the first Portuguese in Japan and a summary of Japanese trade with Portugal under the Shogunate. The last three chapters are translations of articles by C. B. Boxer on Japanese-Portuguese trade and diplomacy during the period 1624-1647.

691. Tsuda Hideo 津田秀郎, "Jōko ōchō-jidai ni okeru Tōkaidō kōtsūro 上古王朝時代に於ける東海道交通路 (The Tōkaidō highway from ancient times through the Heian period)," Chikyū, v. 15 (1931), no. 4, 35-46.

Historically, the Tōkaidō has been Japan's principal land and sea highway. Tsuda traces its importance from ancient times through the Heian period, describing the development of the way-station system, the nature of traffic, shifts in route, and its political and economic significance.

692. Tsukamoto Tsuneo 塚本常雄, "Kyōto shiiki no hensen to sono chirigakuteki kōsatsu 京都市域の變遷と其地理學的考察 (The development of the area of Kyōto city and some geographical considerations)," Chiri ronsō, no. 1 (1932), 155-229.

Kyōto was founded over 1,100 years ago, being laid out according to ancient plans. In this study the writer stresses the influence of the physical environment upon the initial establishment of the city and upon its subsequent growth during this long period. Two-thirds of the paper concerns a survey of physical site factors, i. e., terrain features, geology, drainage, etc. The final third describes the city at three historical stages: the Heian period, the Toyotomi period, and the post-restoration period.

693. Uchida Kan'ichi 内田寛一, "Musashino no keikakuteki kaitaku no ichirei 武藏野の計畫的開拓の一例 (An example of planned colonization in Musashino)," Ōtsuka Chiri Gakkai rombunshū, no. 4 (1934), 81-96; no. 5 (1935), 225-255.

Nishikubo, a part of Musashino-machi in Tōkyō prefecture, was selected as an example of a planned settlement on the Musashino upland. The Tokugawa methods of land allotment and land registration, as well as other Tokugawa controls on colonization, are described and compared with the systems used in recent times.

694. Uchida Kan'ichi 内田寛一, "Tokugawa-jidai nōson ni okeru koguchi zōgen no sōkan-kankei 德川時代農村に於ける戶口增減の相關關係 (Interrelationships between the increase and decrease of families in agricultural villages of the Tokugawa period)," Chiri, v. 4 (1941), no. 1, 13-21.

Essentially a study of population change in a rural area during the last three decades of the Tokugawa period, with some comment on the social implications of such change. Based on data on Nishimata-mura in Tōtōmi province, now part of Shizuoka prefecture, as found in registers of religion in the Tokugawa period.

695. Watanabe Hisao 渡邊久雄, "Chūko ni okeru konden no rekishi chirigakuteki ichikōsatsu--toku ni Tōdaiji-ryō Inanoshō ni tsuite-- 中古に於ける墾田の歷史地理的一考察 — 特に東大寺領猪名庄について — (A study in the historical geography of fields developed in the middle ages, especially at Inanoshō, a manor of the Tōdaiji)," Jimbun chiri, v. 5 (1953), no. 3, 1-16.

Changes in early land utilization and settlement brought on by the incorporation of Inanoshō (in present-day Hyōgo prefecture) into the feudal manorial system of the Tōdaiji temple are analyzed in this monograph, founded on a study of old maps. Comparisons are made between the jōri and shōen land systems.

696. Watanabe Hisao 渡邊久雄, Kōtō-son --hembō-suru toshi kinkōson no kiroku 甲東村 一 變貌する都市近郊村の記録 (Kōtō-son--the record of a changing suburban village), Ōsaka, Katsuragi Shoten, 1942, 3+4+193+4 pp., 2 photos, 36 illus.

A study of a village undergoing urbanization. Kōtō is located between Ōsaka and Kōbe and has become a part of the semi-urbanized belt which has grown up between those cities during the past three decades or so. Watanabe outlines the historical development of this community and discusses the economic and social changes which have affected it. He is particularly interested in recent adaptations in land use as evidenced in a shift from rice farming to truck gardening.

697. Yajima Nikichi 矢嶋仁吉, "Musashino Kodaira-mura ni okeru shinden shūraku no kenkyū 武藏野小平村に於ける新田聚落の研究 (A study of the newer settlements at Kodaira village in the Musashino upland)," Chirigaku hyōron, v. 15 (1939), no. 11, 13-27.

Kodaira-mura is a typical village of the Musashino upland colonized during the middle Tokugawa period. The historical processes of the reclamation of this upland are explained and the areal relationship between "mother" and "daughter" villages is shown with a map. There are references to irrigation, drinking water, land allotment, land use, etc. Two maps of land allotment are added.

698. Yajima Nikichi 矢嶋仁吉, "Tokugawa-jidai Jōshū Annaka-shuku ni okeru sukegō kaeki to nōson jinkō to no kankei 徳川時代上州安中宿に於ける助郷課役と農村人口との關係 (The relationship between the sukegō service and the rural population at Annaka-shuku during Tokugawa era)," Chirigaku hyōron, v. 16 (1940), no. 2, 18-33.

Annaka-shuku, in Kōzuke province, was one of the post stations on the Nakasendō highway. In Tokugawa times villages near stage offices were compelled to furnish, on demand, horses and men for bearer service on the highway. This compulsory labor and supply system was termed sukegō. In this monograph Yajima discusses the social and economic hardships sukegō worked upon the local farming population of some 19 villages in the Annaka area. The system's contributions to farm decay, population stagnation, depopulation, etc., are explained.

699. Yamaguchi Kazuo 山口和雄, Bakumatsu bōekishi 幕末貿易史 (History of foreign trade in the closing years of the Shogunate), Tōkyō, Chūō Kōronsha, 1943, 2+4+380 pp.

An intensive study of foreign trade toward the end of the Tokugawa period and the effect of that trade upon the internal economy of Japan. The policies of the Shogunate government are critically analyzed.

700. Yasuda Hatsuo 安田初雄, "Kōriyama bonchi no keikan henka 郡山盆地の景觀變化 (Changes of landscape in the Kōriyama basin)," Chirigaku hyōron, v. 14 (1938), no. 4, 47-64; no. 5, 58-78.

An historical land use study. The author describes and analyzes changes in the patterns of settlement and cultivation in the Kōriyama basin, with particular emphasis being placed upon alterations in land use brought about by settlers of the samurai class during the late pre-Restoration period. These new groups initiated the first large land reclamation projects in the basin. They began the system of irrigation with water from Lake Inawashiro which quickly attracted to the area more settlers from neighboring regions.

701. Yawata Ichirō 八幡一郎, "Higashi Karorin ni okeru kyoseki iseki (yohō) 東カロリンに於ける巨石遺跡(豫報) (On the megalithic structures of the eastern Caroline islands -- a preliminary report)," Chirigaku hyōron, v. 8 (1932), no. 4, 50-66.

Yawata describes the existing megalithic ruins and related artifacts found along the east coasts of both Kusaie and Ponape islands in the eastern Carolines and suggests, on the basis of comparative analysis, their cultural kinship with ancient Polynesian society.

702. Yonekura Jirō 米倉二郎, "Nōson keikaku to shite no jōrisei — waga kuni chūko no sonraku to sono kōchi— 農村計画としての條里制 — 我國中古の村落と其耕地 (The jōri system as rural land-planning; villages and fields in the middle ages in Japan)," Chiri ronsō, no. 1 (1932), 307-352.

Yonekura briefly explains the historical background for the development of the jōri system of land tenure which came into use following the Taika reforms in 636 A. D. He then discusses in detail jōri settlement in the Kyōto and Nara basins and points out physical factors to account for the local decay or preservation of that system. A map of the original land ownership and field numbering methods employed in the Nara area is reconstructed in large part, and shown superimposed on a map of the recent land-tenure system. A number of more detailed maps of land allotment at selected jōri villages are also included.

703. Yonekura Jirō 米倉二郎, Shūraku no rekishi chiri 聚落の歴史地理 (Historical geography of settlement), Tōkyō, Teikoku Shoin, 1949, 6+229 pp., maps, photos.

Five stages of settlement are considered: prehistoric, ancient, medieval, Tokugawa, and modern. Yonekura is particularly interested in the jōri system.

704. Yoshida Tōgo 吉田東伍, Nihon rekishi-chiri no kenkyū 日本歴史地理の研究 (Studies of Japanese historical geography), Tōkyō, Fuzambō, 1923, 9+6+1011 pp.

Forty-seven of Yoshida's earlier articles are gathered in this volume. They concern a wide variety of historical topics, but the majority pertain to certain aspects of Shogunate rule which affected central Honshū, the Kantō, and the Edo area in particular. An account of the initial development of large-scale engineering works along the Tone and Watarase rivers is an example of content. Difficult or obsolete place names appear in kana.

B. Political geography

705. Akioka Takejirō 秋岡武次郎, "Waga kuni no etaru Kita-Saharen riken 我が國の得たる北サハレン利權 (Rights and interests gained by Japan in north Sakhalin)," Chirigaku hyōron, v. 1 (1925), no. 4, 52-90.

Akioka summarizes the geo-political results of the Japanese-Russian agreement completed in Peking in 1925 and pertaining to north Sakhalin. In addition, he briefly outlines the historical background of that area.

706. Asaka Yukio 浅香幸雄 "*Shimizu-minato Mukōjima no kizoku keisō keika* 清水湊向島の歸屬係爭經過 (The history of the dispute over the reversion of Mukōjima in Shimizu harbor)," *Chirigaku hyōron*, v. 19 (1943), no. 9, 11-27; no. 10, 15-27.
An account of the geographical and political aspects of a long-standing historical dispute among four villages over the ownership or administrative control of Mukōjima, a growing sandbar at the mouth of the Tomoe river in Shimizu harbor, Shizuoka prefecture.

707. Iwata Kōzō 岩田孝三 *Kyōkai seiji chirigaku* 境界政治地理學 (The political geography of boundaries), Tōkyō, Teikoku Shoin, 1953, 3+3+239 pp., 7 maps, 12 photos, and 79 illus.
A discourse on the general significance of feudal and provincial, or *kuni*, boundaries in Japan with an account of their historical development. The author describes the geographical, political, and economic factors affecting boundary disputes and cites a number of specific cases. The final chapter is devoted to the problems of present-day political boundaries, most of which were delimited during the Meiji period.

708. Sasaki Hikoichirō 佐々木彦一郎, "*Mura no kyōkai yori mitaru tochi riyō no hōkō* 村の境界より見たる土地利用の方向 (Trends in land utilization interpreted from the study of village boundaries)," *Chirigaku hyōron*, v. 6 (1930), no. 7, 343-350.
During early periods of colonization when land utilization was relatively extensive, the boundaries of *mura* and lesser administrative units such as *aza* were vague. This is a short study of land use methods and ownership conflicts which arose from the uncertainty of rural boundary demarcation and which became increasingly marked as farming grew more intensive and settlement expanded. A long-standing dispute concerning the political ownership of Lake Towada is cited as a principal example.

709. Sasaki Seiji 佐々木清治, "*Chihō jimusho no chiriteki bumpu* 地方事務所の地理的分布 (The geographical distribution of local governmental offices)," *Chirigaku hyōron*, v. 19 (1943), no. 2, 33-48.
A discussion of the political and geographical significance of the *Chihō jimusho* or Offices of Local Affairs which were established in 1942 as an attempted substitution for the previously abolished *gun* or district offices.

C. Population geography

1. General

710. Asai Tokuichi 浅井得一, "*Hompō toshi no jinkō-chirigakuteki kōsatsu* 本邦都市の人口地理學的考察 (Observations on the population geography of Japanese cities)," *Chiri ronsō*, v. 9 (1937), 1-33.
Population change is analyzed for the period 1898-1935 for all *shi* and *machi* with a 1935 population of over 10,000 and for *mura* with a population density greater than 1,000 per square kilometer. The distribution of these urban units, their growth ratio, and the relationship between change and the size and function of cities is considered. Historical aspects of urban concentration are noted. Occupational composition is also discussed.

711. Asai Tokuichi 浅井得一, "*Kantō chihō shotoshi no jinkō zōgen ni tsuite* 關東地方諸都市の人口增減に就いて (On population changes in various cities of the Kantō region)," *Chiri ronsō*, no. 10 (1939), 1-23.
The 79 *shi* and *machi* in the Kantō area are classified on the basis of selected population criteria and the rates of population change from 1920 through 1935 are computed for individual places and urban categories.

712. Doi Kikukazu 土井喜久一, "*Nihon no kōgyō jinkō bumpu* 日本の工業人口分布 (Distribution of industrial population in Japan)," *Chirigaku hyōron*, v. 14 (1939), no. 11, 26-56.
A study of industrial distribution in Japan and a classification of industrial regions. The author also discusses the pros and cons of further industrial dispersal.

713. Gōda Eisaku 合田榮作, "*Niigata-ken no dekasegi jokō kyōkyū chiiki* 新潟縣の出稼女工供給地域 (The supply regions of factory girls in Niigata prefecture)," *Chirigaku hyōron*, v. 17 (1941), no. 3, 47-63.
Niigata prefecture has been noted as the source of factory help for neighboring regions, particularly for the silk-spinning mills of Nagano and Gumma prefectures. Here Gōda presents his conclusions on the social and economic causes of this labor emigration.

714. Hasegawa Shinji 長谷川新二, "*Waga kuni no jinkō jōtai ni tsuite* 我が國の人口狀態に就いて (On the present aspects of the Japanese population)," *Chirigaku hyōron*, v. 5 (1929), no. 5, 18-30.
A statistical analysis of Japan's population treating the following aspects: density; death, birth, and still-birth rates; natural increase; composition by sex, age, and occupation; rural distribution; and urban concentration.

715. Imamura Gakurō 今村學郎, "*Hompō no sanchi no jinkō wa hatashite ippan ni genshō shitsutsu ari ya* 本邦の山地の人口は果して一般に減少しつつありや (Is the population of mountainous regions in Japan decreasing?)," *Chikyū*, v. 20 (1933), no. 2, 1-10.
Imamura presents data to indicate that, contrary to general belief, Japan's mountain population is not decreasing steadily. He computed population rates of change for 374 mountain settlements scattered throughout Japan and in this paper analyzes his findings.

716. Imamura Takeo 今村丈夫, "*Waga kuni ni okeru daitoshi to chihō no shibōritsu ni tsuite* 吾が國に於ける大都市と地方の死亡率について (Big city and regional death rates in Japan)," *Chirigaku hyōron*, v. 5 (1929), no. 4, 21-36.
A calculation of crude death rates based on 1925 statistics with a comparison of the rural and urban "mortality force."

717. Inoue Shūji 井上修次, "Hompō jinkō zōka keitai ni tsuite 本邦人口増加形態に就いて (On the morphometric analysis of the movement of population in Japan)," Chirigaku hyōron, v. 10 (1934), no. 1, 1-26; no. 2, 51-77.
One of few demographic studies which analyzes population change on a qualitative rather than quantitative basis. Population variation during a 32-year period (1898-1930) is considered, using data from eight census reports compiled during that time which cover all cities and gun in Japan proper, excluding Hokkaidō. Three elements of population change are investigated: rate of increase, mode of movement, and the ratio of the rate of increase. Twelve types of population change are identified and plotted in color on a distribution map (1:6,000,000). Analysis of the map distinguishes clear differences between urban and rural trends of population change.

718. Inoue Shūji 井上修次, "Hompō jinkō zōkaritsu gaikan narabi ni jinkō zōkaritsu to jinkō mitsudo to no kankei ni tsuite (Hompō jinkō zōkaritsu kenkyū daiippō) 本邦人口増加率概観並びに人口増加率と人口密度との關係に就いて（本邦人口増加率研究第一報）(On the increasing rate of population in Japan and on the relationship between increasing rates of population and population density— Studies in the rate of population increase in Japan: first report)," Chirigaku hyōron, v. 8 (1932), no. 11, 1-33.
This article consists of two parts. The first explains the methods used in computing population increase. The second discusses the relationship between the rate of population increase and population density in Iwate-ken, the Aganogawa drainage basin, and the Tōhoku district as a whole, as well as in France and Vienna.

719. Ishii Itsutarō 石井逸太郎, "Karafuto no kikō to jinkō no dōsei 樺太の氣候と人口の動靜 (The climate of Karafuto and changes in its population)," Chirigaku hyōron, v. 3 (1927), no. 2, 36-47.
A report on seasonal migration of population in Karafuto and other relationships between the settlement pattern and the climatic conditions of that land. Two dot maps showing the distribution of summer and winter population, for June and December, 1923, are included. References are also made to climatic influences on economic activities.

720. Ishii Itsutarō 石井逸太郎, "Kisetsu to tomo ni kōhai-suru Noto oki Hekurajima no gyoson 季節と共に興廢する能登沖舳倉島の漁村 (A fishing village that prospers and declines with the seasons on Hekura island, of the Noto peninsula)," Chikyū, v. 5 (1926), no. 4, 70-77.
Hekurajima lies offshore from Wajima-machi on the Noto peninsula. It becomes a fishing center during the summer months but is abandoned during the winter season. Ishii describes the summer fishing activities and fishing communities at Hekura.

721. Itō Hiroshi 伊藤博, "Jinkō yori mitaru shuto saiken no kihon kadai 人口より見たる主都再建の基本課題 (Some basic population problems in the reconstruction of the capital [at Tōkyō])," Shin-chiri, v. 4 (1950), nos. 3, 4, and 5, 72-84.
A consideration of reconstruction and planning problems in Tōkyō in terms of the rate of population increase. Itō surveys Tōkyō's population growth from the Meiji period until 1950.

722. Iwasaki Kenkichi 岩崎健吉, "Kii hantō minami kaigan ni okeru kaigai dekasegi imin no kenkyū (daiippō) 紀伊半島南海岸に於ける海外出稼移民の研究（第1報）(A study of emigrants to foreign countries from the south coast of the Kii peninsula — first report)," Chirigaku hyōron, v. 12 (1936), no. 7, 1-23.
A statistical study of the causes of emigration from the Kii region, based largely on an analysis of occupational data. The principal conclusion reached is that the districts which lead in emigration are areas with high percentages of the total population engaged in fishing as opposed to agriculture, forestry, or other occupations.

723. Iwasaki Kenkichi 岩崎健吉, "Kii hantō minami kaigan ni okeru kaigai dekasegi imin no kenkyū 紀伊半島南海岸に於ける海外出稼移民の研究 (A study of emigrants to foreign countries from the south coast of the Kii peninsula)," Chirigaku hyōron, v. 13 (1937), no. 3, 1-18.
A sequel to an earlier paper on this subject by Iwasaki (see preceding entry). Here he summarizes the motivation of emigration from the south Kii coast as determined from an analysis of more than 400 questionaires returned by the local villagers. He concludes that poverty, contrary to expectation, was not a dominant cause of emigration from that district.

724. Kano Tadao 鹿野忠雄, "Taiwan genjūzoku no jinkō mitsudo bumpu narabi ni kōdo bumpu ni kansuru chōsa 臺灣原住族の人口密度分布並に高度分布に關する調査 (An investigation of the distribution of population density and the altitude of the habitations of the Formosa aborigines)," Chirigaku hyōron, v. 14 (1938), no. 8, 1-19; no. 9, 31-66.
Population characteristics of the several indigenous tribes are discussed in terms of their respective cultures and the physical aspects of their environment. The influence of elevation upon the settlement and demographic peculiarities of these mountain people is emphasized. Numerous maps show general and regional distribution, population density, vertical range of habitation, etc.

725. Kawamoto Chūhei 川本忠平, "Nambu toji no idō han'i to sono kyoriteki seikaku no ichikōsatsu 南部杜氏の移動範圍と其の距離的性格の一考察 (Thoughts on the migration sphere of the sake-making people of the Nambu district and the nature of the distances they travel)," Jimbun chiri, v. 3 (1951), no. 1, 30-43.
Toji are farmers of the Akita-Iwate region of the Tōhoku district who migrate to the brewing centers of Japan during the winter season to work as special technicians in sake making. This paper summarizes the socio-economic influences affecting this traditional class of artisans which in 1950 numbered some 2,800 workers.

726. Kiuchi Shinzō 木内信藏, "Manshū shotoshi no minzoku kōsei (Nihonjin jinkō o chūshin to shite) 満州諸都市の民族構成（日本人人口を中心として）(Racial constitution of various Manchurian cities and towns, with special reference to their Japanese population)," Chirigaku hyōron, v. 16 (1940), no. 3, 38-57.

Gives in map form the total population and the population breakdown by racial percentages for 80 Manchurian cities and towns. The author summarizes demographic characteristics and trends and the relationships between urban function and racial composition. Includes a more detailed analysis of the population of Shinkyō (Changchun).

727. Nagai Tōru 永井亨, Nihon jinkōron 日本人口論 (The Japanese population problem), Tōkyō, Ganshōdō Shoten, 1929, 10+4+424 pp.
Nagai's "social-scientific" doctrine on population is introduced and explained and the population problems of Japan, and of the world as a whole, are discussed on that basis. The writer includes other views on population, principally those held by several prominent Japanese economists. Their arguments are examined critically.

728. Naikaku Tōkeikyoku 内閣統計局 (Statistical Bureau of the Cabinet), Meiji gonen ikō waga kuni no jinkō 明治五年以降我國の人口 (Japan's population since 1872), Tōkyō, Tōkyō Tōkei Kyōkai, 1930, 58+2 pp.
A tabulation of Japan's yearly population, male and female, from 1872 to 1925. Statistics prior to the first national census in 1920 were obtained from local honseki (legal address) registers and cannot be regarded as being highly accurate. Tables of estimated population for representative dates of the Tokugawa period are appended.

729. Nasu Shiroshi 那須皓, Jinkō shokuryō mondai 人口食糧問題 (Population and food problems), Tōkyō, Nihon Hyōronsha, 1927, 2+6+319 pp.
Popular discussions by an outstanding agricultural economist. The book is a collection of Nasu's articles and lectures.

730. Odauchi Michitoshi 小田内通敏, "Nihon no sonraku jinkō narabi ni toshi jinkō no ichikōsatsu 日本の村落人口並都市人口の一考察 (Some considerations on the rural and urban populations of Japan)," Chirigaku hyōron, v. 6 (1930), no. 7, 351-368.
Odauchi made a statistical investigation of the population characteristics of seven selected prefectures; i. e., Iwate, Niigata, Ibaraki, Nara, Kagawa, Hiroshima, and Fukuoka. Some emphasis is placed upon Ibaraki prefecture. The author was concerned with the geographical significance of population from a regional viewpoint.

731. Ogasawara Yoshikatsu 小笠原義勝, "Gyogyōsha no bumpu no yurai 漁業者の分布の由來 (Causes of the distribution of fishermen)," Chirigaku hyōron, v. 19 (1943), no. 5, 33-46.
A short discourse on the relationships between ocean currents, shoals, fishing grounds and fish migration, and their influence upon the size and distribution of Japan's fishing population.

732. Ogasawara Yoshikatsu 小笠原義勝, "Gyomin no bumpu 漁民の分布 (The distribution of fishing folk)," Chirigaku hyōron, v. 19 (1943), no. 4, 11-33, 3 maps.
Differences in the fishing population between east and west Japan are considered in terms of social structure, social psychology, and geographic distribution and density.

733. Ogasawara Yoshikatsu 小笠原義勝, "Nōgyōsha to gyogyōsha to no kankei 農業者と漁業者との關係 (The relationship between farmers and fishermen)," Chirigaku hyōron, v. 18 (1942), no. 12, 1-27.
The ratio between the fishing and farming population of coastal villages is given for Japan as a whole and by regional breakdown. Ogasawara then classifies the coastal settlements of Japan into 20 types on this basis and considers briefly other relationships between agriculture and fishing.

734. Sasaki Hikoichirō 佐々木彦一郎, "Shokugyō jinkō kōsei yori mitaru chiri kubun 職業人口構成より見たる地理區分 (Geographic regions seen from the occupational constitution of population)," Chirigaku hyōron, v. 6 (1930), no. 9, 1-16.
An experiment in regional delimitation based on demographic interpretation. Sasaki used the occupational composition of 52 machi and mura as an exponent in determining five geographical sub-regions or divisions of the Izu peninsula, Shizuoka prefecture.

735. Sasaki Hikoichirō 佐々木彦一郎, "Shokugyō jinkō kōsei yori mitaru Ibaraki-ken no chiriku 職業人口構成より見たる茨城縣の地理區 (Geographic regions of Ibaraki prefecture seen from the occupational constitution of the population)," Chirigaku hyōron, v. 6 (1930), no. 12, 1-11.
A sequel to Sasaki's article on the Izu peninsula in this journal (see preceding entry) that uses the same technique of population analysis in determining the geographic divisions of Ibaraki prefecture.

736. Sekiguchi Takeshi 關口武 and Morifuji Katsumoto 森藤勝元, "Sonraku tsūkonken ni kansuru shomondai 村落通婚圏に關する諸問題 (Various problems on 'the intermarriage circle' of settlements)," Chirigaku hyōron, v. 19 (1943), no. 8, 1-20; no. 11, 48-60.
In this socio-geographic study the writers investigated the spheres of intermarriage between villages for seven mura located north and west of Matsue city in Shimane prefecture. Variations in the size of the marriage sphere are considered in terms of village function and historical change. There is reference to the influence of Matsue city upon local marriage customs.

737. Sonoike Motoki 園池大樹, "Nihon no shigen, jinkō oyobi seikatsu joron -- chirigakuteki shomondai no teiki-- 日本の資源，人口及び生活序論 -- 地理學的諸問題の提起-- (Introduction to the resources, population and living in Japan--the raising of geographical problems)," Tōkyō Daigaku chirigaku kenkyū, no. 1 (1950), 53-60.
Briefly discusses from the geographical standpoint the population and food problems facing Japan.

738. Sumitani Keisuke 炭谷恵副, "Kagawa-ken no jinkō chūshin to seichūten to ni tsuite 香川縣の人口中心と正中點とに就いて (Center and medium point of the population of Kagawa prefecture)," Chirigaku hyōron, v. 9 (1933), no. 4, 37-45.

The distributional center of population in Kagawa prefecture is calculated for the years 1896, 1898, 1902, 1920, 1921, 1925 and 1929. During this 33-year period the distribution of population in Kagawa shows little change.

739. Sumitani Keisuke 炭谷恵副, "Kagawa-ken saikin jūnikanenkan no jinkō zōkaritsu ni tsuite 香川縣最近12ヶ年間の人口増加率に就て (On the increase of population in Kagawa prefecture during the last twelve years)," Chirigaku hyōron, v. 10 (1934), no. 9, 18-32.

Sumitani analyzed the rate of population growth of 174 villages and towns in Kanagawa prefecture on a yearly basis for the period 1919-1931. He concluded in this paper that, although the rate of population increase has some relation to geographic factors, the overall tendency in the prefecture was towards a relatively uniform rate, regardless of peculiar environmental conditions or settlement type.

740. Tachi Minoru 館稔, Nihon jinkō no shōrai 日本人口の將來 (The future of Japan's population), Tōkyō, Sekai Keizai Chōsakai, 1947, 162 pp.

An authoritative demographic study based on an analysis of reliable statistical data compiled and tabulated by the official Jinkō Mondai Kenkyūsho 人口問題研究所 (Population Problem Research Institute). Tachi summarizes the influence of the war upon Japan's population, forecasts the population increase for the near future, and discusses the population problem in relation to national resources, food supply, industrial production, employment, and other social and economic factors. With four graphs and 37 tables.

741. Takano Iwasaburō 高野岩三郎, Hompō jinkō no genzai oyobi shōrai 本邦人口の現在及び將來 (The present and future of Japan's population), Tōkyō, Dōjinsha, 1916, 2+3+145 pp.

A pioneer population study founded on statistical analysis. The author draws conclusions on future trends and makes some interesting suggestions on social policy.

742. Takemi Yoshiji 武見芳二, "Karafuto nyūimin no keizai-chirigakuteki kōsatsu 樺太入移民の經濟地理學的考察 (An economic-geographical study of immigrants in Karafuto)," Chirigaku hyōron, v. 4 (1928), no. 9, 37-55; no. 10, 34-59.

Made at the height of emigration from Japan proper to Karafuto, this is an investigation of the demographic characteristics of the immigrants, their settlement distribution, and their occupational breakdown and economic absorption. Takemi identifies four principal settlement areas in Karafuto and notes the nature of four general periods of immigration in relation to the development of extractive industries on the island.

743. Takemi Yoshiji 武見芳二, "Okinawatō shutsuimin no keizai-chirigakuteki kōsatsu 沖繩島出移民の經濟地理學的考察 (An economic and geographical study of Okinawa emigrants)," Chirigaku hyōron, v. 4 (1928), no. 2, 1-22; no. 3, 12-47.

Takemi examines the economic reasons for continuing population pressure in the Ryūkyū islands in spite of decreasing population caused by emigration. The location and status of Okinawan immigrants is treated in the second part of this study.

744. Takemi Yoshiji 武見芳二, "Waga shokuminchi ni okeru naichijin nyūimin 我が植民地に於ける內地人入移民 (Immigrants in Japanese territories from Japan proper)," Chirigaku hyōron, v. 5 (1929), no. 2, 29-34.

Using 1920 and 1925 census data, Takemi studies the distribution of "permanent" Japanese settlements in Karafuto, Korea, Formosa, the Kwantung leased territories, and the mandated islands. He analyzes the place of origin of the overseas settlers in order to summarize recent immigration trends and forecast future rates.

745. Takeuchi Jōgyō 竹内常行, "Niigata, Toyama, Ishikawa sanken ni okeru chōson menseki to chikei narabi ni jinkōmitsudo to no kankei 新潟・富山・石川三縣に於ける町村面積と地形並びに人口密度との關係 (The relationship between topography and density of population and the area of towns and villages in Toyama, Ishikawa, and Niigata prefectures)," Chirigaku hyōron, v. 8 (1932), no. 7, 18-31.

All machi and mura in the three named Hokuriku prefectures were classified according to area and also divided into three general landform categories. Population density was then computed and correlated in terms of these two factors and the resulting generalizations were indicated.

746. Tanaka Keiji 田中啓爾, "Chūō Nihon ni okeru sanroku no jimbun chishigakuteki kenkyū gaihō 中央日本に於ける山麓の人文地誌學的研究概報 (A study of human geography in the piedmont regions of central Japan: an outline report)," Chirigaku hyōron, v. 6 (1930), no. 7, 594-633.

Colonization, land use, transportation, and communications in the marginal zones of mountain ranges are considered in terms of problems involved in interpreting the geographical patterns of such districts.

747. Tanaka Keiji 田中啓爾 and Kōda Kiyoki 幸田清喜, "Hakusan sanroku ni okeru dezukuri chitai 白山山麓に於ける出作地帶 (The district of visiting farms in the peidmont of Mt. Hakusan)," Chirigaku hyōron, v. 3 (1927), no. 4, 19-36; no. 5, 12-26.

The authors present a descriptive survey of temporary and migratory agriculture in the Hakusan district of Fukui and Ishikawa prefectures. The term dezukuri is used for both shifting cultivation in the mountains as well as for isolated permanent farms far detached from the owners' villages. Both types are common along the western foot of Mt. Hakusan. The crops, cultivation methods, marketing, labor migrations, subsidiary industries, etc., connected with this form of agriculture are analyzed in detail and the reasons for the development of this practice are summarized.

748. Tanakadate Hidezō 田中館秀三, "Kantō chihō no jinkō mitsudo no chikeiteki bunseki 関東地方の人口密度の地形的分析 (Topographical analysis of the density of population in the Kantō district)," Chigaku zasshi, v. 50, no. 598 (1938), 11-25.

Geomorphologic divisions of the Kantō district are considered as the base for interpreting the significance of the region's population and land use patterns. With a table of shi, machi, and mura, and four maps.

749. Tanakadate Hidezō 田中館秀三, "Kyūshū chihō no jinkō mitsudo no chikeiteki bunseki 九州地方の人口密度の地形的分析 (Topographical analysis of the density of population in the Kyūshū district)," Chigaku zasshi, v. 51, no. 605 (1939), 16-35.

Similar in presentation to earlier articles on topography and population by this author.

750. Tanakadate Hidezō 田中館秀三, "Tōhoku chihō jinkō mitsudo no chikeiteki bunseki 東北地方人口密度の地形的分析 (Topographical analysis of the density of population in the Tōhoku district)," Chigaku zasshi, v. 42, no. 494 (1930), 6-17; no. 495, 30-41; no. 496, 43-50.

The first in a series of articles by Tanakadate on population distribution and density in relationship to geomorphologic controls. General categories of land use are considered in conjunction with his identification of landform and population divisions. With one map.

751. Tanakadate Hidezō 田中館秀三 and Yamaguchi Yaichirō 山口彌一郎, "Tōhoku chihō ni okeru dezukuri oyobi dekasegi shūraku no keizai chiri 東北地方に於ける出作り及び出稼聚落の經濟地理 (The economic geography of detached farms and the emigration of labor from villages in the Tōhoku region)," Chirigaku hyōron, v. 12 (1936), no. 3, 26-55.

Two examples are described in detail. Hinoemata-mura, Minamiaizu-gun, in Fukushima prefecture was selected to illustrate a mountain community where the land farmed is scattered at some distance from the village. Tayama-mura, Ninohe-gun, in Iwate prefecture is considered to be representative of an agricultural mountain village where winter emigration of labor is necessary.

752. Ueda Masao 上田正夫, "Chiikiteki ni mita jinkō genshō no hendō 地域的にみた人口現象の變動 (Changes in population phenomena seen regionally)," Shin-chiri, v. 3 (1949), no. 5, 19-32.

Comparison of pre-war and post-war population statistics on the prefectural level. Changes are analyzed on a regional basis.

753. Ueda Teijirō 上田貞次郎, Nihon jinkō mondai kenkyū 日本人口問題研究 (Studies on Japanese population problems), Tōkyō, Kyōchōkai, 1933-34, 2 v.

Among the earliest surveys of the population question based on reliable statistical data, these two volumes constitute a major collaborative work. Accomplished under the guidance of T. Ueda, a pioneer Japanese demographer, with the backing of the Conciliation Society of Tōkyō (Kyōchōkai), the voluminous study contains 45 separate monographs on various aspects of the population problem facing Japan and several reviews of pertinent foreign literature. Techniques used to compute birth and death ratios, and other factors in population forecasting are analyzed. Considerable attention is devoted to the serious unemployment situation existing at the time of publication and to the migration of labor within Japan. The studies in volume one are relatively general, those in volume two more intensive. The research is of a high standard and the papers are written in an objective manner. With bibliography.

754. Yamamoto Kumatarō 山本熊太郎, "Waga kuni no shuyō 'kisetsuteki dekasegi' no idō bumpu 我國の主要「季節的出稼」の移動分布 (Distribution of the migration of 'seasonal emigrant laborers' in Japan)," Chirigaku, v. 1 (1933), no. 2, 112-120; no. 3, 111-118; v. 2 (1934), no. 1, 274-257.

In the cold and snowy regions of Japan many farm and mountain people migrate in winter to seek additional income. In this paper the geographic, social, and economic factors behind this seasonal movement are discussed and the types and locations of the supplementary employment are surveyed.

755. Yazawa Taiji 矢澤大二, "Manshū ni okeru waga nōgyō iminson no genjō ni kansuru jakkan no kōsatsu 滿洲に於ける我が農業移民村の現狀に關する若干の考察 (Some thoughts on the present state of Japanese colonized villages in Manchuria)," Chirigaku hyōron, v. 15 (1939), no. 7, 32-57.

The history of Japanese agricultural colonization in northern Manchuria is outlined briefly and the various forms of early Japanese settlement there are explained in general, followed by a more detailed description of the morphologic and economic characteristics of selected villages.

756. Yoshimura Shinkichi 吉村信吉, "Chimei ni yoru jinkō idō no ichikōsatsu 地名による人口移動の一考察 (A study on population movement based on place names)," Chirigaku hyōron, v. 6 (1930), no. 2, 61-76.

Yoshimura uses place names as indicators of population movement on the Kantō plain. Explanation of settlement expansion and shift by this method does not appear to be altogether satisfactory.

2. Medical geography

757. Horiguchi Tomoichi 堀口友一, "Nihon ni okeru fūdobyō no shippei chirigakuteki kenkyū (daiippō)--Shinanogawa, Aganogawa ryūiki no tsutsugamushibyō ni tsuite-- 日本に於ける風土病の疾病地理學的研究(第1報) 信濃川, 阿賀ノ川流域の恙虫病に就いて (A medical geographical study of endemic diseases in Japan: The tsutsugamushi disease of the Shinano and Agano river districts)," Tōkyō Daigaku chirigaku kenkyū, no. 2 (1952), 180-187.

Horiguchi investigates the physical and social environmental factors related to localizing the virus-borne tsutsugamushi disease, or Japanese river fever, to several river deltas along the Japan Sea coast of central and northern Honshū. A semi-technical discussion of the pathogenic complex causing the spread of this disease is included, with regard to the geographical influences affecting it.

758. Horiguchi Tomoichi 堀口友一, Shippei chirigaku 疾病地理學 (The geography of disease), Tōkyō, Kokon Shoin, 1952, 2+231 pp., 16 photos, 82 illus.

This work is an attempt to systematize the branch of geography dealing with the areal analysis of medical phenomena. The author explains the general nature of the subject with reference to the ideas of Max Sorre and J. M. May. The distribution of certain diseases in Japan is discussed in relation to environmental influences and the country is divided, in this respect, into eight "medical" regions. Well illustrated and with 16 statistical tables.

759. Nakamatsu Iyahide 仲松彌秀, "Ryūkyū-rettō ni okeru mararia-byō no chirigakuteki kenkyū 琉球列島に於けるマラリア病の地理學的研究 (A geographical study of malaria in the Ryūkyū islands)," Chirigaku hyōron, v. 18 (1942), no. 4, 49-73.

A scientific inquiry into the geographical influences affecting the localization and spread of tropical malaria in the various island groups of the Ryūkyū archipelago.

D. Transportation and communications

760. Fuji Tokujirō 富士徳治郎, "Waga kuni shuyō kōwan no kaijō bumpaiken ni tsukite 我が國主要港灣の海上分配圏につきて (A division of the spheres of ocean influence of the main ports of Japan)," Chirigaku hyōron, v. 1 (1925), no. 5, 1-11.

The principal Japanese ports are classified on the basis of their predominant function as either trans-shipment points or as ocean terminals.

761. Furushima Toshio 古島敏雄, Shinshū chūma no kenkyū 信州中馬の研究 (Study of chūma in Shinshū), Tōkyō, Itō Shoten, 1944, 5+407+4 pp., 12 tables.

Subtitled, "One case in the history of land transportation in Shogunate Japan." Originally, chūma was a system of privately-operated pack-horse transportation authorized by the Shogunate in areas where the official governmental system, shukueki, was not organized. Furushima gives an excellent account of the expansion of chūma and of the subsequent decline of official transport. He also touches upon other clashes between private and governmental interests brought on by the rise of the merchant class.

762. Hatanaka Daisuke 畠中大輔, "Waga kuni ni okeru bōeki hatten no keika 我が國に於ける貿易發展の經過 (Development of Japanese foreign trade)," Chiri kyōiku, v. 26 (1937), no. 5, 191-200.

On the type and amount of imports and exports according to pre-1937 data, with comment on future trading prospects and the effect of overseas trade upon home industries.

763. Hosoi Ichiroku 細井一六, "Dōro no bumpu genshō ni kansuru kōsatsu 道路の分布現象に關する考察 (A consideration of phenomena relating to the distribution of roads)," Chirigaku hyōron, v. 7 (1931), no. 3, 22-37.

The distribution of trunk highways in Japan is considered, followed by a morphometric analysis of the road-net in the drainage area of the Toyokawa in Aichi prefecture. A technique advanced by Jefferson is used.

764. Iiyama Toshiharu 飯山敏春 and Sasaki Hikoichirō 佐々木彦一郎, "Kantō no kōtsūmō 關東の交通網 (Traffic nets in Kantō)," Chirigaku hyōron, v. 9 (1933), no. 3, 54-69; no. 4, 31-36.

The first part of this paper concerns the development of the Kantō rail network after 1893 and the accompanying expansion of passenger and freight service. The last part deals with the growth of established truck and bus routes and their relationship to the railway system. The influence of the transportation net upon the size of the hinterland of various Kantō cities is discussed. With map.

765. Iwasaki Kenkichi 岩崎健吉 and Tanaka Uichirō 田中卯一郎, "Futatabi Ōsaka-shi no shōgyō seiryoku han'i ni tsuite 再び大阪市の商業勢力範圍に就いて (Second report on the sphere of commercial influence of the city of Ōsaka)," Chirigaku hyōron, v. 6 (1930), no. 8, 75-81; no. 9, 52-55.

See J. Yamagiwa's article on this subject published in the Chirigaku hyōron in 1925 (entry 783). This is a sequel to that initial research. It treats changes which developed during the subsequent 15-year period in the respective commercial spheres of Tōkyō and Ōsaka. The writers followed Yamagiwa's methods of analysis.

766. Kawada Shirō 河田四郎, "Kamotsu unsōro to shite no Tōkyō-shinai kasen unga 貨物運送路としての東京市内河川運河 (Rivers and canals in Tōkyō city as routes of freight transportation)," Chirigaku hyōron, v. 7 (1931), no. 4, 40-49.

The distribution of canals and river routes is studied in relation to their functional influence upon the various districts of Tōkyō, and the types of water-borne freight are described. The canal system is shown in map form.

767. Kōda Kiyoki 幸田清喜, "Edogawabashijō ni okeru kōtsū chōsa ni tsuite 江戸川橋上に於ける交通調査に就いて (On an investigation of traffic over Edogawa bridges)," Chirigaku hyōron, v. 12 (1936), no. 8, 45-67.

An analytical survey of the nature of the highway traffic entering Tōkyō across the Ichikawa and Katsushika bridges, the principal eastern gateways to the metropolis. Reported are Kōda's detailed observations on the types of vehicle and freight, their destination and origin, etc., and the significance of this traffic as concerns Tōkyō's sphere of commercial attraction. Numerous maps and tables are appended.

768. Koyama Kazue 小山嘉壽榮, "Jawatō e no Nihon shikki yushutsu ni tsuite ジャワ島への日本漆器輸出に就いて (On the export of Japanese lacquerware to Java)," Chirigaku hyōron, v. 14 (1938), no. 10, 49-69; no. 11, 16-25.

An account of the pre-war Japanese lacquerware trade with Java emphasizing technical problems in manufacturing and shipping imposed by climatic differences. With suggestions on production and marketing improvements.

769. Nagai Masatarō 長井政太郎, "Mogamigawa no suiun oyobi yusō busshi ni tsuite 最上川の水運及び輸送物資に就て (On river transportation on the Mogamigawa and the goods transported)," Chirigaku hyōron, v. 6 (1930), no. 10, 30-48.

Treats historical change in the characteristics of freight transported on the main stream draining Yamagata prefecture. Three periods are considered: pre-Meiji, post-Meiji, and after the opening of rail service. With maps and selected statistical data from the 1890s onward.

770. Ōhira Tatsuaki 大平辰秋, "Kitakami sanchi no dōro 北上山地の道路 (The roads of the Kitakami mountains)," Chirigaku hyōron, v. 16 (1940), no. 4, 12-30.

The road network of this portion of eastern Tōhoku is analyzed in terms of terrain and livestock farming. Regional differences in the characteristics of the road pattern are explained.

771. Ōkubo Takehiko 大久保武彦, "Dōromō no keitaigakuteki kenkyū 道路網の形態學的研究 (Observations on the morphology of roads)," Chirigaku hyōron, v. 10 (1934), no. 11, 80-101.

Road networks reflect the nature of the local terrain and the stage of geomorphologic evolution. Ōkubo analyzed the road patterns of a variety of landscape types in Japan, such as river flood plains, alluvial fans, coastal dunes, dissected uplands, etc., in order to clarify the relationship between road development and settlement form in landscape interpretation.

772. Ozaki Torashirō 尾崎虎四郎, "Kamogawa-chō ni oyoboseru tetsudō no eikyō ni tsuite 鴨川町に及ぼせる鐵道の影響に就いて (On the influence of the railroad on the town of Kamogawa)," Chirigaku hyōron, v. 7 (1931), no. 2, 34-47.

Compares the functional and morphological aspects of Kamogawa in Chiba prefecture while it was the railhead on the Pacific coast of the peninsula with the changes which occurred after the railway was extended around the whole peninsula. Analysis is largely based on passenger and freight statistics and data from commercial establishments such as hotels and restaurants.

773. Sasaki Hikoichirō 佐々木彦一郎, "Shijōken no chiriteki genkai 市場圏の地理的限界 (The geographic limits of market spheres)," Chirigaku hyōron, v. 4 (1928), 26-35.

Analysis of the geographic, economic, and political significance of regularly or seasonally scheduled markets.

774. Suhara Iyo 須原伊豫, "Nihon no bōeki 日本の貿易 (Japan's foreign trade)," Chiri kyōiku, v. 6 (1927), no. 1, 27-36.

Several aspects of Japanese international trade are summarized on the basis of a study of import and export statistics for 1926.

775. Tanaka Kaoru 田中薫, "Shinsai chokugo ni okeru Tōkyō-shi no kōtsū 震災直後に於ける東京市の交通 (Communication in Tōkyō city just after the 1923 earthquake)," Chirigaku hyōron, v. 1 (1925), no. 2, 56-70; no. 3, 54-67; no. 5, 34-44.

A study of the disruption of Tōkyō's transportation and communication services by the great Kantō earthquake and of the subsequent reconstruction and improvement of those systems. Post-1923 suburban growth and the attendant rapid expansion of the city's rail-net to serve the increased commuter traffic are treated in some detail.

776. Tanaka Keiji 田中啓爾, "Bōeki no kanō chiiki 貿易の可能地域 (Regions with trade possibilities)," Shin-chiri, v. 3 (1949), no. 6, 1-8.

Potential increases in Japan's current foreign trade and future expansion into new overseas markets are considered from a variety of viewpoints: geographic, financial, political, economic, and technical.

777. Tanakadate Hidezō 田中館秀三 and Takahashi Mikio 高橋幹雄, "Yoneshirogawa ryūiki ni okeru ichiba no kenkyū 米代川流域に於ける市場の研究 (Markets in the drainage area of the river Yoneshiro)," Chirigaku hyōron, v. 13 (1937), no. 12, 1-24.

On the historical origin and distribution of local marketing centers in the Yoneshiro region of northern Akita prefecture. This paper is part of a broader geographical study, by the same authors, of the markets of the Tōhoku district. Selected markets are described in detail.

778. Teishinshō 遞信省 (Postal and Communications Ministry), Teishin jigyōshi 遞信事業史 (History of postal and communications services), Tōkyō, Teishin Kyōkai, 1940, 7 v.

A long and detailed official history of postal service and telecommunications in Japan. The best reference work upon the subject. Each volume consists of about 800 pages and much of the material is in table and statistical form.

779. Tetsudōin 鐵道院 (Railroad Institute), Hompō tetsudō no shakai oyobi keizai ni oyoboseru eikyō 本邦鐵道の社會及經濟に及ぼせる影響 (Social and economic effects of the railroad in Japan), Tōkyō, Tetsudōin, 1916, 136 maps.

Poorly titled, this is a series of traffic flow maps indicating the quantitative movement of passengers and principal commodities by rail throughout Japan. The maps show general trends adequately but the base scale is too small for any local detail.

780. Tetsudōshō 鐵道省 (Ministry of Railways), Nihon tetsudōshi 日本鐵道史 (History of Japanese railways), Tōkyō, Tetsudōshō, 1921, 3 v.

In three volumes, this is a detailed history of the Japanese railways prior to the establishment of the Ministry of Railways in 1920. The first volume deals with the initial period of railroad development that lasted until 1892, when a law controlling construction came into effect. The second volume treats the period from 1892 until government ownership of the railroads was established in 1906, and the third volume covers the early years of nationalization and the building of Japan's modern railway network. Roughly 1,000 pages per volume with a chronological table included at the end of the set. Of use to the transportation geographer requiring background information.

781. Tetsudōshō Atami Kensetsu Jimusho 鐵道省熱海建設事務所 (The Atami Construction Office of the Ministry of Railways), Tanna zuidō kōjishi 丹那隧道工事誌 (History of the engineering of the Tanna tunnel), Tōkyō, Tetsudōshō Atami Kensetsu Jimusho, 1936, 12+602 pp.

The construction of the Tanna tunnel, north of the Izu peninsula, is unique in the history of railroad engineering in Japan. Sixteen years in a struggle against unfavorable conditions were spent on this project. This is a complete collection of all the documents concerning the work, with details on planning, equipment, engineering methods, accidents, labor, finance, etc.

782. Unno Kazutaka 海野一隆, "Seinō suiun no chiikiteki kōzō 西濃水運の地域的構造 (Regional structure of water transport in western Mino)," Jimbun chiri, v. 2 (1950), no. 4, 27-39.

Describes the post-Meiji decline in the use of river transport on two water routes, the Kuise and Suimon rivers, along the west side of the Nōbi plain, principally in Gifu prefecture.

783. Yamagiwa Jirō 山極二郎, "Ōsaka-shi no shōgyō no seiryoku han'i 大阪市の商業の勢力範圍 (The sphere of commercial influence of the city of Ōsaka)," Chirigaku hyōron, v. 1 (1925), no. 1, 50-61; no. 2, 45-55.

Yamagiwa analyzes telegram, telephone, freight transportation, and related commercial data for 1921 and 1922 to determine the relative zones of business attraction for Tōkyō and Ōsaka. He draws a demarcation line across central Honshū between the respective spheres of influence, indicating that Ōsaka overshadows Tōkyō in areal domination.

784. Yamaguchi Heishirō 山口平四郎, "Shinano no kōtsū chiri 信濃の交通地理 (Geography of communications in Shinano)," Jimbun chiri, v. 5 (1935), no. 4, 1-18.

The mountainous nature of Shinano province, now Nagano prefecture, has been the most dominant factor in shaping the development of the routes of communication in that region. Environmental control is emphasized in Yamaguchi's discussion of the historical growth of local transport. The paper treats three aspects of this subject: 1) the early establishment of routes and their lasting character or relatively unchanged alignment down to the present; 2) the modernization of communications, stressing the economic influence of the introduction of railroads upon such matters as the rapid rise of the silk industry; and 3) external commercial relationships as analyzed from rail freight statistics.

785 Yamamoto Fumio 山本文雄, Nihon shimbunshi 日本新聞史 (History of Japanese newspapers), Ōsaka, Kokusai Shuppan Kabushiki Kaisha, 1948, 2+6+286 pp.

This is a modification and refinement of Yamamoto's earlier history on the development of the Japanese newspaper (Nihon shimbun hattatsushi). In this work he has restricted the time period to the Meiji era and has made a more detailed and theoretical analysis, avoiding pure description as much as possible. The author attempts to show the growth of the newspaper in relationship to Japan's political development and the emergence of the modern social structure.

786. Yasuda Hatsuo 安田初雄, "Shōken kenkyū no ichihōhō — Kōriyama-shi no shōken 商圏研究の一方法—郡山市の商圏 (A method for the study of commercial spheres— the commercial sphere of Kōriyama city)," Chirigaku, v. 5 (1937), no. 4, 77-89.

Yasuda discusses various economic and geographic methods for determining the commercial spheres of urban or local centers. He bases the discussion upon his study of the problem at Kōriyama city in Fukushima prefecture.

787. Zōsen Kyōkai 造船協會 (Shipbuilding Association), Nihon kinsei zōsenshi (Taishō jidai) 日本近世造船史(大正時代) (History of recent Japanese shipbuilding: the Taishō period), Tōkyō, Zōsen Kyōkai, 1935, 4+17+7+676 pp., 86 photos and 53 illus.

An excellent historical survey of shipbuilding during the Taishō era, the period when Japan emerged as a leading naval power. This book is valuable to those interested in Japan's rise to a position of world significance in naval and shipping affairs and is of particular use to the geographer concerned with industrial patterns and commercial development in modern Japan. In addition to treating both naval and merchant marine shipping from the construction standpoint, the work discusses legislation affecting shipping control, port regulations, shipbuilding organizations, etc.

E. Settlement geography

1. Rural

788. Aono Hisao 青野壽郎, "Kujūkurihama kaigan heiya no chishigakuteki kenkyū sono ni 九十九里濱海岸平野の地誌學的研究その二 (Geographical studies of the plains along the Kujūkurihama coast)," Chirigaku hyōron, v. 8 (1932), no. 9, 50-73; no. 10, 35-55.

Settlement characteristics of Kujūkurihama beach in Chiba prefecture are covered in this description of the fishing villages there. The distribution of naya villages which originated in the use of storage sheds for fishing tools, other historical aspects, and the development of purse seining are detailed.

789. Aono Hisao 青野壽郎, "Kujūkurihama kaigan heiya ni okeru shūraku no idō 九十九里濱海岸平野に於ける聚落の移動 (Development of settlements in the coastal plain of Kujūkurihama)," Chirigaku hyōron, v. 7(1931), no. 1, 1-25.

This article concerns the differentiation of settlement types in the Kujūkurihama district of eastern Chiba prefecture. The present development clearly reflects chronological occupancy in terms of site factors, land use and other functional aspects. The author distinguishes and describes three classes of village: oka or mother settlements, shinden of the Tokugawa period, and recent naya communities.

790. Bekki Atsuhiko 別技篤彦, "Seinō heiya ni okeru wajū no chirigakuteki kenkyū 西濃平野に於ける輪中の地理學的研究 (A geographical study of reclaimed swamps in the plain of western Mino province)," Chiri ronsō, no. 1 (1932), 230-282.

Wajū, similar in many respects to the polders of Holland, are reclaimed and dyked areas characteristic of the delta plains bordering Ise bay in Aichi and Gumma prefectures. After an historical foreword treating the origin and development of wajū reclamation projects, the author presents detailed case examples of this type of land use and settlement, discussing irrigation and drainage problems, dyke construction, house types, and agricultural practices.

791. Hosoi Ichiroku 細井一六, "Hida Takayama no shūraku to kōtsūro 飛騨高山の聚落と交通路 (Settlements and roads of Takayama, Hida province)," Chirigaku hyōron, v. 7 (1931), no. 5, 49-69.

Population distribution and change during the period 1890-1905 are explained and followed by a general description of settlement forms and the related road pattern. The rather unusual static condition of cultural features in this district of Gifu prefecture is pointed out.

792. Hosoi Ichiroku 細井一六, "Toyokawa ryūiki ni okeru den'en senkyo no keitai 豊川流域に於ける田園占居の形態 (The form of agricultural occupance in the drainage area of the Toyokawa)," Chirigaku hyōron, v. 5 (1930), no. 6, 32-41.

Different stages of farm settlement in the Toyokawa lowland of Aichi prefecture are discerned through an analysis of present forms.

793. Ichikawa Wataru 市川渡, "Etchū Kurehayama itō no shūraku ni tsuite 越中呉羽山以東の聚落について (On the rural settlements of Etchū to the east of Mt. Kureha)," Chirigaku hyōron, v. 10 (1934), no. 4, 1-24.

An interesting paper on settlement geography, this article concerns the disseminated hamlets and farm dwellings of the Etchū district in Toyama prefecture. Ichikawa discusses the origin of the scattered settlements in that region and makes comparisons with similar settlement forms in other parts of Japan.

794. Imori Rikuhei 井森陸平, Nōson no shakai to seikatsu 農村の社會と生活 (Society and life in farm villages), Tōkyō, Jidaisha, 1943, 458 pp.

Socio-anthropological studies of rural life in the Japanese farming community. Comprises a collection of the author's lectures and published monographs. The treatment is not systematic. A few of the studies based on Imori's field work are fairly detailed but the majority of the material is rather general and is largely presented from the sociological point of view.

795. Inoue Shūji 井上修次, "Hompō saikō chiiki no kyojū jōtai to kyojū jōgenkai no bumpu to ni tsuite— kōchi kyojū chōsa daiippō 本邦最高地域の居住状態と居住上限界の分布とについて, 高地居住調査(第一報) (On the state of habitation in the highest areas in Japan and on the distribution of height limits to habitation— Studies of highland habitation, first report)," Chirigaku hyōron, v. 10 (1934), no. 8, 27-52; no. 9, 45-61.

High level settlement in the mountains of central Japan, including farms, individual habitations, and temporarily used buildings, was located and identified largely from Inoue's study of 1:50,000 topographic maps. Such habitation is classified functionally, the various types described, and specific representative examples cited. The writer reaches some conclusions as to trends in mountain occupancy and to the height limits affecting mountain dwellings.

796. Ishii Itsutarō 石井逸太郎, "Kyojū chirigaku-jō yori mitaru Etchū heiya 居住地理學上より見たる越中平野 (The Etchū plain viewed from the standpoint of habitation geography)," Chirigaku hyōron, v. 6 (1930), no. 7, 1-12.

What is known of prehistoric and ancient forms of habitation on the Toyama plain is summarized and contrasted with a description of the settlement pattern of recent times, stressing the nature of disseminated villages in the western part of the lowland. Ishii's dot map of population distribution is included.

797. Itō Ryūkichi 伊藤隆吉, "Tōkyō-shi seikō ni okeru yashikimori no keitai to kinō 東京市西郊に於ける屋敷森の形態と機能 (Types and functions of yashikimori on the upland in the western suburbs of Tōkyō)," Chirigaku hyōron, v. 15 (1939), no. 8, 54-72; no. 9, 18-31.

The types and uses of farm woodlots at two representative villages, Mitaka and Hōya, in Tōkyō prefecture are studied in detail with reference to the general nature and distribution of such forest land in other sectors of Japan.

798. Iwasaki Kenkichi 岩崎健吉, "Tosa Murotozaki fukin engan shūraku ni kansuru keitai keisoku kekka no bumpu ni tsuite 土佐室戸岬附近沿岸聚落に關する形態計測結果の分布に就て (On the distribution of measurements of the morphology of houses and stonewalls near Murotozaki in Tosa province)," Chirigaku hyōron, v. 17 (1941), no. 4, 30-50.

An interpretation of regionality as determined by a survey of house types and enclosing walls in a coastal district of southeastern Shikoku. The characteristics of roofs, eaves, and stone walls are noted in great detail, and the writer attempts to account for local variations.

799. Kagami Kanji 鏡味完二, "Nōbi heiya ni bumpu-suru matsubayashi no chirigakuteki igi 濃尾平野に分布する松林の地理學的意義 (Geographical significance of the distribution of pine forests in the Nōbi plain)," Chirigaku hyōron, v. 11 (1935), no. 2, 13-40.
The distribution of pine forests is selected as one aid in interpreting the cultural landscape of the Nōbi plain in Aichi and Gifu prefectures. Forest density is calculated and its relationship to economic activity described.

800. Kagami Kanji 鏡味完二, "Okinawatō no shūraku 沖縄島の聚落 (Settlements in the Okinawa islands)," Chirigaku hyōron, v. 18 (1942), no. 9, 33-56; no. 10, 17-43.
The urban geography and cultural origin of the two most important Okinawan cities, Naha and Shuri, are discussed and followed by a general summary of rural settlement characteristics. Towns and villages are classified according to their morphology and physical site, and their distribution is mapped. In conclusion, the type of town classed geographically as Akropolis-lage is compared theoretically with the gusuku settlements of Okinawa.

801. Kawada Shirō 河田四郎, "Kōfu bonchi ni kansuru nisan no chirigakuteki kōsatsu 甲府盆地に關する二、三の地理學的考察 (Some geographical considerations concerning the Kōfu basin)," Chirigaku hyōron, v. 7 (1931), no. 12, 19-30.
Observations on settlement, land use, and general aspects of the natural and cultural landscape of this mountain basin in Yamanashi prefecture. Climatic factors in regard to the vineyards and cherry orchards of the district are particularly considered.

802. Kawaguchi Takeo 川口丈夫, "Hokkaidō no kako jinkō to kaitakuchiiki 北海道の過去人口と開拓地域 (The past population of Hokkaidō and reclamation areas)," Chirigaku hyōron, v. 13 (1934), no. 7, 63-82; no. 8, 1-17.
Population density maps of Hokkaidō in 1887, 1895, 1907, 1918, and 1935 are the major materials in these articles. By comparing these maps with one another, the reclamation areas within Hokkaidō are studied.

803. Kikuchi Toshio 菊地利夫, "Kujūkuriura naya shūraku no seiritsu 九十九里浦納屋聚落の成立 (Formation of naya settlements on the Kujūkuri coast)," Shin-chiri, v. 3 (1949), no. 1, 14-22.
A paper on one type of land use peculiar to the fishing communities along the Kujūkuri beaches, on the Pacific coast of Chiba prefecture.

804. Kiuchi Shinzō 木内信藏 and Ōtsuki Takahiko 大槻高彦, "Nambu magariya no bumpu 南部曲家の分布 (Distribution of the southern magariya dwellings)," Chirigaku hyōron, v. 18 (1942), no. 11, 73-79.
Magariya are L-shaped farm houses peculiar to the Tōhoku district. Here, the distribution and general character of those located in the Morioka area is explained. With a diagrammatic plan of this house-type.

805. Kodera Kanekichi 小寺廉吉 and Iwamoto Hideo 岩本秀夫, "Ōigawa karyū no sankyosei sonraku 大井川河流の散居制村落 (On the scattered settlements on the alluvial fan of the Ōigawa)," Chirigaku hyōron, v. 15 (1939), no. 9, 32-56; no. 10, 20-43.
A two-part study of the disseminated villages in the lower valley of the Ōi river and on the adjacent coastal plain in Shizuoka prefecture. The reasons for Strendorf are examined in terms of land use characteristics and the general historical development of this region. Population problems are summarized in this regard.

806. Komaki Saneshige 小牧實繁, "Nihon ni okeru shūraku no kōkyo gendo (yohō) 日本に於ける聚落の高距限度(豫報) (A preliminary report on the height limits of settlements in Japan)," Chiri ronsō, no. 1 (1932), 127-154.
All communities situated above 1,300 meters, or approximately 4,300 feet in elevation, are identified from a study of 1:50,000 topographic maps and the primary function of each is determined and listed. Settlements in Hokkaidō are not included.

807. Kon Wajirō 今和次郎, "Ogasawara guntō no kenkyū— shūraku no keitai oyobi kaoku— 小笠原群島の研究 聚落の形態及家屋 (A study of the Ogasawara islands: settlement types and houses)," Jimbun chiri, v. 1 (1926), no. 1, 36-46.
A study of dispersed settlement and land utilization in the Ogasawara (Bonin) islands. Settlement forms rather unique to these islands are explained and the characteristic types of farm dwelling and out-building are described in some detail.

808. Kusamitsu Shigeru 草光繁, "Daichisonraku no keitai 臺地村落の形態 (Forms of rural settlements in some uplands of Japan)," Chirigaku hyōron, v. 8 (1932), no. 5, 1-18.
A descriptive classification of upland settlement forms based on a selective study of three upland districts, i. e., Musashino, Shimoosa, and Mikata-ga-hara. Morphologic characteristics and the type of site topography constitute the principal classification criteria employed.

809. Kusamitsu Shigeru 草光繁, "Hinokawa heiya no sonrakukei ni kansuru keitaigakuteki kenkyū (yohō) 簸川平野の村落景に關する形態學的研究(豫報) (Studies in the morphology of rural settlement types on Hinokawa plain -- a preliminary report)," Chirigaku hyōron, v. 6 (1930), no. 8, 45-61.
In this article Kusamitsu classifies morphologically the types of disseminated village found in the small Hikawa lowland of northern Shimane prefecture.

810. Maeda Toraichirō 前田虎一郎, "Gaison-gata shūraku no hassei narabi ni hensen no ichikōsatsu: Saitama-ken Sōka-machi 街村型聚落の發生並に變遷の一考察 埼玉縣草加町 (A study of the origin and development of a Strassendorf: Sōka in Saitama prefecture)," Chirigaku hyōron, v. 12 (1936), no. 8, 1-23.

Developmental changes of Sōka in Saitama prefecture since its initial growth as a stage town on the Nakasendō highway are explained. The town is considered as representative of the Strassendorf type. Sōka's present form is then analyzed in detail. Information is given on house density, the distribution and types of shop, etc. The data were gathered during a field survey by the author.

811. Matsui Isamu 松井勇, "Tonami heiya no ichibu ni okeru sanson no bumpu jōtai ni kansuru tōkeiteki ichikōsatsu 礪波平野の一部に於ける散村の分布狀態に關する統計的一考察 (A statistical survey of the distribution of disseminated villages in a part of the Tonami plain)," Chirigaku hyōron, v. 7 (1931), no. 6, 13-30.
Map interpretation and theoretical considerations of the settlement pattern of the Tonami lowland in Toyama prefecture where perhaps the best examples of disseminated villages in Japan occur. The district to the east of Fukuno-machi comprises the study area.

812. Misawa Katsue 三澤勝衛, "Taniguchi shūraku no kenkyū 谷口聚落の研究 (A study of settlements at the mouths of valleys)," Chirigaku, v. 2 (1934), no. 1, 69-99.
An inquiry into the structural and functional aspects of Nishi-Nagano, a commercial suburb of Nagano city in north-central Honshū, with general references to the characteristics of similar towns in other areas. Misawa feels that Nishi-Nagano is representative of many communities geographically situated at valley mouths.

813. Mitomo Kunigorō 三友國五郎, "Tankō shūraku no nisan 炭坑聚落の二三 (Some examples of coal-mining settlements)," Chirigaku, v. 3 (1935), no. 10, 121-124.
Mining communities along the lower Ongagawa in northern Kyūshū are described, with comment on land use and social problems. Mitomo stresses the need for settlement planning in the coal fields.

814. Miyamoto Michihiko 宮本道彦, for the Kyōchōkai 協調會 (Conciliation Society), Nōson chōsa oboegaki 農村調査覺書 (Notes on village surveys), Tōkyō, Kyōchōkai, 1943, 2+2+168 pp.
Miscellaneous observations on rural communities, mainly mountain and fishing villages, with suggestions for the field researcher planning sociological surveys. Certain topics may interest the settlement geographer.

815. Mochizuki Katsumi 望月勝海 and Saitō Gaiji 齊藤外二, "Tetorigawa senjōchi no shūraku keikan 手取川扇狀地の聚落景觀 (View of settlement on the Tetorigawa fan)," Chirigaku hyōron, v. 13 (1937), no. 7, 21-36.
In this alluvial fan area, the most important in rice production of Ishikawa prefecture, many of the villages are agglomerate. The geographical and social reasons for this are analyzed and the landforms, the development of irrigation canals, and flood damage are described. Some reference is made to architecture, the irimoya type of house predominating here.

816. Mori Sumie 森壽美衛, "Ōmura senjōchi no reki 大村扇狀地の礫 (Gravel of the Ōmura alluvial fan)," Chirigaku hyōron, v. 8 (1932), no. 6, 1-21.
In this paper Mori presents a study of agricultural settlement in an alluvial area where the land must be cleared of coarse gravel and stones before cultivation is possible. The district concerned is on the east side of Ōmura bay in Nagasaki prefecture. Farming problems and the use of cleared stone for construction purposes are described. With brief historical notes on the initial colonization of the Ōmura district.

817. Mori Sumie 森壽美衛, "Ukujima oyobi Ojikajima 宇久島及び小値賀島 (Uku island and Ojika island)," Chirigaku hyōron, v. 7 (1931), no. 6, 31-55.
These two islands, together with minor islets, form the northern group of the Gotō chain located west of Kyūshū. This is a general survey of the physical and cultural geography of the islands, with added notations on emigration and population problems.

818. Muramatsu Shigeki 村松繁樹, "Kaitokō 垣内考 (A study of kaito)," Jimbun chiri, v. 4 (1952), no. 2, 1-10.
Kaito are rural social groups, forming distinct settlement units in the mura and ōaza of the Yamato basin and vicinity. The types of kaito are described by citing specific examples. Marked differences in the nature and characteristics of the groups are examined in terms of possible geographical causation.

819. Muramatsu Shigeki 村松繁樹, "Nihon shūraku chirijō no ichimondai—iwayuru kangō buraku ni tsuite—日本聚落地理上の一問題 所謂環濠部落に就いて (A geographical problem on settlements in Japan - on the so-called water-encircled settlements of the Yamato basin)," Chirigaku hyōron, v. 18 (1942), no. 8, 1-8.
A short investigation into the nature of kangō villages, or moat-surrounded settlements, in the Yamato basin. Their appearance, distribution, and origin are considered.

820. Murata Teizō 村田貞藏, "Isawa no keikan ni kansuru jakkan no kiroku 膽澤の景觀に關する若干の記錄 (Some notes on the landscape of the Isawa district)," Chirigaku hyōron, v. 15 (1939), no. 9, 57-74.
Concerning settlement forms in a triangular shaped hill and valley section of south-central Iwate prefecture, bordered on the east by the Kitakami river.

821. Murata Teizō 村田貞藏, "Sanson no bunsando o shiru ichihōhō 散村の分散度を知る一方法 (A method of measuring the degree of dispersal of disseminated villages)," Chirigaku hyōron, v. 6 (1930), no. 12, 48-57.
Murata introduces a grid system for classifying settlement dissemination on the basis of rural house density as determined from map analysis. He illustrates the application of his technique to the settlement pattern of the Tonami plain in Toyama prefecture.

822. Murata Teizō 村田貞藏 and Yoshimura Shinkichi 吉村信吉, "Shūraku no jinkō to sono kōsaku menseki no rironteki kōsatsu 聚落の人口とその耕作面積の理論的考察 (A theoretical consideration of village population and the area of cultivated fields belonging to it)," Chirigaku hyōron, v. 6 (1930), no. 5, 1-31.

The researchers devised a formula for calculating the areal limits of agricultural land which can be reasonably maintained from one village, considering the following factors: size of population, distance, labor time, type of cultivation. On the basis of this formula, a comparison of the field limits of Strassendorf and agglomerated villages is made.

823. Nakamura Keisaburō 中村慶三郎, "Yamakuzure chihō ni okeru shūraku ni tsuite 山崩れ地方に於ける聚落に就いて (On settlement in landslide regions)," Chirigaku hyōron, v. 8 (1932), no. 2, 30-38.
Nakamura investigated the reasons for persistent settlement in certain areas in Niigata prefecture subject to the danger and damage of frequent landslides. His conclusions reported here relate to topographic and other advantages for initial settlement afforded by such sites in spite of their unsafe features.

824. Nishida Masao 西田正夫, "Hitachi Kashima chihō no bunka chirigakuteki kenkyū 常陸鹿島地方の文化地理學的研究 (A cultural geographical study of the settlements in the Kashima region of Hitachi province)," Chirigaku hyōron, v. 10 (1934), no. 3, 23-44; no. 4, 41-58.
An historical and geographical study of the establishment and growth of settlement in a sand dune area on the Pacific coast in Ibaragi prefecture, just north of the lower Tonegawa. Nishida classifies several types of village and traces the old travel routes along which settlement developed.

825. Nishiki Masao 西龜正夫, "Kita-Nihon no shūraku 北日本の聚落 (Settlements of northern Japan)," Chikyū, v. 9 (1928), no. 3, 33-38; no. 4, 54-60.
A brief paper on the settlement geography of Tōhoku and Hokkaidō. The article stresses the significance of variations in building construction. Nishiki compares the types of northern settlement with representative forms in Kyūshū and Chūgoku.

826. Nishiki Masao 西亀正夫, "Shūraku no seitai ni tsuite 聚落の生態について (On the ecology of settlements)," Chikyū, v. 5 (1926), no. 4, 120-128.
On the origin, growth, developmental fluctuations, and decay of Japanese settlements. A rather general discussion with no attempt at regional interpretation.

827. Nishida Yoshirō 西田與四郎, "Suzuka-tōge no shūraku 鈴鹿峠の聚落 (Settlements of Suzuka pass)," Chirigaku hyōron, v. 2 (1926), no. 6, 21-31.
Suzuka-tōge is a principal pass on the old Tōkaidō highway. This paper concerns the historical growth and decay of the former stage towns located on either side of that pass. The towns described are Seki, Tsuchiyama, and Bange, all of the Strassendorf type.

828. Nishida Yoshirō 西田與四郎, "Shūraku no kenkyūhō 聚落の研究法 (The methodology of settlement geography)," Chiri kyōiku, v. 4 (1926), no. 5, 24-34.
An outline of the geographical approach to the study of rural and urban settlement founded on an analysis of Japanese forms, with consideration of both physical and cultural elements. The majority of the examples cited pertain to the morphological development of Japanese cities.

829. Nō Toshio 能登志雄, "Musashino daichi gaison ni kansuru kenkyū 武藏野臺地街村に關する研究 (The Strassendorf villages of the Musashino upland)," Chirigaku hyōron, v. 11 (1936), no. 1, 48-65.
The German term Strassendorf is usually transliterated into Japanese as gaison, and the expression is frequently used by Japanese geographers in spite of some discrepancies in its usage. In this paper, Nō attempts to clarify the definition of gaison as a compound concept of form and function, but with emphasis upon the structure of this type of "shoe-string" settlement. Examples cited are chiefly in the Musashino upland west of Tōkyō.

830. Odauchi Michitoshi 小田内通敏, "Chōsen no jimbun chirigakuteki shomondai 朝鮮の人文地理學的諸問題 (Some problems of human geography in Korea)," Chirigaku hyōron, v. 1 (1925), no. 6, 1-19.
In this monograph Odauchi regards Korea as a transitional region between Japan and China, both naturally and culturally. Observations on social stratification, kinship settlements, periodic fairs, and various social and economic aspects of Korean life are cited, together with remarks on the development of Japanese influences and interests.

831. Odauchi Michitoshi 小田内通敏, Kyōdochiri kenkyū 郷土地理研究 (Geographical study of the home provinces), Tōkyō, Tōkō Shoin, 6+5+4+325 pp.
A discussion of the theory and methodology of regional settlement geography with reference to examples in Japan.

832. Odauchi Michitoshi 小田内通敏, Shūraku to chiri 聚落と地理 (Settlement and geography), Tōkyō, Kokon Shoin, 1927, 8+5+311 pp.
A highly regarded series of monographs on Japanese settlement geography, including studies of villages in Korea and Manchuria. Odauchi was a pioneer in this field.

833. Odera Renkichi 小寺廉吉, "Etchū Gokayama no uchi no Nashidani shūraku no kenkyū 越中五ケ山の内の梨谷聚落の研究 (A study of the settlement of Nashidani in Gokayama, Etchū province)," Chirigaku hyōron, v. 10 (1934), no. 5, 1-30.
The Etchū villages of Gokayama in Toyama prefecture are secluded in deep mountain valleys and remain relatively isolated from settlements in the nearby lowlands. This is an historical and social interpretation of these valley communities based on the writer's examination of old documents and records at Gokayama.

834. Ogasawara Yoshikatsu 小笠原義勝, "*Setonaikai no gyoson to nōson* 瀬戸内海の漁村と農村 (Agricultural and fishing villages of the Inland Sea)," *Chirigaku hyōron*, v. 16 (1940), no. 7, 1-16.
A study in the historical geography of agricultural and fishing villages in the islands of the eastern Inland Sea, with specific reference to the Geiyō and Shiaku island groups. Ogasawara is primarily interested in the economic base of these settlements.

835. Ogawa Takuji 小川琢治, "*Jimbun chirigaku-jō yori mitaru Nihon no sonraku* 人文地理學上より觀たる日本の村落 (Japanese villages seen from the standpoint of human geography)," *Chikyū*, v. 5 (1926), no. 4, 1-26.
In this monograph the structure and origin of house types are especially emphasized, but the writer also devotes considerable attention to disseminated and *kaito* settlement. Also described in some detail is the development of stage and ferry towns. Ogawa was a recognized authority on both Japanese settlement geography and history and this study is a classic of its type.

836. Ōnishi Seiji 大西青二, "*Tōgeshita shūraku* 峠下集落 (Settlement below passes)," *Jimbun chiri*, v. 3 (1951), no. 1, 43-56.
An examination of the cultural phenomena accompanying the decline of Futatsuya and Itadori, once important pass settlements on routes connecting the Kinai and Hokuriku regions. The article gives special regard to the social and population changes that followed the opening of the railroad which passed these communities.

837. Sakurai Shizuka 櫻井靜, "*Gaison ni tsuite no kenkyū* 街村についての研究 (A study of *Strassendorf*)," *Chikyū*, v. 23 (1935), no. 4, 51-61; no. 5, 46-55.
The characteristics of *gaison* or "shoe-string" villages are studied in order to determine some of the reasons for their formation.

838. Sasaki Hikoichirō 佐々木彦一郎, "*Bonchi shūraku no kikō ni tsuite* 盆地聚落の機構に就いて (On the nature of settlements in a mountain basin)," *Chirigaku hyōron*, v. 2 (1926), no. 9, 1-12.
The writer selected the Kazuno basin in Akita prefecture as being a fairly representative mountain basin in terms of both geographic and human considerations. This is a study of the origin and growth of settlement in that basin.

839. Sasaki Hikoichirō 佐々木彦一郎, "*Kantō no mura no oitachi* 關東の村の生ひ立ち (The growth of villages in the Kantō)," *Chiri kyōiku*, v. 17 (1933), no. 4, 10-18.
A description and classification of the towns and villages of the Kantō district with comments on their origin, adaptation to topographic conditions, and development.

840. Sasaki Hikoichirō 佐々木彦一郎, "*Setouchi-shotō no dōro to jūka to no ritchi kankei* 瀬戸内諸島の道路と住家との立地關係 (Relationships between roads and houses in the landscape of various islands of the Inland Sea)," *Chirigaku hyōron*, v. 8 (1932), no. 1, 38-47.
Contains the writer's general observations on the human geography of the Setonaikai district. Several geographical problems are posed.

841. Sasaki Seiji 佐々木清治, "*Sankakusu shinden — toku ni sono bumpu ni tsuite —* 三角洲新田 — 特にその分布に就いて — (Reclaimed rice-fields on deltas with special reference to their distribution)," *Jimbun chiri*, v. 5 (1953), no. 5, 1-18.
Shinden or reclaimed villages in the deltas of the Fuji and Ōi rivers in Shizuoka prefecture are studied. Two types of development are cited: land reclaimed in the midst of established settlements and new fields added to the fringe of old settlements.

842. Satō Hiroshi 佐藤弘, "*Hompō ni okeru gaison no bumpu* 本邦に於ける街村の分布 (Distribution of *Strassendorf* settlements in Japan)," *Chirigaku hyōron*, v. 6 (1930), no. 7, 550-556.
A paper describing representative types of *Strassendorf* or "shoe-string" settlements and summarizing locational factors.

843. Shima Yukio 島之夫, "*Hokkaidō Asahikawa fukin no tondenhei kaikon shūraku ni tsuite* 北海道旭川附近の屯田兵開墾聚落に就いて (On settlements begun by the colonial militia in the Asahigawa area of Hokkaidō)," *Chikyū*, v. 17 (1932), no. 1, 44-50.
A short account of the colonization of certain areas in Hokkaidō at the beginning of the Meiji era by the soldier-farmers of the *tondenhei*, or colonial militia. After tracing the historical background of this movement, Shima briefly describes the present-day condition of three of the villages which were founded by military immigrants.

844. Shōyama Toshio 庄山俊夫, "*Shimōsa daichi ni okeru shūraku no bumpu* 下總臺地に於ける聚落の分布 (Distribution of settlements on the Shimōsa upland)," *Chirigaku hyōron*, v. 10 (1934), no. 5, 44-62.
The author bases this article on the morphology and distribution of settlement in an upland section of Chiba prefecture on his morphometric analysis of topographic maps correlated with a study of aerial photographs covering the district.

845. Takeuchi Jōgyō 竹内常行, "*Karafuto nōgyō shūraku no bumpu, hattatsu, keitai no kenkyū* 樺太農業聚落の分布, 發達, 形態の研究 (A study of the distribution, growth, and form of rural settlements in Karafuto)," *Chirigaku hyōron*, v. 11 (1935), no. 11, 26-39.
This is a comprehensive description of farm villages in Karafuto (Sakhalin), devoting most detail to their morphologic character. The historical background of settlement is outlined.

846. Takeuchi Jōgyō 竹内常行, "Kurobegawa senjōchi no shūraku keitai ni kansuru nisan no kōsatsu 黒部川扇狀地の聚落形態に關する2.3の考察 (Some thoughts on the form of settlement on the fan of the Kurobe river)," Chirigaku hyōron, v. 8 (1932), no. 2, 16-29.

Geographical site factors are chiefly considered in the winter's analysis and classification of settlement forms in the alluvial area at the mouth of the Kurobe river in eastern Toyama prefecture. The predominant dispersed and isolated villages of the Kurobe region are compared with types characteristic of the western portion of the prefecture.

847. Tanakadate Hidezō 田中館秀三 and Yamaguchi Yaichirō 山口彌一郎, "Tōhoku chihō ni okeru shijōken, shijōgun oyobi shijō shūraku no hattatsu 東北地方に於ける市場圏 市場群及び市場聚落の發達 (Market areas of the Tōhoku district and their grouping, together with the development of marketing settlements)," Chirigaku hyōron, v. 18 (1942), no. 5, 18-40; no. 6, 48-69.

In Tōhoku there are still preserved some of the periodic markets which were once common throughout Japan. The authors classify these markets by type and by the size of their spheres of attraction. They argue that these market spheres constitute economic regions which can be interpreted as significant cultural-geographic units within which the populace can live with relative commercial self-sufficiency.

848. Tanioka Takeo 谷岡武雄 "Jōri shūraku 條里集落 (Jōri villages)," Jimbun chiri, v. 1 (1948), no. 2, 72-75.

An elementary explanation of the nature and significance of the jōri land system is followed by descriptions of the physical form of selected jōri villages.

849. Toda Hideo 戸田秀雄, "Keikan kubun ni kansuru ichikōsatsu 景觀區分に關する一考察 (On morphological divisions of landscape)," Chirigaku hyōron, v. 8 (1932), no. 3, 44-71.

Elements of the cultural landscape in the Mt. Fuji and Mt. Aso districts are studied to discover interpretive patterns of use in classifying homogeneous areal units.

850. Tomita Yoshirō 富田芳郎, "Nambu Taiwan no ichibu ni okeru shūkyogata nōson shūraku to sono keieikei 南部臺灣の一部に於ける集居型農村聚落とその經營景 (Agglomerate rural settlements and their planning in a part of southern Formosa)," Chirigaku hyōron, v. 9 (1933), no. 7, 7-38.

The origin and morphology of settlements on the Toroku alluvial fan and the adjacent coastal plain are explained and the form and construction of houses described. The agglomerate settlement in this district is in contrast to disseminated types prevailing in the northern sections of the island.

851. Tomita Yoshirō 富田芳郎, "Taiwan ni okeru gōsei shūraku to shite no Matō oyobi Kari 臺灣に於ける合成聚落としての麻豆及佳里 (Matō and Kari as examples of compound settlements in Formosa)," Chirigaku hyōron, v. 11 (1935), no. 6, 17-27; no. 7, 18-33.

The preservation of distinct agricultural and commercial sections in these villages in Tainan province, southwestern Formosa, led the author to consider them as a form of "compound" agglomerate settlement. In this connection, their characteristics are described.

852. Tsujii Kōtarō 辻井浩太郎, "Futamigaura torii-mae-machi no kōsatsu 二見浦鳥居前町の考察 (A survey of Futamigaura, a town developed in front of Ise shrine)," Chiri kyōiku, v. 17 (1933), no. 5, 51-54; no. 6, 51-54.

This paper outlines the functional development of Futamigaura, initially a recreation place for shrine visitors, later a hostelry, and finally a local commercial center.

853. Tsujimoto Yoshirō 辻本芳郎, "Chikasui yori mitaru Sagamino no shūraku 地下水より見たる相模野の聚落 (Settlement in the Sagami plain considered in relation to undergound water)," Chirigaku hyōron, v. 13 (1937), no. 11, 53-68.

This is a discussion of the problem of settlement location based on a survey of sub-surface water conditions in the Sagami plain. It takes note of distribution, inclination of the ground water surface, and depths of wells. With map.

854. Ueno Fukuo 上野福男, "Bandokoro ni okeru shūraku to kōchi, dezukuri shūraku, sono teichaku oyobi teichaku to kōsaku no kōkyo no mondai o chūshin ni 番所に於ける聚落と耕地、出作聚落、其の定著及び定著と耕作の高距の問題を中心に (Settlement and cultivation at Bandokoro: on the problems of a temporary settlement for farmers, its becoming a fixed settlement, and the upper limit of fixed settlements and cultivation)," Chirigaku hyōron, v. 15 (1939), no. 12, 15-40.

This paper investigates the process of establishing Bandokoro, at the eastern foot of Mt. Norikura, as a permanent farm settlement. This community had long been a temporary summer encampment for farmers from the nearby village of Onogawa. Farming problems and the upper limits of cultivation in this mountainous area are also discussed.

855. Watanuki Isahiko 綿貫勇彦, "Setouchi no shūraku 瀬戸内の聚落 (Villages in the Setouchi)," Chirigaku hyōron, v. 7 (1931), no. 12, 59-70.

Watanuki classifies villages in the Inland Sea region into four general categories determined by the proportionate emphasis in each upon agricultural or fishing activities. He then describes a village selected as being representative of the purely agricultural type, the settlement of Mori in Ochi district, Ehime prefecture. Geographic, social, and economic aspects of this village are considered.

856. Watanuki Isahiko 綿貫勇彦, Shūraku chirigaku zōhoban 聚落地理學 增補版 (Settlement geography, enlarged edition), Tōkyō, Yajima Shobō, 1948, 2+2+4+286 pp., 125 illus.

This is a monumental work by the late Professor Watanuki, one of Japan's leading geographers. General geographical characteristics of settlement are considered first, followed by detailed studies of eight Japanese type examples determined by topographical site conditions, e. g., a mountain village and settlements on uplands, alluvial fans, rocky coasts, dunes, etc. Similar to the first edition of 1933 except for the addition of four supplementary monographs.

857. Yajima Nikichi 矢嶋仁吉, Jōshū Annaka-machi no shūraku keikan 上州安中町の聚落景観 (A view of the settlement of Annaka town, Kōzuke province)," Chirigaku hyōron, v. 16 (1940), no. 8, 45-55; no. 9, 1-9.
The historical development of this former stage and castle town on the Nakasendō highway, located on the Kantō side of the important Usui pass, is outlined with particular reference to the several changes in its functional role.

858. Yajima Nikichi 矢嶋仁吉, "Musashino shinden shūraku no kenkyū 武蔵野新田聚落の研究 (A study of shinden villages in the Musashino upland)," Chiri kyōiku, v. 27 (1937), no. 2, 50-55; no. 3, 49-54.
A portion of Yajima's long-term research on shinden, "new" settlements established by colonization in the Edo period. This paper treats Onumada (in Kodaira, Kitatama district, Tōkyō prefecture) as a typical shinden, and is a descriptive and historical sketch of that community and its land holdings.

859. Yajima Nikichi 矢嶋仁吉, "Tama kyūryō seihokubu Nanao-mura fukin no chikasui to shūraku ritchi 多摩丘陵西北部七生村附近の地下水と聚落立地 (Groundwater and settlements in the neighborhood of Nanao-mura in the northwest part of the Tama hills)," Chirigaku hyōron, v. 19 (1943), no. 12, 8-30.
Nanao-mura in Tōkyō prefecture is employed as a part of the author's long-continued study of shinden villages in the Musashino upland. The condition of ground water in this vicinity is surveyed in relation to the development of villages and other cultural elements. Comparisons of the Tama hills with the Musashino and Akitome uplands are made.

860. Yajima Nikichi 矢嶋仁吉, "Tōkyō-fu Kurume-mura fukin no chikasui to shūraku ritchi no kenkyū 東京府久留米村附近の地下水と聚落立地の研究 (A study of the groundwater near Kurume village, Tōkyō prefecture, and the development of settlement)," Chirigaku hyōron, v. 17 (1941), no. 11, 1-24.
Kurume-mura in Tōkyō prefecture is in a region of valley heads dissecting the margin of the Musashino upland, where there are many springs. It provides the area for this study of groundwater in relation to settlement. Reference is made to the heavy rainfall in July, 1941.

861. Yamaguchi Sadao 山口貞夫, "Izu-shotō no shūrakukei 伊豆諸島の聚落型 (Settlement types on the Izu islands)," Chirigaku hyōron, v. 11 (1935), no. 12, 1-29.
The so-called Izu Shichitō, or seven isles of Izu, actually consist of eleven islands supporting some twenty-five villages. This is a comparative study and general classification of these settlements, mainly in terms of their form, function, physical site, and other environmental factors. Comments on population, historical development, and social and economic conditions are included.

862. Yamaguchi Sadao 山口貞夫, "Sanson shūraku no seitai 山村聚落の生態 (Ecology of a mountain village)," Chirigaku, v. 4 (1936), no. 4, 183-191; no. 6, 35-44.
Anrakujō village in Mogami district, Yamagata prefecture, was selected by Yamaguchi as an example of a typical mountain village in Tōhoku. He studied sociological and economic conditions there in an effort to increase the understanding of mountain life. Of interest to the geographer are his observations on land use, land ownership, and agricultural methods.

863. Yamaguchi Sadao 山口貞夫, "Tanegashima gaihō 種子島概報 (A reconnaissance report on Tanegashima)," Chirigaku hyōron, v. 13 (1937), no. 5, 54-72.
Tanegashima, situated to the south of Kyūshū, is under the jurisdiction of Kagoshima prefecture. The cultural landscape of this island is described in regard to topographic influences. The ecology of settlement is discussed. Based on a field survey made in 1936.

864. Yamaguchi Yaichirō 山口彌一郎, "Rikuchū Isawa senjōchi ni okeru sankyo to sono seikatsu 陸中膽澤扇状地に於ける散居とその生活 (On disseminated settlements on the alluvial fan of Isawa and on life there)," Chirigaku hyōron, v. 17 (1941), no. 5, 1-21.
This article provides a comprehensive description of the form of disseminated settlement, land use, and village life in the Isawa district of Iwate prefecture. The author includes background material on the early colonization and agricultural reclamation of the area.

865. Yamaguchi Yaichirō 山口彌一郎, Tankō shūraku 炭鑛聚落 (Coal mining settlements), Tōkyō, Kokon Shoin, 1942, 4+2+4+287 pp., 44 illus.
Reasons for the growth and decline of mining towns, their functional and demographic composition, and daily aspects of the miners' lives are covered in this work. Frequent reference is made to Taira, the author's home city in the Jōban coal district.

866. Yamaguchi Yaichirō 山口彌一郎, "Tankō shūraku no zen'isei 炭鑛聚落の漸移性 (The migration of coal-mining settlements in Japan)," Chirigaku hyōron, v. 10 (1934), no. 7, 1-14.
Discusses the shifting of settlement, peculiar to mining communities which move with the mines along coal seams following local exhaustion of deposits. Examples are cited from the Jōban, Ishikari, Ube, and other Japanese coal fields and from Manchuria.

867. Yamaguchi Yaichirō 山口彌一郎, "Tōhoku-chihō ni okeru matagi shūraku no kikō to sono hensen 東北地方に於けるマタギ聚落の機構と其の變遷 (The form of, and changes in, matagi settlements in the Tōhoku district)," Chirigaku hyōron, v. 18 (1942), no. 2, 1-29.

Matagi is the name of a group of hunters living in the deep mountains of Tōhoku. Nowadays there are no pure communities of matagi, since agriculture and forestry have been introduced. Nevertheless, many special aspects of matagi life are still maintained. This is a study of the distribution of these settlements and their culture.

868. Yamaguchi Yaichirō 山口彌一郎, "Yagō ni yoru shūraku no ichikōsatsu. Iwate-ken Kunohe-gun Ube-mura Kuki 屋號に依る聚落の一考察. 岩手縣九戸郡宇部村久喜 (Investigation of a settlement through its 'by-names.' Kuki, Ube village, Kunohe district, Iwate prefecture)," Chirigaku hyōron, v. 13 (1937), no. 6, 1-11.

In secluded fishing and mountain villages the custom of yagō or the use of family "by-names" is still practiced, providing excellent material for the study of settlement geography. This is true of Kuki, a fishing village in northern Iwate prefecture, where the yagō relate to the development of the village and to changes enforced by the occurrence of tsunami or seismic tidal wave disasters.

869. Yamaoka Masaki 山岡政喜, "Enkai shūraku no imi 沿海聚落の意味 (The significance of coastal villages)," Chiri, v. 4 (1941), no. 2, 90-104.

Concerning the blending or intermingling of the fishing and farming aspects of life typical of the Japanese coastal community.

870. Yamazaki Teiichi 山崎禎一, "Shūrakuichi no chirigakuteki kenkyū 聚落位置の地理學的研究 (Geographical study on the situation of settlements)," Chirigaku hyōron, v. 10 (1934), no. 6, 25-54.

The locations of settlements in Japan are classified into 22 types according to topographical considerations. Each of the 22 types is explained and a map of Shizuoka prefecture illustrates the distribution of the 22 types which were actually studied there.

871. Yanagida Kunio 柳田國男, Kaison seikatsu no kenkyū 海村生活の研究 (Studies of life in seashore villages), Tōkyō, Nihon Minzoku Gakkai, 1949, 3+472 pp.

A companion volume to Yanagida's earlier work, Sanson seikatsu no kenkyū. He led eleven folklore specialists in detailed investigations of seashore communities, not all of which were gyoson or fishing villages. Recorded are observations and opinions on social and economic activities, beliefs, family life, fishing operations, organizations, etc.

872. Yanagida Kunio 柳田國男, Sanson seikatsu no kenkyū 山村生活の研究 (Studies of life in mountain villages), Tōkyō, Iwanami Shoten, 1937, 2+5+593pp.

The result of three years of intensive research in the field by thirteen folklore specialists under the direction of Dr. Yanagida. Contents are classified under some 65 headings, e. g., history, labor, dwellings, clothing, traditions, etc. Of interest to the anthropo-geographer.

873. Yoshida Keiichi 吉田敬市, "Pa-si-pheong kō 波市坪考 (Thoughts on the traveling fishing villages [in Korea])," Jimbun chiri, v. 4 (1952), no. 5, 1-13.

The broad coastal banks off western Korea are the scene of brief seasonal fishing by large numbers of fishermen for Sciaena schlegelii. Pa-si-pheong is the name given to the temporary fishing bases. Yoshida here considers the nature of these seasonal encampments, their history, and their distribution. References in old documents are cited.

2. Urban

874. Andō Masuo 安藤萬壽男, "Ōgaki-shi no Meiji ishin igo ni okeru toshiteki hatten 大垣市の明治維新以後に於ける都市的發展 (Urban development of Ōgaki since the Meiji Restoration)," Chiri, v. 2 (1939), no. 4, 36-55.

During the Tokugawa period, Ōgaki was the political center of Mino province, now a part of Gifu prefecture. In this article, Andō studies the geographical changes the city has undergone since the Meiji Restoration and attempts to determine the adaptations the city has had to make in its functional shift from castle town to industrial center.

875. Doi Kikuichi 土井喜久一, "Naichi shichōson-betsu toshishisū no bumpu to chiiki kubun 內地市町村別都市指數の分布と地域區分 (Distribution of vocations in the population of urban types in Japan and the geographical divisions that result from it)," Chirigaku hyōron, v. 15 (1939), no. 2, 11-27; no. 4, 237-248.

Toshishisū, which literally means " city index figures," refers in this article to the percentage of the population of a town-type in a particular vocation in comparison with the total population. It is considered an indication of the degree of urbanization. Calculating this figure for cities, towns and villages of Japan , geographical differences are studied and these studies result in a regional division of Japan. This article has often been referred to by geographers.

876. Fujioka Kenjirō 藤岡謙二郎, "Kinki chihō no machi 近畿地方の町 (Towns in the Kinki district)," Jimbun chiri, v. 3 (1951), no. 3, 12-27.

Utilizing 1948 data, Fujioka analyzed the demographic characteristics of 241 machi in the Kinki district in order to establish a functional classification system based on predominant occupational categories. Reporting on his results in this paper, he identifies and discusses six types of town, namely agricultural, forestry, fishing, industrial, commercial, and "white collar." Remarks are made on the historical origins of Kinki towns, changes in their nature, and the significance of their distribution.

877. Hoyanagi Mutsumi 保柳睦美, "Kaoku tōmitsudosen yori mitaru dai-Tōkyō seibu 家屋等密度線より見たる大東京西部 (The western part of greater Tōkyō seen from iso-density lines for houses)," Chirigaku hyōron, v. 9 (1933), no. 5, 57-76.
A short scholarly discourse on building density as an index of urban expansion. Radial growth along main transportation routes in western Tōkyō is considered in this regard and iso-density maps are utilized in an effort to determine growth boundaries in that area. Several theories on the limits of urbanization are discussed and the contribution of density mapping to this problem is pointed out.

878. Iiyama Toshiharu 飯山敏春, "Kantō chihō toshi mitsudozu 関東地方都市密度圖 (A distribution map of cities and towns in the Kantō plain)," Chirigaku hyōron, v. 8 (1932), no. 9, 33-49.
Iiyama divides the Kantō lowland into three sub-regions and analyzes the distribution and density of urban settlement from topographic maps. He discusses the strong influence of terrain upon the settlement pattern, a factor which is clearly indicated in his distribution map of cities and towns.

879. Kagami Kanji 鏡味完二, "Jūjigai no keikan -- Nagoya-shi no rei 十字街の景観 -- 名古屋市の例. -- (The landscape of street crossings: examples from Nagoya)," Chirigaku hyōron, v. 14 (1938), no. 8, 45-58.
An interpretation of the character of street intersections and their influence upon building density as one form of morphological analysis.

880. Kagami Kanji 鏡味完二, "Seisan to hambai to kara mita Nagoya-shi (toshi kinō no kenkyū) 生産と販賣とから見た名古屋市 (都市機能の研究) (Nagoya, viewed in the light of production and sales; a study of urban function)," Chirigaku hyōron, v. 15 (1939), no. 1, 35-63.
Coefficients of production and sale are calculated for a variety of merchandise categories such as food and drink, clothing, chemical products, machinery and tools, etc. The results are plotted and a map compiled indicating the functional breakdown of Nagoya.

881. Kagami Kanji 鏡味完二, "Toshi no seichō kōzō 都市の成長構造 (Structure of city development)," Chirigaku hyōron, v. 13 (1937), no. 5, 1-16.
Gradational structural changes from the core area outward to the urban periphery are investigated in order to clarify city growth factors, using Nagoya as a case study.

882. Kagawa Kan'ichi 香川幹一, "Chihō shōgyō toshi no botsuraku 地方商業都市の没落 (The decline of commercial cities in the provinces)," Chirigaku, v. 5 (1937), no. 1, 107-179.
The historical development of coastal cities along the head of Sagami bay is surveyed in order to determine the causes for the commercial decay of such former centers as Odawara, Fujisawa, and Kamakura. Special attention is devoted to Fujisawa, which is now largely a resort and residential city in contrast to its chief function in the past as a marketing town.

883. Kanomata Yoshihide 鹿又義英, "Shokugyō kōsei kara mita hompō shotoshi no tokushoku 職業構成から見た本邦諸都市の特色 (The special characteristics of cities in Japan, seen from their occupational constitution)," Chirigaku, v. 3 (1935), no. 10, 253-258.
Selected Japanese cities are classified according to occupational breakdown. Population percentages by employment categories are listed for each city.

884. Kiuchi Shinzō 木内信藏, "Gendai toshi no bunrui 現代都市の分類 (Classification of modern cities)," Jimbun chiri, v. 1 (1948), no. 2, 7-17.
Kiuchi reviews the several most common theories and methods used for classifying Japanese cities and then presents his own variation based on economic considerations, primarily per capita production value, which for this study he derives from 1940 statistics.

885. Kiuchi Shinzō 木内信藏, "Hokushi Sansei-shō no chihō toshi ni tsuite 北支山西省の地方都市に就て (On the cities of Shansi province, north China)," Chirigaku hyōron, v. 19 (1943), no. 3, 11-35.
The form, function, and population of cities in the north China province of Shansi are studied. Urban distribution and the general commercial and industrial patterns of the province are presented with maps. Some broader reference to the cultural development of Shansi and its population growth as a whole is made and examples of city plans are also shown.

886. Kiuchi Shinzō 木内信藏, "Pekin no toshi keitai gaihō 北京の都市形態概報 (A general morphological study of Peking)," Chirigaku hyōron, v. 15 (1939), no. 3, 48-69.
An outline of the historical growth of the old Chinese capital is followed by a concise but comprehensive description of Peking's modern appearance and layout. With a city plan, detail maps, statistical information, and photographs.

887. Kiuchi Shinzō 木内信藏, "Physiognomie tōri. Tōkyō-shi ni okeru toshi keikan chiri Physiognomie 通り. 東京市に於ける都市景観地理 (The landscape of a city as seen in Tōkyō: physiognomy and streets)," Chirigaku hyōron, v. 12 (1936), no. 2, 12-41.
The term "physiognomy" is defined and explained as applied to urban geography. The definition is followed by an interpretation of the morphology and functional role of streets in selected commercial sections of Tōkyō.

888. Kiuchi Shinzō 木内信藏, Toshi chirigaku kenkyū 都市地理學研究 (Studies in urban geography), Tōkyō, Kokon Shoin, 1951, 20+435 pp., 27 photos, 34 tables, and 116 illus.

This is not a fully comprehensive treatment of the principles of urban geography but it is the best recent Japanese work on this subject. The book summarizes most concepts of the field applicable in the study of the Oriental city and offers a great deal of information valuable to the urban geographer. Following a general discussion of the growth of urban geography and its several schools, Kiuchi concentrates upon the distribution and areal structure of cities, referring to both Western and Eastern examples. Emphasis, however, is upon Japanese cities and, secondarily, upon Chinese cities. This is particularly the case in the final section on morphology. With an excellent bibliography in English of approximately 1,000 references covering all phases of geography relative to urban interpretation.

889. Kiuchi Shinzō 木内信藏, "Toshi hatten ni okeru ten'i genshō to keikan chiiki no bunka 都市發展に於ける轉移現象と景觀地域の分化 (Migrations of functions and differentiation of landscape regions in urban development)," Chirigaku hyōron, v. 13 (1937), no. 12, 64-81.
Changes in the internal functional patterns of cities resulting from urban growth are investigated and explained, with Tōkyō used as a case study.

890. Kiuchi Shinzō 木内信藏, "Toshi misshū jūtaku chiku no chirigakuteki kenkyū joron. Toku ni Tōkyō-shi ni okeru bumpu ni tsuite 都市密集住宅地區の地理學的研究序論.特に東京市に於ける分布に就いて (A foreword to geographical study of urban slums. With particular reference to their distribution in Tōkyō city)," Chirigaku hyōron, v. 17 (1941), no. 7, 33-54.
Part of a joint study made by a number of teachers, doctors, social scientists, architects, and planners in an effort to analyze the physical, social, and economic causes of slum development. With some reference to examples of housing blight in Kyōto and Ōsaka.

891. Kiuchi Shinzō 木内信藏, "Toshi seichō no chikara to shigai no keitai 都市成長の力と市街の形態 (Forces of city growth and the forms of streets)," Shakai chiri, no. 13 (1949), 6-10.
Remarks on urban planning, particularly in regard to the post-war rehabilitation of Tōkyō. Building density and population maps for each six-month interval from 1945 through 1947 are analyzed and recommendations made concerning needed adjustments in street pattern. The problem of dispersal of civic and commercial centers is briefly discussed, as well as the need for thorough physical inventory as a preliminary procedure in appraising urban requirements so as to avoid the dangers of overly idealistic city planning.

892. "Matsuri to ichiba 祭と市場 (Festivals and markets)," Chirigaku, v. 4 (1936), no. 6, 7-11; no. 7, 92-97.
The origin of specialized markets in connection with shrine and temple festivals is related and five specific examples described, e. g., the agricultural-tool market at the Hachiman shrine in Chiba prefecture.

893. Miura Kin'ichi 三浦欣一, "Akashi-shi no toshichiriteki ichisokumen 明石市の都市地理的一側面 (One view of the urban geography of Akashi city)," Chiri kyōiku, v. 19 (1933), no. 6, 46-52.
Interpretation of the regional character of this city in Hyōgo prefecture and a summary of its historical and geographical development.

894. Miyata Toyo 宮田とよ and Watanabe Yoshie 渡邊芳枝, "Futagomachi to shite no Mito-shi 双子町としての水戸市 (Mito as a twin city)," Chirigaku, v. 3 (1935), no. 10, 280-285.
A former castle town, the city of Mito in Ibaragi prefecture still clearly reflects its feudal divisions. The Kami-ichi section was mainly the residence of the samurai and the Shimo-ichi district was, and yet remains, the commercial center. The authors investigate from the geographical viewpoint the economic and social ramifications of these inherited Tokugawa divisions.

895. Murata Teizō 村田貞藏, "Banchikai no keitai oyobi banchikaizu ni yoru shinkyū dōro no shikibetsu 番地界の形態及び番地界圖に依る新舊道路の識別 (The forms of land allotments and the distinction between old and new roads as shown on cadastral maps)," Chirigaku hyōron, v. 8 (1932), no. 8, 57-79.
Patterns of land allotment and functional utilization are correlated with landforms and the alignment of roads in an attempt to evaluate the relative age of through routes by detecting harmonious or dissimilar morphological relationships. Historical changes in the Tōkaidō in the Ōmori district are discussed in this regard.

896. Nagai Masatarō 長井政太郎, "Yamagata-shi 山形市 (Yamagata city)," Chirigaku, v. 4 (1936), no. 6, 23-31; no. 7, 106-113.
Nagai compares recent and old maps of Yamagata and describes the physical changes which have occurred. He also inquires into the economic base of that city, past and present, and remarks on some of the social and political developments which have marked its growth.

897. Nishida Yoshirō 西田與四郎, "Jōkamachi no kenkyū 城下町の研究 (A study of castle towns)," Chiri kyōiku, v. 7 (1927), no. 5, 8-14.
A brief survey of the historical geography of castle towns, the political centers of feudal Japan, with emphasis upon their current development. Factors of site, situation, type classification, and morphology are touched upon.

898. Nishimura Mutsuo 西村睦男, "Taihoku-shi no chirigakuteki kenkyū 臺北市の地理學的研究 (A geographical study of the city of Taihoku)," Chiri ronsō, no. 10 (1939), 169-222.
A study in urban geography portraying Taihoku, at that time the Japanese capital of Formosa, as a "representative" colonial city of the empire. The nature of the city is delineated in terms of functional analysis and description.

899. Ogasawara Yoshikatsu 小笠原義勝, "Chihō toshi no toshido no kenkyū 地方都市の都市度の研究 (A study of the urbanization of provincial cities)," Chirigaku hyōron, v. 13 (1937), no. 6, 12-27.
The degree of urbanization and the peculiar structural and functional characteristics of Shimoda, in Shizuoka prefecture, are depicted in some detail. Shimoda is then compared with Gifu, selected as another representative type of local center, and their functional similarities and differences are considered.

900. Ogasawara Yoshikatsu 小笠原義勝, "Toshido to toshi-genkai 都市度と都市限界 (Degree of urbanization and city boundaries)," Chirigaku hyōron, v. 12 (1936), no. 7, 41-72.
The degree and density of urbanization along main thoroughfares in Tōkyō and Gifu city are investigated in terms of commercial specialization and the spacing of buildings, in an effort to establish a general, standardized index for interpreting urban development and limits.

901. Ogasawara Yoshikatsu 小笠原義勝, Nakayama Masayoshi 中山正嘉, and Kashimura Ichirō 柏村一郎, "Toshido no kenkyū (Kishū Tanabe-machi ni tsuite) 都市度の研究(紀州田邊町について) (A study of degree of urbanization, on the town of Tanabe of Kii province)," Chirigaku hyōron, v. 17 (1941), no. 8, 19-39.
An interesting study founded on the assumption that retail specialization provides an index of the degree of urbanization. The town of Tanabe in Wakayama prefecture is taken as a case study, followed by a comparison of Tanabe, Gifu, and Shimoda as illustrations of commercial and morphological transition from town to city status.

902. Ogawa Takuji 小川琢治, "Jimbunchirigaku-jō yori mita Nihon no toshi 人文地理學上より見た日本の都市 (Japanese cities from the viewpoint of human geography)," Chikyū, v. 5 (1926), no. 5, 1-13; no. 6, 1-12.
The significance of place names in the origin of Japanese cities is discussed. There follows a brief historical summary of the development of urban centers. The summary is founded in part on an analysis of population growth and distribution as determined by map interpretation. Ogawa outlines some of the causes of urban agglomeration and morphological difference and touches upon the standard methods of classifying Japanese cities.

903. Ōkubo Takehiko 大久保武彦, "Pattern: toshi kōgai kenkyū josetsu (1) Pattern: 都市郊外研究序説 (Pattern: introduction to the study of city environs; part one)," Chirigaku hyōron, v. 13 (1937), no. 1, 25-40.
An introductory explanation of the author's conception of pattern as an approach to the study of urban landscape. "Pattern" in Ōkubo's usage refers to the higher order of organic elements in the cultural scene as contrasted with fragmented elements identified through detailed analysis. Selection of the broader grain of a pattern often provides a better basis for creating a useful interpretive synthesis, the writer believes.

904. Ono Tadahiro 小野忠凞, "Suō chihō ni okeru seikatsu chiiki kōzō 周防地方に於ける生活地域構造 (Structure of life regions in the Suō district)," Jimbun chiri, v. 3 (1951), no. 3, 40-49.
Certain spatial aspects of human ecology are explored in this paper. The writer's "life regions" are spheres of activity and influence affecting the daily life of individuals. Examples of such real relationships in regard to week end recreation, amusement, shopping, etc. are selected from the Suō area in eastern Yamaguchi prefecture.

905. Sasaki Hikoichirō 佐々木彦一郎, "Toshi no kōzō keitai no ichibunrui 都市の構造形態の一分類 (A classification of urban structural patterns)," Chirigaku hyōron, v. 10 (1934), no. 10, 36-45.
A distributional approach to urban classification. This is an attempt to classify Japanese cities on the basis of the differing internal location patterns of their major institutions and installations: *i. e.*, government offices, shrines, factories, etc. Study of 117 cities by this technique resulted in Sasaki distinguishing three general structural categories which he defines and compares.

906. Sasaki Hikoichirō 佐々木彦一郎, "Tōkyō no toshin chōsa 東京の都心調査 (Reconnaissance of centers in Tōkyō)," Chirigaku hyōron, v. 9 (1933), no. 9, 9-19.
Sasaki made a comparative study of three of Tōkyō's major commercial centers in an effort to promote further understanding of the economic and physical structure of that metropolis. He selected the Ginza, Shinjuku, and Asakusa districts and compared such varied factors as the occupational breakdown, the height and type of construction of buildings, and the number of families per dwelling unit.

907. Sasaki Seiji 佐々木清治, "Shukubamachi no keizai keitai 宿場町の經濟形態 (The economic landscape of a stage town)," Chiri to keizai, v. 1 (1936), no. 3, 83-89; no. 4, 101-107.
Fukuori-machi, located on the Tōkaidō in Shizuoka prefecture, is discussed as an example of a stage town. In outlining its history, Sasaki stresses those geographic, economic, and social factors peculiar to the development of such towns.

908. Tanaka Keiji 田中啓爾, "Shōtengai no henshitsu sayō 商店街の變質作用 (The transformation of a shopping thoroughfare)," Jimbun chiri, v. 4 (1952), no. 1, 1-5.
The urban section surrounding the Ikegami railway station in Tōkyō is studied as an example of the development of a shopping district in that metropolis. Tanaka traces the transformation of Ikegami from a monzen-machi or "temple town" at the Hommonji temple, formerly outside the city, to its present status as an important outlying business center.

909. Tanaka Keiji 田中啓爾, "Yokohama no chirigakuteki kōsatsu 横濱の地理學的考察 (A geographical study of Yokohama)," Chirigaku hyōron, v. 1 (1925), no. 1, 15-31; no. 2, 24-33.
This paper provides a good summary of the historical geography of the Yokohama area, tracing the growth of the city from a cluster of fishing villages to principal Japanese port and important segment in the urbanized Keihin industrial zone. A fairly detailed description of the city just prior to its destruction by the 1923 earthquake is included.

910. Tanakadate Hidezō 田中館秀三 and Takahashi Mikio 高橋幹雄, "Hachirōgata engan ni okeru ichiba no kenkyū 八郎潟沿岸に於ける市場の研究 (A study of the markets on the coast of Hachirōgata lake)," Chirigaku hyōron, v. 15 (1939), no. 4, 32-46.

The frequency of market days and the distribution of market sites in this district of Akita prefecture are explained, with a description of the type and form of each market. Marketing trends in this vicinity are also considered.

911. Tanakadate Hidezō 田中館秀三 and Takahashi Mikio 高橋幹雄, "Yokote bonchi ni okeru ichiba no kenkyū 横手盆地に於ける市場の研究 (A study of market places in the Yokote basin)," Chirigaku hyōron, v. 13 (1937), no. 2, 49-74; no. 3, 26-44.

The distribution and characteristics of the markets in the Yokote mountain basin of Akita prefecture are examined in detail to determine their spheres of influence and their effect upon local settlement patterns.

912. Yajima Nikichi 矢嶋仁吉, "Tōkyō-shi seihembu no toshika to tochi riyō no henka 東京市西縁部の都市化と土地利用の變化 (Urban development and change of land utilization in the western part of Tōkyō city)," Chirigaku hyōron, v. 18 (1942), no. 4, 22-48.

Changes in social and economic conditions accompanying the development of a metropolis are analyzed through shifts in land use in the marginal area of a great city. Suginami in the western part of Tōkyō, a transitional zone between city and farm villages, provides the study area.

913. Yamaguchi Yaichirō 山口彌一郎, "Gaisongata chihō toshi chōsa no ichihōhō--rei: Miyagi-ken Watari-machi 街村型地方都市調査の一方法 一例. 宮城縣亘理町 (A method for the study of Strassendorf towns in the provinces— example: the town Watari in Miyagi prefecture)," Chirigaku hyōron, v. 11 (1935), no. 5, 1-11.

Morphometric analysis of building density in reference to an explanation of overall layout and the identification of the core area and secondary centers. The typical arrangement of houses and shops and functional specialization are noted and discussed.

914. Yamaguchi Yaichirō 山口彌一郎, "Mitsudo keisoku ni yoru chihō toshi kōsatsu no ichirei 密度計測による地方都市考察の一例 (The consideration of local cities in terms of population density - one example)," Chirigaku hyōron, v. 10 (1934), no. 2, 41-50.

Yamaguchi investigated settlement forms in the Jōban coastal district of Fukushima prefecture. Here he is principally interested in urban and village development and activities along the main roads in that region. He believes that the basic economic and social characteristics of an area can be ascertained, relatively easily, in a reconnaissance of local thoroughfares.

915. Yamaguchi Yaichiō 山口彌一郎, "Tōhoku chihō no ichiba 東北地方の市場 (Markets in the Tōhoku district)," Chirigaku, v. 9 (1941), no. 1, 179-185.

Discusses the characteristics of open street markets still commonly found in Tōhoku towns although declining both in numbers and frequency. The writer distinguished two types, one informally operated, the other run on a more official auction system. His description principally concerns the former.

916. Yasuda Hatsuo 安田初雄, "Takada-shi no keikan 高田市の景觀 (The landscape of the city of Takada)," Chirigaku hyōron, v. 15 (1940), no. 5, 17-31.

A study of a city in Niigata prefecture, located in the center of a region noted for its extremely heavy snowfall. The theme concerns cultural adaptation to the physical environment, and the author describes several urban characteristics which reflect the severe winter conditions of this district of Honshū. Yasuda stresses the development of certain construction features and building types which are peculiar to Takada and its neighboring communities.

F. Linguistic geography

1. Dialects

917. Chiba Tokuji 千葉徳爾, "Hōgen no bumpu to chiikisei 方言の分布と地域性 (The distribution of dialects and their regionality)," Chirigaku, v. 9 (1941), no. 10, 135-143.

Regional differences in terminology concerning the rice harvest, and particularly variation in terms for the stacking and drying of rice prior to threshing, are classified into seven groups. A map shows the distribution of the dialect groups and the author explains some of the language differences between east and west Japan in relation to local customs.

918. Miyanaga Masamori 宮良當壮, Nihon no sumizumi 日本の隅々 (Nooks and corners of Japan), Nara, Yōtokusha, 1947, 196 pp.

The author is an anthropologist who specializes in linguistics, particularly the study of Japanese dialects. However, this book is not restricted to language study, being instead a general social and ethnological survey of a number of interesting communities in back-country Japan. Observations are made on village life at selected places in the following scattered areas: the Shimokita, Oga, Noto, and Sada peninsulas; Tsushima island and islands in the northern Ryūkyūs; in the Echigo mountains; and in Tsugaru and Dewa provinces.

2. Place name studies

919. Akioka Takejirō 秋岡武次郎, "Gaikokujin no meimei-seru Nihon no chimei no jakkan 外國人の命名せる日本の地名の若干 (Some Japanese place names originated by foreigners)," Chirigaku hyōron, v. 6 (1930), no. 7, 577-593.
Concerned with the western names given to Japanese geographical features by English and other European map makers, in particular those derived from the names of members of Captain James Cook's third expedition to the Pacific in 1779, e. g., Cape King, Cape Gore, etc.

920. Ishida Ryūjirō 石田龍次郎, "Shinden shinkai to sono chimei 新田·新開とその地名 (Shinden and shinkai: a study of their place names)," Chiri kyōiku, v. 24 (1936), no. 1, 42-48.
Ishida investigates some aspects of the relationship between oya-mura and ko-mura, "mother" and "daughter" villages, as determined by an analysis of place names on maps and in documents. Shinden and shinkai are villages of relatively recent establishment or settlement.

921. Kagami Kanji 鏡味完二, "Kaigan no chimei 海岸の地名 (Place names on the coast of Japan)," Chirigaku hyōron, v. 16 (1940), no. 11, 1-19; v. 17 (1941), no. 7, 15-32; no. 11, 50-65; v. 18 (1942), no. 3, 43-65; no. 5, 41-66; no. 7, 40-60; no. 8, 34-49.
Geographical names and place names taken from 1:50,000 and 1:200,000 topographic maps are listed and classified according to their frequency of occurrence, by suffix, etc. They are shown in table and map form. This is a thorough survey and indexing of coastal names.

922. Kagami Kanji 鏡味完二, "Kasen no chimei 河川の地名 (Names of rivers)," Chirigaku hyōron, v. 20 (1944), no. 2, 40-62.
River names are analyzed for their cultural and geographical significance. Kagami also surveys regional differences in the use of suffixes which mean "stream" or "river."

923. Kagami Kanji 鏡味完二, "Ryūkyū-rettō ni okeru chimei no bumpu 琉球列島に於ける地名の分布 (The distribution of geographical names in the Ryūkyū archipelago)," Chirigaku hyōron, v. 18 (1942), no. 11, 52-68; no. 12, 28-44.
The origin, meaning, and distribution of some sixteen commonly used geographical names in the Ryūkyū islands are investigated to establish their source regions; i. e., southern Japan, Formosa, etc. The author finds little linguistic connection between Okinawan names and those of other southern regions. His conclusions as to their Japanese origin are presented.

924. Komatsu Saburō 小松三郎, "Mura no chimei 村の地名 (Place names of villages)," Chikyū, v. 20 (1933), no. 2, 11-32.
A collection and classification by meaning of the place names of the northern Ina valley in Nagano prefecture. Names are first listed by aza according to the official readings registered at the mura offices.

925. Matsuo Toshio 松尾俊郎, "Hatano bonchi no chimei 秦野盆地の地名 (Place names in the Hatano basin)," Shakai chiri, no. 8 (1948), 2-6; no. 9, 8-11.
Some of the place names common to this district are cited and their origin is investigated with reference to the geographical conditions of the surroundings and to the legends told by aged people. Interwoven are descriptions of the ways of local living that will help in the understanding of this district.

926. Mochizuki Katsumi 望月勝海, "Jōban chihō no kōsen to chimei to no kankei 常磐地方の鑛泉と地名との關係 (The relation between mineral springs and place names in the Jōban district)," Chirigaku hyōron, v. 1 (1925), no. 9, 50-64.
The linear distribution of mineral springs in the Jōban district of Ibaraki and Fukushima prefectures is investigated in relation to tectonic lines and landforms. Place names of the region, many of which contain the character 塩, meaning "salt," are correlated with the local existence of salt-water and other types of spring.

927. Nagai Masatarō 長井政太郎, "Yamagata-ken no chimei kenkyū 山形縣の地名研究 (A study of the place names in Yamagata prefecture)," Chirigaku hyōron, v. 15 (1940), no. 11, 28-58.
Some 40,000 place names in Yamagata prefecture were collected by the Yamagata Kyōdo Kenkyūkai 山形鄕土研究會 (The Yamagata Society for Prefectural Study). The geographical, historical, and social significance of these place names is analyzed in this article and the names are grouped into certain categories, such as names derived from rice farming, from geographic features, from religious beliefs, monuments, customs, etc.

928. Nakamura Keisaburō 中村慶三郎, "Yamakuzure chiiki ni okeru oizukuri to chimei 山崩れ地域に於ける追作りと地名 (Oizukuri and place names in landslide areas)," Chirigaku hyōron, v. 8 (1932), no. 11, 34-45.
Oizukuri refers to the use of former field boundaries when these are disturbed by landslides. The present article is a study of land use and ownership problems arising from landslides or land slippage, particularly in regard to the dislocation of property lines.

929. Nakano Hiroshi 中野弘, "Waga kuni no yama no na no bumpu 我が國の山の名の分布 (The distribution of mountain names in Japan)," Chirigaku hyōron, v. 16 (1940), no. 10, 25-55; no. 11, 20-30.
A linguistic and geographical study of the names of mountains and volcanoes in Japan, including a survey of prefixes and suffixes which are commonly used in mountain names. Nakano classifies the names into several groups based mainly on frequency of occurrence. The location of the mountains whose names are selected for study is indicated on accompanying maps as are certain distribution patterns based on usage.

930. Nishizuru Sadayoshi 西鶴定嘉, "Kita-Karafuto oyobi Ajia tairiku ni okeru Ainu-go chimei no bumpu ni tsuite 北樺太及亞細亞大陸に於ける アイヌ語地名の分布に就いて (On the distribution of Ainu place names in north Karafuto and the Asiatic continent)," Chirigaku, v. 5 (1937), no. 6, 59-63; no. 8, 125-130; no. 10, 33-37; no. 11, 56-60.

The writer identifies names of Ainu origin used on old maps and in documents for places in north Karafuto, eastern Siberia, and neighboring areas. He interprets their meaning, and summarizes the extent of their distribution and the significance of their usage. Pronunciations are given in both syllabic characters and romanization.

931. Satō Hisashi 佐藤久, "Tōhoku chihō ni okeru chimei 'tai' no bumpu 東北地方に於ける地名タイの分布 (Distribution of the geographical name 'tai' in the Tōhoku district)," Chirigaku hyōron, v. 18 (1942), no. 12, 68-72.

In the Tōhoku district there are numerous place names ending in the suffix "tai" or "dai." The kanji for these suffixes is sometimes 平, sometimes 台, 岱, etc. Satō identifies these names and shows their distribution on a map. He presents preliminary conjectures as to whether or not the name endings are of significance in reflecting actual conditions of local topography.

932. Sugaya Yasumasa 菅谷泰昌, "Sanson goshō 山村語抄 (Select vocabulary of mountain villages)," Chikyū, v. 20 (1933), no. 2, 54-68.

This paper was written as a supplement to Yanagida Kunio's work, Sanson goi (The vocabulary of mountain villages), Tōkyō, Dai-Nihon Sanrinkai, 1932. It adds about 100 additional words and phrases peculiar to mountain dialects, notes the names of the places where they were collected, and interprets their meanings.

933. Yamaguchi Shigeichi 山口茂一, "Chimei zakkō 地名雜考 (Miscelleneous thoughts on place names), Shakai chiri, no. 23 (1950), 2-3.

Memorandum on place names which are difficult to read. Examples are shown and pronunciations given in syllabic charaters.

934. Yamaguchi Yaichirō 山口彌一郎, "Chimei yori mita kaikon shūraku no hattatsu 地名より見た開墾聚落の發達 (Development of reclaimed settlements studied from place names)," Chirigaku hyōron, v. 19 (1943), no. 11, 13-31.

A general study of the nature and significance of names given to agricultural settlements during times of extensive land reclamation and expanded cultivation, principally during the Edo period. Historical and geographical interpretations in terms of place-name analysis are made with reference to all parts of Japan but the establishment and growth of "new" farm communities in northern Japan are particularly investigated.

935. Yanagida Kunio 柳田國男, Chimei no kenkyū 地名の研究 (Studies on place names), Tōkyō, Kokon Shoin, 1936, 4+5+368+3 pp.

A collection of some fifty of the author's monographs, previously published in Chirigaku hyōron and other leading journals, synthesizing the approaches of geography, philology, history, and folklore in the interpretation of place names. Regarded as one of the best and most valuable Japanese works on this subject.

G. Religious geography

936. Mochizuki Katsumi 望月勝海, "Chirigakuteki ni mitaru Nihon no shūkyō 地理學的に見たる日本の宗教 (Religions in Japan seen geographically)," Chirigaku hyōron, v. 6 (1930), no. 1, 65-74.

A map study of the three major Japanese religions, Buddhism, Shintoism, and Christianity, presenting the relative number and distribution of their respective adherents and the location of their edifices and institutions by prefecture. With special reference to the Shinshū sect in the Hokuriku district of Honshū.

937. Murakami Hideo 村上英雄, "Hompō shūkyō bumpu no kenkyū (Shintō kakuha) fu shūkyō dempa no chiriteki seiyaku ni tsuite 本邦宗教分布の研究（神統各派）附宗教傳播の地理的制約に就いて (On the distribution of the religious sects in Japan - of each sect of Shintō - and the geographical conditions of their propagation)," Chirigaku hyōron, v. 11 (1935), no. 4, 1-25; no. 5, 12-30.

The relative following of the various Shintō sects is indicated and the distribution of their adherents throughout Japan is shown, followed by a historical discussion of the origin and geographical expansion of each sect. The most detailed treatment is given to the Tenri sect.

938. Yanagida Kunio 柳田國男, Nihon no matsuri 日本の祭 (Festivals of Japan), Tōkyō, Kōbundō, 1942, 4+280 pp.

A general description and interpretation of Japanese festivals by an outstanding folklore authority. Numerous examples of specific local matsuri are cited, but the author does not here attempt a systematic regional or provincial description.

H. Geography of folklore and folk songs

939. Masuzawa Kiyoshi 増澤淑, Nihonjin no seikatsushi 日本人の生活史 (Life history of the Japanese), Tōkyō, Nihon Shobō, 1942, 6+14+436+11 pp.

Written during the Pacific War, this book is not without the influence of the war-time ideology. Its scope is broad and the treatment is general and somewhat popular. Coverage ranges from such topics as clothing, diet, and housing, to religious practices and customs. Of some interest to the social geographer and anthropologist.

940. Oikawa Hiroshi 及川周, *Yukiguni no seikatsu* 雪國の生活 (Life in snowy country), Kyōto, Usui Shobō, 1949, 184 pp., 1 map, and several illus.

An account of the author's life in the Uonuma district of Niigata prefecture, the heart of Japan's heavy snow country. Oikawa's professional field is hygienics, so observations on that subject occur from time to time, including discussion of certain health experiments conducted at the Niigata Ika Daigaku 新潟医科大學 (Niigata Medical College). However, on the whole, the book is concerned with descriptions of the various phases of everyday life in an area where the heavy snowfall greatly influences social and cultural forms. Oikawa's observations on housing and clothing are particularly interesting.

941. Tamura Eitarō 田村榮太郎, *Nihon fūzokushi* 日本風俗史 (History of Japanese manners and customs), Tōkyō, Mikasa Shobō, 1936, 1+8+307 pp., 15 illus.

A history of Japanese manners and customs from primitive times through the Edo period, written by a specialist. Analysis is on the basis of social class. After the primitive stage of social development, Tamura distinguishes two historical eras: a) rule of the *kuge* or court nobles, subdivided into the Nara and Heian periods; and b) domination by the samurai, subdivided into the Kamakura, Muromachi, and Edo periods.

942. Yanagida Kunio 柳田國男, *Nihon minzokugaku kenkyū* 日本民俗學研究 (Folklore studies in Japan), Tōkyō, Iwanami Shoten, 1935, 5+2+585 pp.

A collection of fourteen lectures, delivered at the first grand meeting of the Nihon Minzoku Gakkai 日本民俗學會 (Japanese Folklore Association) in 1935, treating various Japanese traditions, legends, provincial ceremonies, local religious customs, Ainu characteristics, etc. Of interest to the anthropo-geographer.

I. Culture traits and other cultural distributions

1. General

943. Tōkyō Teikoku Daigaku Nōgakubu Nōseigaku Kenkyūshitsu 東京帝國大學農學部農政學研究室 (Institute of Agricultural Politics, College of Agriculture, Tōkyō Imperial University), *Nōson seikatsu no chōsa* 農村生活の調査 (Researches on life in farm villages), Tōkyō, Iwanami Shoten, 1933, 2+92 pp.

This is a survey made of the domestic expenditures of over 650 households in order to establish a consumers' index for typical rural communities. Statistics were gathered at 18 *buraku* in three prefectures— Niigata, Yamagata, and Saitama. Conclusions reached on farm expense and the classification of farm purchases will be of interest to the economic geographer.

944. Tsujimura Tarō 辻村太郎, *Keikan chirigaku kōwa* 景觀地理學講話 (Lectures on the geography of landscape), Tōkyō, Chijin Shokan, 1937, 362 pp., 146 photos.

A pioneer work on the geography of landscape in Japan. Previously quite a few separate articles by various authors were published on the theory of landscape, but this is the first consolidated, systematic treatment by a Japanese scholar. Tsujimura sums up the principal concepts and theories advanced by foreign geographers, interprets this thinking in terms of the Japanese scene, and contributes his own ideas. He also presents the viewpoints of other Japanese. With a bibliography, mainly of foreign works. Many excellent photographs, most of them taken by Japanese geographers.

945. Yamashina Yoshimasa 山階芳正, "*Tōshosei ni kansuru kōsatsu* 島嶼性に關する考察 (A study concerning insularity)," *Tōkyō Daigaku chirigaku kenkyū*, no. 2 (1952), 147-161.

A summary of Yamashina's concept of insular geography that recognizes the problems of interpretation inherent in the peculiar physical and social isolation of islands. Various factors of the natural and cultural environment are considered from this point of view and numerous specific examples are cited in respect to minor Japanese islands, *e. g.*, Gotō, Tsushima, Hatsushima.

2. Specific

946. Adachi Isamu 足立勇, *Kaitei Nihon shokumotsushi* 改訂日本食物史 (History of Japanese food, revised), Tōkyō, 1950, Yūzankaku, 2 v.

A highly-detailed historical survey of Japanese food production, processing, and cooking from ancient times into the Meiji period, with comment on social and economic factors which determined food habits. One chapter is devoted to brief descriptions of the present staple foods and traditional meals in each region. Several interesting woodcuts.

947. Chūō Shokuryō Kyōryokukai 中央食糧協力會 (Central Cooperative Association on Food), *Hompō kyōdoshoku no kenkyū* 本邦郷土食の研究 (Study of traditional foods in Japan), Tōkyō, 1944, Tōyō Shokan, 2+4+3+706 pp.

Prompted by the war time rice shortage, this is a detailed investigation of staple foods and traditional Japanese foods in an effort to find and promote adequate substitutes. The use of selected staples was studied on a regional basis.

948. Fujita Motoharu 藤田元春, *Nihon minkashi* 日本民家史 (History of house types in Japan), Tōkyō, Tōkō Shoin, 1927, 2+4+4+26+604 pp., 252 photos and illus.

Rather than being a historical study, this is a detailed and systematic description of the common forms of existing Japanese dwellings as determined by a national survey. Emphasis is upon the architectural point of view, although Fujita includes a brief historical summary of the evolution of Japanese houses and takes into account geographical, social, and other influences. Considered are construction materials and methods, roof types, house shape and room layout, the nature of house lots, etc. Well illustrated, this is a useful reference work for the cultural geographer.

949. Kon Wajirō 今和次郎, *Kusayane* 草屋根 (Thatched roofs), Tōkyō, Kangensha, 1946, 238 pp., 85 illus.
A reprinting of twenty-eight essays by an expert in the study of Japanese rural house types and clothing. The majority of the essays relate to houses of central Honshū and Tōhoku and four concern varying regional styles of farm dress.

950. Kon Wajirō 今和次郎, *Nihon no minka* 日本の民家 (Houses of Japan), Kyōto, Kōseikaku Shoten, 1927, 8+8+336 pp., 6 photos, 100 illus.
A description of Japanese rural houses from the structural standpoint, with explanations of the influence of folklore upon architectural forms. Of interest to the social geographer, the work consists of three parts. The first deals with rural houses in general and specifically describes 48 selected house types representative of farming, fishing, and mountain communities. Discussion of the examples is accompanied by sketches, and the types are presented in geographical order, from north to south, although Hokkaidō types are not included. The third part is concerned with the houses of two typical mountain villages.

951. Ogawa Tōru 小川徹, "*Tōkyō-fu oyobi Yamanashi-kenka ni okeru yane no keishiki o chūshin to shita minka no chōsa* 東京府及び山梨縣下に於ける屋根の型式を中心とした民家の調査 (A survey of house types in Tōkyō and Yamanashi prefectures with special reference to roof shapes)," *Chirigaku hyōron*, v. 14 (1938), no. 2, 48-73.
Miscellaneous data and information pertaining to rural house types, emphasizing the origin and meaning of differing roof structure found in seven separate areas of the prefectures named. Illustrations, including photographs, accompany the text.

952. Okagami Tadao 岡上忠雄, *Shokuryō keizai no rekishi* 食糧經濟の歴史 (A history of food economy), Tōkyō, Yūzankaku, 1948, 182 pp.
A small book packed with facts and data on food and agriculture in Japan, past and present. Beginning with a discussion of the diet and food sources of Japan's primitive inhabitants, Okagami traces the development of agriculture into recent times. His section on modern agriculture offers valuable information to the student of Japanese land utilization. It includes such topics as types of crop, crop improvement, development of farming techniques and agricultural tools, insect and blight control, and harvest and storage problems.

953. Ozawa Shigeru 小澤滋, *Nihon no shokumotsu-bunka* 日本の食物文化 (Cultural study of Japanese foods), Tōkyō, Dai-Nihon Shuppansha Hōbunsō, 1940, 2+2+257 pp.
A popular study of the significance and place in Japanese society and culture of traditional food and drink. Also considered is the use of some Western introductions such as bread and coffee. The book unfortunately contains no index.

954. Sasaki Hikoichirō 佐々木彦一郎, "*Minka no keitai ni arawareta chiiki no tokusei* 民家の形態に現はれた地域の特性 (Regional characteristics as revealed in house types)," *Chirigaku hyōron*, v. 7 (1931), no. 9, 67-74.
Based on field observation, this is an explanation of how regional characteristics are expressed in rural house types. Examples of houses located on the alluvial fan of the Ōi river, in a *monzen* town at the Tsukuba shrine, and at various communities representing former stage towns along the Tōkaidō highway are described in this regard.

955. Shima Yukio 島之夫, *Nihon min'oku chiri* 日本民屋地理 (Geography of Japanese rural houses), Tōkyō, Kokon Shoin, 1937, 2+7+194 pp., 28 photos, 76 illus.
In part one of this book representative house types of east Asia are described. In part two rural houses in Japan are considered with special reference to those characteristic of overseas areas. A third part deals with a discussion of house structure and climate, particularly roof styles in relation to rain and snowfall conditions. A supplement provides detail for selected house types in rural areas of Japan proper, *e. g.*, on the Noto peninsula and in the Nara basin.

956. Tōki Ryūshichi 東木龍七, "*Jūkyo keiei no gensoku to chiiki tokuchō no seiin* 住居經營の原則と地域特徴の成因 (Basic principles for the maintenance of habitations and the origins of regional characteristics)," *Chirigaku hyōron*, v. 7, 1931, no. 10, 14-30.
The author identifies nine motivating principles or basic considerations in the selection of habitation sites and in the maintenance of residence and discusses the regional characteristics of settlement in Japan in terms of local variation in the relative emphasis of these fundamental needs.

957. Yoshizaka Takamasa 吉阪隆正, *Jūkyogaku hanron* 住居學汎論 (An introduction to housing science), Tōkyō, Sagami Shobō, 1950, 228 pp., many illus.
Written by an architect with broad experience at home and in Europe, this is a somewhat unique approach to housing in Japan as an element of human ecology. Prior to investigating the present housing shortage and other housing problems, the author discusses the morphology of Japanese houses and outlines their historical evolution. He then surveys progressive housing ideas, including his own original concepts, in terms of improving the Japanese home of the future. With many photographs, diagrams, and sketches as well as a bibliography and a French translation of chapter headings.

Chapter X

Economic Geography

The majority of geographical studies in Japan contain some economic geographical data. This is because most studies are concerned with small areas and encompass all or most aspects of the geography of those areas. On the other hand, it is often difficult to get a picture of the national whole in the field of economic geography.

Under this section of the bibliography are entered only those studies which are primarily economic in character.

There has been a relative neglect of the industrial field, and the changes brought about by the war and reconstruction have not yet been studied in detail. Especially numerous are studies of agricultural production and problems. This is a natural development in a country where agriculture is the most important occupation and dominates the national landscape.

In more recent times, studies in land reclamation and hydro-electricity have been undertaken with enthusiasm. Regional planning and the conservation of natural resources have also more and more drawn the interest of Japanese geographers. Again most of these studies cover limited and small areas. Land utilization is being studied on a national scale by the Chiri Chōsajo or Geographical Survey Institute and a sequence of excellent maps and some essays have resulted.

A. General

958. Iki Tsunenaka 伊木常誠, "Hompō yuden no jōkyō 本邦油田の状況 (The status of oil fields in Japan)," Chigaku zasshi, v. 21, no. 242 (1909), 1-15; no. 243, 17-26; no. 244, 9-25; no. 245, 21-27; no. 246, 23-33, 1 map.
Written mainly from the geological point of view this is a description of the oil fields of Japan. A general discussion of petroleum deposits is followed by a consideration of each producing field, including its geological structure, history, operations in 1909, and future prospects. Outdated, but nonetheless a valuable article for the geologist or geographer interested in Japan's power resources.

959. Kimura Yasuji 木村靖二, Nihon nōmin sōtōshi 日本農民争闘史 (History of farmers' uprisings in Japan), Tōkyō, Hakuyōsha, 1930, 3+13+544 pp.
A history of Japanese agrarian revolts (ikki) and agricultural disputes from feudal times through the Tokugawa and Meiji periods.

960. Misawa Katsue 三澤勝衛, "Yatsugatake kazan sanroku no keikankei 八ヶ岳火山々麓の景観型 (Patterns of landscape in the foot-slopes of Yatsugatake volcano)," Chirigaku hyōron, v. 5 (1929), no. 9, 36-67.
A map study of the cultural and physical landscape of the Yatsugatake district in the border between Yamanashi and Nagano prefectures. Relationships between the man-made and natural elements are analyzed. Fifteen maps treat such factors as settlement, population, land use, landforms, etc.

961. Satō Hiroshi 佐藤弘, Keizai chiri 經濟地理(Economic geography), Tōkyō, Shinkigensha, 1951, 3+8+386 pp., 36 illus.
Although written as a college text on economic geography, this book contains considerable original theory and opinion. Arranged in three parts, the first treats methodology, the second discusses the role of resources, and the third is a presentation of Satō's theory of the Standoertlichkeit of industries. Numerous examples pertaining to Japanese industry are cited. There is much post-war data and information on the Japanese metal and chemical industries and hydro-electric power generation.

962. Ueda Shinzō 上田信三, "Nihon sanson keizai chiri kenkyū josetsu 日本山村經濟地理研究序說(Introduction to an economic geographical study of Japanese mountain villages)," Chirigaku hyōron, v. 15 (1939), no. 1, 1-16.
Largely limited to a discussion of economic factors affecting mountain communities in southwestern Japan. In the main, this study consists of general observations on agriculture, forestry, home industry, trade, and communications and refers to topographic and climatic influences. Some comparisons with lowland villages are made.

963. Ueda Shinzō 上田信三, "Yamagata-ken Kita-Oguni-mura no keizai chirigakuteki kōsatsu — hekienchi kenkyū no ichirei to shite 山形縣北小國村の經濟地理學的考察 — 僻遠地研究の一例として — (Observations on the economic geography of Kita-Oguni village, Yamagata prefecture, as an example of a remote place)," Chirigaku hyōron, v. 14 (1938), no. 4, 1-27.
Problems confronting the economic geographer concerned with the study of isolated districts are discussed in this study of the economic and social environment of Kita-Oguni village, secluded in the mountainous backcountry of Yamagata prefecture. Detailed observations on village life constitute the latter half of this paper.

B. Land use

964. Ishida Ryūjirō 石田龍次郎, "Iōjima no sangyōteki shinka 硫黄島の産業的進化 (Industrial development of Iōjima)," Chirigaku hyōron, v. 6 (1930), no. 7, 523-537.
The economic development of Iōjima in the Volcano Islands (Kazan rettō) is outlined as an example of an economic geographical study of an isolated environment. Efforts to maintain colonization on the basis of sulphur mining, fishing, sugar cane cultivation, and the growing of cocoa for cocaine production are touched upon.

965. Kagami Kanji 鏡味完二, "Owari no kuni Tan'yō-mura no tochiwari 尾張國丹陽村の土地割 (Land-allocation in Tan'yō village in the province of Owari)," Jimbun chiri, v. 4 (1952), no. 1, 20-29.

The village concerned is located about eight kilometers north of Nagoya city in Aichi prefecture. It was selected as being representative of communities where mixed rice farming and truck gardening are practiced simultaneously by growing vegetables on small artificially-raised earth platforms in the midst of the paddy fields. The reasons for this intensive form of land use are explained and the various problems of cultivation, irrigation, labor, etc., connected with it are described.

966. Kagose Yoshiaki 籠瀬良明, Kōchi no chiriteki kenkyū— sono kaihatsu to kairyō 耕地の地理的研究——その開發と改良 (A geographical study of cultivated fields— their reclamation and improvement of land), Tōkyō, Surugadai Shuppansha, 1953, 8+150 pp., 18 photos, 24 tables and 45 illus.

A study based upon field observations of land use problems and an examination of several land reclamation and improvement projects with emphasis upon geographical aspects. The author attempts to determine the contributions geography can make to the planning phases and execution of such programs. The first chapter reports on improvement work in progress in the Kurobe alluvial fan, Toyama prefecture, where the main problem is how to raise the temperature of the water for irrigation. The second chapter summarizes research on an area in the Kantō plain where rice is sown directly on the fields without transplanting. Subsequent chapters investigate irrigation problems in the Kantō and Echigo regions and trace the historical development of irrigation in these districts.

967. Kawada Saburō 川田三郎, "Taiwan ni okeru tochi riyō 臺灣に於ける土地利用 (Land utilization in Formosa)," Chirigaku hyōron, v. 19 (1943), no. 6, 23-38.

A comprehensive summary of the crop systems and farming methods employed on Formosa during the closing years of Japanese administration. Agricultural data is interpreted on a gun or district basis and a land utilization map is included, showing predominant crops by gun.

968. Matsui Isamu 松井勇, "Bumpu ni kansuru nisan no mondai, toku ni Tama kyūryō no ichibu ni tsuite 分布に關する二三の問題，特に多摩丘陵の一部に就いて (Some problems in spatial distribution, especially in the Tama hill region)," Chirigaku hyōron, v. 9 (1933), no. 5, 1-22.

Patterns of the cultural landscape are analyzed in terms of areal inter-relationships for this district of Kanagawa prefecture lying north and west of Yokohama city. The influence of relief surfaces and slope upon the spacing of cultural elements is stressed.

969. Matsui Isamu 松井勇, "Tama kyūryō ni okeru shinrin hatachi suiden no senkyo mensekihi no bumpu jōtai ni kansuru hitotsu no shiryō 多摩丘陵に於ける森林畑地水田の占居面積比の分布狀態に關する一つの資料 (Data for the distribution and ratio of forest, dry field and rice field in the Tama hills)," Chirigaku hyōron, v. 11 (1935), no. 12, 30-46.

A quantitative map study of the landscape of the Tama hills in Kanagawa prefecture based on an analysis of 1:25,000 topographic sheets.

970. Matsui Isamu 松井勇, "Tama kyūryō oyobi Sagamihara ni okeru suidendo, futsūbatado, kajubatado to no bumpu ni kansuru hitotsu no shiryō 多摩丘陵及び相模原に於ける水田度，普通畑度，果樹畑度との分布に關する一つの資料 (Data on the distribution and percentage acreage of paddy fields, farm sites, and orchards in the Tama hills and the Sagami plain)," Chirigaku hyōron, v. 13 (1937), no. 1, 1-24.

Using agricultural statistics for 1929, Matsui calculated percentage acreage by crop and other land use types for this section of Kanagawa prefecture. Distributional characteristics are discussed and shown in map form at scales of 1:300,000.

971. Misawa Katsue 三澤勝衛, Fūdo sangyō, atarashiki Nihon no tame ni 風土産業，新しき日本のために (Fūdo industries for a new Japan), Okaya, Tateshima Shobō, 1947, 4+10+385 pp., 24 photos, 23 illus.

Misawa is a long-time student of fūdo or topographic, climatic, and biological environment in his native province Shinano (Nagano prefecture). This is a collection of his lectures on the subject of regionality and environmental adjustment. He advocates the promotion of national prosperity by developing new industires in harmony with local environment.

972. Misawa Katsue 三澤勝衛, "Kiso Yamaguchi-mura ni okeru sangyō keikan no ryōteki kenkyū 木曽山口村に於ける産業景觀の量的研究 (A quantitative study of the cultural scene at Yamaguchi village, Kiso)," Chirigaku hyōron, v. 9 (1933), no. 11, 1-18.

An experiment in the quantitative representation of the cultural landscape. Misawa selected certain land use features, analyzed them statistically, and showed their numerical distribution on maps.

973. Miyazaki Kenzō 宮崎健三, "Hōjōzugata (Koshinogata) fukin no keikan 放生津潟（越の潟）附近の景觀 (The landscape of the Hōjōzugata or Koshinogata region)," Chirigaku hyōron, v. 9 (1933), no. 12, 21-32.

Hōjōzugata is a large lagoon in the northern sector of the Toyama plain. This is a geomorphological study of the lagoon and its adjacent marshy terrain together with an interpretation of the cultural landscape of that region.

974. Morita Keiichi 森田敬一, "Teishitsu chitai no chiiki kōzō 低濕地帶の地域構造 (The regional structure of a swampy lowland)," Jimbun chiri, v. 4 (1952), no. 1,29-40.

Not a study in geomorphology, but a description of land use along the marshy lower reaches of the Yodogawa in Ōsaka prefecture, with principal reference to the cultivation of wicker-willow and related land reclamation efforts in this district.

975. Ogasawara Yoshikatsu 小笠原義勝, "Nihon no tochiriyōku 日本の土地利用區 (The land-use regions of Japan)," Chigaku zasshi, v. 60 (1951), no. 2, 21-29.

A classification of Japanese land utilization founded on a regional interpretation of differing agricultural systems. Considers such criteria as a)types of crop, b) intensity of farming, c) degree of land use in relation to surface features, and, d) the distribution and use of pasture and other grasslands. Sub-divisions of Japan are made in terms of this classification.

976. Sugai Shirō 酉水孜郎, "Kazan chiiki ni okeru tochiriyō 火山地域に於ける土地利用 (Land utilization on volcanic areas)," Chirigaku hyōron, v. 15 (1939), no. 6, 16-41.
Land use in volcanic districts is discussed, notably on the foot slopes of Mt. Fuji and in central and southern Kyūshū. The unstable aspects of agriculture in these regions are noted. Includes observations on wet rice farming and the hindrances to the greater development of this form of cultivation in volcanic areas.

977. Sugai Shirō 酉水孜郎, "Kazan chiiki ni okeru tochiriyō no kenkyū 火山地域に於ける土地利用の研究 (A study of land utilization in volcanic regions)," Chirigaku hyōron, v. 18 (1942), no. 9, 1-18.
A sequel to a paper published by Sugai in this journal in 1939 (see preceding entry). Portrays the fundamental characteristics of agricultural crops and land management in volcanic districts throughout Japan.

978. Sugai Shirō 酉水孜郎, Kokudo no kaihatsu 國土の開發 (Evolution of the national land), Tōkyō, Kokon Shoin, 1950, 2+2+192 pp.
Sugai appraises regional and national land planning in Japan and advances his own ideas on this subject. He presents a synthesis of concepts from the social and natural sciences. Included is a summary of legislation pertaining to land development and planning, especially in reference to Hokkaidō.

979. Sugai Shirō 酉水孜郎, "Tōkyō-shi kinkō ni okeru tochiriyō 東京市近郊に於ける土地利用 (Land use in the suburban areas of Tōkyō)," Chirigaku hyōron, v. 5 (1929), no. 12, 51-72; v. 6 (1930), no. 1, 21-48.
An able description of agricultural land use on the western fringe of metropolitan Tōkyō. Of interest both to the urban geographer and the land utilization specialist.

980. Tanaka Kaoru 田中薫, "Taiwan ni okeru Taiyaru zoku no tochi riyō no keitai ni tsuite 臺灣に於けるタイヤル族の土地利用の形態に就いて (Types of land utilization of the Taiyal tribe in Formosa), "Chirigaku hyōron, v. 13 (1937), no. 2, 31-48; no. 5, 17-26.
The crop system, cultivation techniques, and migratory farming practices of the Taiyal aborigines in the mountains of Formosa are described in detail. The text is accompanied by a number of photographs on Taiyal land use and related ethnographic topics. Representative views of the highland terrain are included.

981. Tanakadate Hidezō 田中館秀三 and Yamaguchi Yaichirō 山口彌一郎, Tōhoku chihō no keizai chiri kenkyū 東北地方の経済地理研究 (Studies in the economic geography of the Tōhoku district), Tōkyō, Kokon Shoin, 1953, 344 pp., 14 illus.
A collection of twelve previously published monographs on the economic and settlement geography of Tōhoku. Papers deal with various aspects of the following general topics: reclamation, agriculture and crop failure, markets, rural settlement, and population. Included is a bibliography of more than 100 papers concerning Tōhoku written by Tanakadate and Yamaguchi and a ten-page index.

982. Ueno Fukuo 上野福男, "Keishachi riyō no kadai 傾斜地利用の課題 (Problems of the utilization of slope lands)," Chigaku zasshi, v. 62 (1953), no. 1, 1-12.
In this article Ueno surveys the problems and methods of farming and related forms of land use on steeply sloping land, particularly in reference to those regions in Japan where cultivation on mountain and hill slopes is intensively developed. Steep fields are considered as a special form of "upland" fields. The distribution of this type of farm land is noted and utilization trends discussed.

983. Yazawa Taiji 矢澤大二, "Suwa bonchi ni hattatsu-seru 'tsuka-zukuri' ni tsuite 諏訪盆地に發達せる「塚作り」に就いて (Concerning 'mound cultivation,' a development in the Suwa basin)," Chirigaku hyōron, v. 14 (1938), no. 6, 26-42.
Tsuka-zukuri, literally translated "mound cultivation," involves the building of small hillocks or barrows upon which fruit trees are grown in the midst of low-lying paddy areas. This form of land use, characteristic of the Suwa district of Nagano prefecture, is explained in relation to climatic influences and with reference to its combination with other local agricultural practices.

984. Yoneda Masatake 米田正武, "Serebesu Minahasashū ni okeru tochiriyō no jōtai セレベス，ミナハサ州に於ける土地利用の狀態 (The state of land utilization in the Minahassa district, Celebes)," Chirigaku hyōron, v. 16 (1940), no. 1, 15-37.
A general survey of land use in the region at the end of the northern-most peninsula of the Celebes, treating four zones: a) coastal mangroves, b) coconut plantations in the coastal lowlands, c) spice cultivation in the mountains above 300 meters elevation, and d) wet rice farming inside old volcanic calderas. With a discussion of the numbers, racial divisions, and cultural composition of the population.

985. Yoshimura Shinkichi 吉村信吉, "Chiiki keitai sokutei no ichihōhō to sono rei— bunka keikan to chikei to no mensekiteki sōkan 地域形態測定の一方法と其の例 — 文化景觀と地形との面積的相關 (A method and an example of the morphometry of cultural landscape— the spatial relationship between landforms and the cultural landscape)," Chirigaku hyōron, v. 6 (1930), no. 11, 1-30; no. 12, 12-47.
The writer advances a technique of quantitative map analysis useful in relating cultural features to terrain types and in correlating various landscape elements to reach an interpretive synthesis. Land use in the Shimōsa upland and in the Tama hills is studied.

C. Forestry

986. Matsuda Takashi 松田孝, "Tokushima-ken nambu no ringyō ni tsuite 徳島縣南部の林業に就いて (On forestry in the southern region of Tokushima prefecture)," Tōkyō Daigaku chirigaku kenkyū, no. 2 (1925), 162-179.
Economic and social aspects of lumbering and charcoal-making in the mountain forests of southwestern Shikoku are surveyed in this article. Methods of timber extraction and charcoal manufacturing are noted along with factors of labor supply, financing, transportation, and marketing. The nature of forest and terrain in the region is described briefly.

987. Matsui Isamu 松井勇, "Tama kyūryō ni okeru shinrindo no bumpu 多摩丘陵に於ける森林度の分布 (The distribution of forest in the Tama hills)," Chirigaku hyōron, v. 16 (1940), no. 12, 18-41.
A morphometrical map-study aimed at determining a uniform method for assessing regional characteristics through distributional analysis. The percentage area of forest land in this sector of Kanagawa prefecture is indicated and the distribution of forest area is shown in map form.

988. Mizuyama Takayuki 水山高幸, "Kyōto bonchi seien ni okeru chikurin no bumpu 京都盆地西縁に於ける竹林の分布 (The distribution of bamboo cultivation in the western border area of the Kyōto basin)," Jimbun chiri, v. 5 (1953), no. 4, 18-37.
Land, water, labor, marketing, and other factors related to the commercial production of bamboo sprouts (Phyllostchys edulis, or mōsōchiku in Japanese) in this district are considered in conjunction with an explanation of the localization and economic base supporting this specialized agricultural industry.

989. Ueno Fukuo 上野福男, "Chiriteki ni mitaru Shimpi kokkyō fukin sanson no rōdō no kisetsuteki bungyō 地理的に見たる信飛國境附近山村の勞働の季節的分業 (The seasonal cycle of labor in the mountain villages at the border of Shinano and Hida provinces, seen from the viewpoint of geography)," Chirigaku hyōron, v. 16 (1940), no. 8, 16-33.
Variations in the seasonal demand for labor in three different types of mountain village on the border zone of Nagano and Gifu prefectures are compared and interpreted. Labor problems are discussed in connection with specific forms and combinations of land use, i. e., farming, forestry, livestock raising, forestry and farming, etc.

D. Agriculture

1. Crops and crop systems

a. General

990. Arai Hiroshi 新井浩, "Kita-Kambara sakyūretsu no nashibata ni tsuite --kantan naru keisoku no ichirei to shite-- 北蒲原砂丘列の梨畑に就いて —簡単なる計測の一例として— (On the pear-gardens on the sanddunes at Kita-Kambara, as an example of a simple choromorphometrical study)," Chirigaku hyōron, v. 11 (1935), no. 4, 51-65.
A map study of orchard development on dune topography in Niigata prefecture between 1911 and 1931 to illustrate choromorphometrical techniques.

991. Awaji Shōzō 淡路正三, "Ashitaka-Hakone sanroku ni okeru kōchi to keisha to no kankei 愛鷹箱根山麓に於ける耕地と傾斜との關係 (The relationship between cultivated land and slope in the Ashitaka-Hakone mountains)," Chirigaku hyōron, v. 8 (1932), no. 9, 19-32.
The area, height, and steepness of cultivated land on the western slopes of Ashitaka and Hakone volcanoes, located to the south and southeast of Mt. Fuji, were calculated in this map study aimed at determing the slope limitations of crop land. Results are shown partially in map form. Awaji notes the complete absence of terraced fields in this mountain area. 1:25,000 topographic sheets were analyzed.

992. Hoyanagi Mutsumi 保柳睦美, "Kagan kōsakukei no keitaigakuteki kenkyū yohō 河岸耕作景の形態學的研究豫報 (A preliminary report on the morphology of the cultivation landscape of river valleys)," Chirigaku hyōron, v. 6 (1930), no. 3, 1-23.
An examination of varying land use patterns in Japanese river valleys where significant agricultural differences can be observed in connection with river control features and flood protection measures. The general effect of flood danger upon the cultural landscape is noted and specific crop systems and farming practices, employed on land within river levees and beyond them, are contrasted.

993. Hoyanagi Mutsumi 保柳睦美, "Keisha to suiden no kaidan to no kankei 傾斜と水田の階段との關係 (Relationships between rice terraces and angle of slope)," Chirigaku hyōron, v. 13 (1937), no. 8, 49-65.
The results of a statistical study of the relationship between the degree of slope and the width of rice terraces are reported upon. Terraces were studied on the alluvial fan of the Asakawa and on the slopes at Obasute, in Nagano prefecture, and on slopes of river terraces near Numata in Gumma prefecture. Observations on the average and maximum heights of individual terraces are included.

994. Ichikawa Wataru 市川渡, "Etchū Takamineyama fukin no chiriteki tokushoku (yohō) 越中高峯山附近の地理的特色(豫報) (Geographical features in the vicinity of Mt. Takamine, Etchū province — a preliminary report)," Chirigaku hyōron, v. 11 (1935), no. 3, 1-14.
The terrain of Mt. Takamine, in Toyama prefecture, and the land forms of the Tane basin below its north flank are described in conjunction with a discussion of land use in that region. The local communal system of forest ownership is given attention.

995. Iseki Hirotarō 井関弘太郎, "Nōgyō koyō rōdō ni kansuru chiriteki kenkyū— Tōkyō kinkō suiden chiiki no baai— 農業雇傭労働に関する地理的研究 — 東京近郊水田地域の場合 —(Geographical studies of agricultural employment with reference to a paddy-field area in the outskirts of Tōkyō)," Jimbun chiri, v. 2 (1950), no. 2, 11-26.

Differences in the management of farm land and field labor between similar rice-growing areas in Tōkyō and Saitama prefectures are compared on the basis of agricultural census statistics for 1936 and 1947.

996. Ishida Ryūjirō 石田龍次郎, "Oki Dōzen no makibata 隠岐島前の牧畑 (The pastures of Dōzen island in the Oki group)," Chirigaku hyōron, v. 5 (1929), no. 2, 1-20.

A broad examination of the social and physical aspects of the land system of Oki island located north of Shimane prefecture in the Sea of Japan. Particular attention is given to the peculiar pasture allotment custom practiced there.

997. Ishida Ryūjirō 石田龍次郎, "Oki no makibata soshiki no jizoku 隠岐の牧畑組織の持續 (A sequel on the system of pasture fields in the Oki islands)," Chirigaku hyōron, v. 5 (1929), no. 6, 76-85.

A sequel to the preceding paper. Analyzes the reasons for historical change in the makibata system and cites case studies made at four villages. The effects of population growth on the traditional land system are considered.

998. Iwasaki Kenkichi 岩崎健吉, "Kii hantō seinan engan ni okeru nōgyō keikan chiiki ni tsuite 紀伊半島西南沿岸に於ける農業景観地域に就いて (On the agricultural regions on the southeast coast of the Kii peninsula)," Chirigaku hyōron, v. 10 (1937), no. 12, 1-32.

Selected elements of the agricultural landscape of the southern coastal district of the Kii peninsula are analyzed in order to identify local homogeneous units or microregions. Describes land use in six villages in the Hidaka area of Wakayama prefecture chosen as the most representative of the agricultural divisions.

999. Koike Yōichi 小池洋一, "Chōsen kademmin no hassei 朝鮮火田民の發生 (Origin of the burning cultivators in Korea)," Jimbun chiri, v. 5 (1953), no. 2, 21-32.

An investigation of the causes of the rapid growth of the kademmin, or the migratory fire-field farm population, in the mountains of Korea at the beginning of the Shōwa era. The government's agricultural policy, designed to stimulate rice production for export to Japan, actually promoted tenancy and migration to the mountains, the writer believes, and here cites evidence to support his conclusion.

1000. Kondō Yasuo 近藤康男, Sengo no nōgyō mondai 戰後の農業問題 (Post-war agricultural problems), Tōkyō, Nihon Hyōronsha, 1947, 7+219 pp.

A critical analysis of the Japanese agricultural system and farming methods in the light of the land reforms instituted by the American occupation and in terms of post-war conditions generally. This book is a collection and rearrangement of Kondō's earlier newspaper and magazine articles and radio lectures.

1001. Kurihara Hakuju 栗原百壽, Nihon nōgyō no kiso kōzō 日本農業の基礎構造 (The basic structure of Japanese agriculture), Tōkyō, Chūō Kōronsha, 1948, 3+8+2+434+6 pp.

Four phases of the recent agrarian situation are analyzed critically in this study which views farming as a part of capitalistic development. Chapter headings are: 1) farm households, 2) ownership of farm land, 3) agricultural production, and 4) farming techniques. With a 143-page statistical supplement.

1002. Maruyama Shizuo 丸山静雄, "Nōson ni okeru rōdō-jikan no kisetsuteki haibun 農村に於ける労働時間の季節的配分 (Seasonal variations in working hours at farm villages)," Chiri to keizai, v. 1 (1936), no. 1, 98-101.

Maruyama reports his observations on rural labor practices in Gifu prefecture. He investigated farm operations from the viewpoint of labor efficiency at Kasagi-mura in Ena-gun and charted the monthly variations in working hours etc.

1003. Masuda Kazuji 桝田一二, "Hatano bonchi no jimbun chirigakuteki kansatsu 秦野盆地の人文地理學的觀察 (View of the human geography of the Hatano basin)," Chirigaku hyōron, v. 6 (1930), no. 7, 478-495.

The Hatano basin in Kanagawa prefecture is known for tobacco cultivation and with this in mind its landforms and land use are described.

1004. Matsui Isamu 松井勇, "Nasu senjōchi nōgyō-chiri (dainihō): suiden chiiki to hata chiiki to no taihi (1) 那須扇状地農業地理（第二報）水田地域と畑地域との対比 (1) (The agricultural geography of the Nasu-fan district, second report: comparison of the paddy-field areas with the upland-field areas, part 1)," Tōkyō Daigaku chirigaku kenkyū, no. 1 (1950), 2-15.

Makes a comparison of cropping and farm management techniques practiced on the paddy-field and upland-field sections of the Nasu alluvial fan as delimited in Matsui's previous study.

1005. Matsui Isamu 松井勇, "Nōgyō keiei soshiki ni yoru gun no bunrui 農業経営組織による郡の分類 (A classification of the districts of Japan by agricultural management types)," Chirigaku hyōron, v. 19 (1943), no. 1, 1-16.

A systematic classification of gun in terms of their predominant type of farm management, using 1938 statistical information. Major categories are as follows: 1) agriculture, 2) agriculture and sericulture, 3) agriculture and livestock raising, and 4) agriculture, sericulture, and livestock raising. Regional characteristics are discussed on the basis of this breakdown.

1006. Matsui Isamu 松井勇, "Nōgyō keiei soshiki yori mitaru waga naichi no chiri kubun 農業経営組織より見たる我が内地の地理区分 (A geographical division of Japan seen through agricultural management types)," Chirigaku hyōron, v. 19 (1943), no. 6, 1-22; no. 7, 32-50; no. 8, 21-39.

A sequel to the author's article in vol. 19, no. 1 of this journal (see preceding entry) regarding the classification of gun by agricultural type. This is an extension of his classification system to include five major agricultural regions and 46 sub-regions.

1007. Matsui Isamu 松井勇, "Rikuchū hokubu kaigan (Taneichi, Nakano, Samuraihama) sanson no nōgyō keiei no gaikan 陸中北部海岸.種市.中野.侍濱3村の農業經營の概觀 (General view of agricultural management at three villages along the northern coast of Rikuchū: Taneichi, Nakano, and Samuraihama)," Chirigaku hyōron, v. 17 (1941), no. 12, 1-28; v. 18 (1942), no. 1, 42-73.

Observations on farming, livestock raising, and supplementary employment at three agricultural villages located on the coastal terraces of northern Iwate prefecture.

1008. Matsui Isamu 松井勇, "Tama, Yurugi nikyūryō no chihyō keisha to kōsakukei 多摩 淘綾二丘陵の地表傾斜と耕作景 (Surface inclination and cultivation phenomena in the Tama and Yurugi hills)," Chirigaku hyōron, v. 8 (1932), no. 7, 1-17; no. 8, 21-33.

Matsui employed a grid method in this statistical analysis of areal relationships between cultivation features and terrain elements of the landscape in the Yurugi-Tama district of Kanagawa prefecture. The influence of degree of slope on cultivation is the primary consideration of this study.

1009. Miura Yasutoshi 三浦保壽, "Kyūshū sanchi ni okeru yakihata keiei kakuzetsu sanson no kenkyū 九州山地に於ける焼畑經營隔絶山村の研究 (A study of isolated villages managing fire farming in a mountainous region of Kyūshū)," Jimbun chiri, v. 4 (1953), no. 6, 25-37.

The ownership, labor, and other economic and social factors involved in fire-field agriculture (yakihata) at Itsugi-mura in Kumamoto prefecture are analyzed as a case study in an effort to explain the general causes for the maintenance of this primitive farming system in Kyūshū and in other parts of Japan.

1010. Mori Sumie 森壽美衛, "Taradake nanroku no kōsaku keikan 多良嶽南麓の耕作景觀 (Agricultural landscape of the southern slopes of Taradake)," Chirigaku hyōron, v. 6 (1930), no. 11, 60-90.

The volcanic area considered in this monograph on rural land use is located on the border of Nagasaki and Saga prefectures in Kyūshū. Taradake is a dissected conical mountain, the lower slopes of which are rather extensively cultivated. Slope and elevation and their effect upon the development and density of farm plots are investigated and problems of farming on the lower, wet fields and the upper, dry fields are compared.

1011. Muramoto Tatsurō 村本達郎, "Saitama-ken no nōgyō chiri 埼玉縣の農業地理 (The agricultural geography of Saitama prefecture)," Chirigaku hyōron, v. 16 (1940), no. 6, 1-11.

Physical and cultural characteristics of Saitama are summarized and followed by an explanation of the prefecture's agricultural patterns. Farming differences in three generalized sub-divisions of the prefecture are discussed, with the emphasis upon human rather than natural influences.

1012. Nagai Masatarō 長井政太郎, "Yamagata-ken Shōnai kaigan Hamanaka no keizai chiriteki kōsatsu 山形縣庄内海岸濱中の經濟地理的考察 (A view of the economic geography of Hamanaka village on the Shōnai coast of Yamagata prefecture)," Chirigaku hyōron, v. 17 (1941), no. 9, 1-16; no. 10, 39-49.

This fishing village on the coast of the Japan Sea, situated in a depression between sand dunes, has long carried on afforestation for providing firewood. This and other characteristic features and activities of village life are noted, e. g., the making of salt from sea water, fishing, agriculture, and the local family systems.

1013. Nagai Masatarō 長井政太郎, Nitō Tsuneo 仁藤庸雄, and Naganuma Katsumi 永沼桂美, "Zaōsan seiroku no nōson 藏王山西麓の農村 (Agricultural villages on the western foot of Mount Zaō)," Chirigaku hyōron, v. 13 (1937), no. 9, 1-25; no. 10, 1-21.

After a general description of the geology and geomorphology of the Zaō area, the distribution of cultivated fields, rice farming, sericulture, livestock raising and forestry are discussed in detail, using data by ōaza and koaza. The origin and growth of these settlements and the seasonal and permanent emigration of the farm population are related.

1014. Nihon Rekishi Chiri Gakkai 日本歴史地理學會 (Association of Japanese Historical Geography), Nihon nōminshi 日本農民史 (History of the Japanese peasantry), Tōkyō, Nihon Gakujutsu Fukyūkai, 1925, 2+4+373 pp.

Fifteen geographers and historians collaborated on this study of Japanese agriculture and peasant society. Rural oppression is highlighted in several of the articles which deal with agrarian revolts from early times through the uprising of 1873. The final two chapters cover agrarian movements and related developments during and after the Meiji period.

1015. Ogawa Tōru 小川徹, "Shizuoka-ken Suntō-gun chihō no kōsakukei ni tsuite 靜岡縣駿東郡地方の耕作景に就いて (On cultural landscape in the Suntō district, Shizuoka prefecture)," Chirigaku hyōron, v. 11 (1935), no. 9, 20-33.

A picture of the general agricultural landscape of this region is presented with the aid of statistical data, and transitional aspects of land use at three representative mura are interpreted.

1016. Sasaki Hikoichirō 佐々木彦一郎, "Kazuno bonchi no keizai chiri kōsei 鹿角盆地の經濟地理構成 (Structure of economic geography in the Kazuno basin)," Chirigaku hyōron, v. 2 (1926), no. 8, 1-26.

This mountain basin in Tōhoku constitutes most of the upper drainage area of the Noshiro river, where a serious pollution problem has existed because of contamination from the Kosaka and Osarizawa copper mines. The effect of this upon local agriculture is discussed and mining, forestry, manufacturing, and other local enterprises are outlined to create an overall picture of the economic pattern of this district situated on the boundary between two historic regions, the former provinces of Rikuchū and Ugo.

1017. Sasaki Hikoichirō 佐々木彦一郎, "*Sanson no keizai chiri I: Saitama-ken Chichibu-gun Urayama-mura no chōsa* 山村の經濟地理I・埼玉縣秩父郡浦山村の調査 (Economic geography of mountain villages, I: Reconnaissance of Urayama village, in Chichibu district, Saitama prefecture)," *Chirigaku hyōron*, v. 11 (1935), no. 6, 28-48; no. 7, 49-75.
Sasaki depicts the nature of life in this mountain village which has been little affected by outside influences in spite of its relative proximity to the Tōkyō metropolitan area. With some reference to burned-field mountain agriculture in Japan as a whole.

1018. Sasaki Seiji 佐々木清治, "*Kōson no nōgyō keitai* 郊村の農業形態 (Agricultural forms of suburban villages)," *Chirigaku hyōron*, v. 8 (1932), no. 4, 1-18; no. 5, 19-44.
From the standpoint of reciprocal economic relations between urban centers and surrounding rural settlement, Sasaki identifies various types of suburban communities and explains their agricultural production and their role as food suppliers to the city markets. Examples of several truck-farming villages serving Tōkyō are noted and the horticultural role of Yoshikawa-mura near Hamamatsu, Shizuoka-ken, is detailed.

1019. Sugai Shirō 酉水孜郎, *Nihon no nōgyō* 日本の農業 (Japanese agriculture), Tōkyō, Kokon Shoin, 1949, 520 pp., maps, illus.
This is an economic geography of Japanese agriculture. Research deals with the basis of agriculture, descriptions of arable land and its distribution, urban influence upon the agricultural landscape, etc. Emphasis is placed on rice culture, land reclamation in mountain regions, and agriculture as a foundation for regional planning. With tables, figures, photos, numerous distribution maps, and index.

1020. Sugimoto Hisashi 杉本壽, *Nōsanson keizai no kisoteki kenkyū* 農山村經濟の基礎的研究 (Basic study of the economy of mountain farm villages), Ōsaka, Kōbunsha, 1944, 9+2+7+831 pp., 17 illus.
A critical analysis of the economic and social structure of highland agricultural settlements in Japan with comment on historical change. Numerous references and old village documents are cited. Included are two monographs on *dezukuri* (migratory farming) and *maki* (pasture systems) in Okinawa.

1021. Tada Fumio 多田文男, "*Nōchi no chirigaku* 農地の地理學 (Geography of agricultural lands)," *Jimbun chiri*, v. 3 (1951), no. 4, 1-9.
Remarks briefly upon miscellaneous aspects of agricultural geography including land use problems, flood control and erosion prevention, irrigation, climatic influences, and the height limitations of rice cultivation.

1022. Tanabe Ken'ichi 田邊健一, "*Noto hantō (1) kōchi no keitaigakuteki kenkyū* 能登半島 (1) 耕地の形態學的研究 (The Noto peninsula, 1: a morphological study of cultivated land)," *Chirigaku hyōron*, v. 16 (1940), no. 6, 34-49.
Geomorphological controls over land use on this hilly peninsula are interpreted from the agricultural pattern of the area. Ten types of farm land are identified.

1023. Tanaka Keiji 田中啓爾, "*Chūō Nihon ni okeru kōchi no jimbun chishigakuteki kenkyū gaihō* 中央日本に於ける高地の人文地誌學的研究概報 (General view of the human geography of highlands in central Japan)," *Chirigaku hyōron*, v. 6 (1930), no. 8, 62-94.
A description of the highland agriculture and industry typical of settlement in the mountainous districts of central Honshū. Height and climatic limitations of crops such as the various grains, vegetables, fruits, and mulberry are considered. The author refers to specific communities throughout the article.

1024. Tanaka Toyoji 田中豊治, "*Nōchi kaihō to sanson keizai* 農地開放と山村經濟 (The redistribution of farm land and the economics of mountain villages)," *Jimbun chiri*, v. 2 (1950), no. 3, 12-25.
The inadequacy of the post-war land reform program as regards mountain communities is discussed and the economic and social factors related to land problems are summarized. The author cites specific case studies, selected on a regional basis.

1025. Toda Shintarō 戸田愼太郎, *Nihon shihon shugi to Nihon nōgyō no hatten* 日本資本主義と日本農業の發展 (Development of Japanese capitalism and agriculture), Tōkyō, Minshu Hyōronsha, 1947, 8+4+271 pp.
Essentially a revised an expanded edition of the author's *Nihon nōgyōron* (Study of Japanese agriculture), published by Sōbunkaku, Tōkyō, in 1936. A study of land ownership and agricultural economics since the Meiji revolution.

1026. Tōhata Seiichi 東畑精一, ed., *Nōson no jittai* 農村の實態 (The real state of farm villages), Tōkyō, Sangensha, 1950, 2v.
Compiled by the Nōgyō Fukkō Kaigi 農業復興會議 (Council for Agricultural Rehabilitation). A survey of representative rural economics. Volume one is primarily concerned with crops and production methods. The second volume deals with assorted efforts to improve the productivity of farming and other farm industry. It cites cases of farm failure, tenant dispossession, and peasant migration because of low productivity. A rare source of more than usual value to the settlement geographer and land use specialist since it deals specifically with present-day conditions.

1027. Ueno Fukuo 上野福男, "*Chūbu Nihon ni okeru kōrei chiiki no nōgyō* 中部日本に於ける高冷地域の農業 (The agriculture of the regions with high elevation and cool climate in central Japan)," *Chirigaku hyōron*, v. 18 (1942), no. 1, 1-28; no. 2, 44-63.
A report on highland agriculture in the drainage basins of the Nishino and Na rivers, in Kiso province, and of the Masuda river in Hida province where the upper limit of farm settlement ranges between 840 and 1,500 meters. The effect of altitude, climatic conditions, and insolation upon crops is investigated and the methods of cultivation of paddy rice and selected upland crops are described.

1028. Ukita Tsuneyoshi 浮田典良, "Shiga-ken no chagyō 滋賀縣の茶業 (The tea industry in Shiga prefecture)," Jimbun chiri, v. 4 (1952), no. 5, 32-48.

A brief outline of the history of tea production in Japan and a description of growing and processing methods and management techniques in the three principal tea producing districts. The causes promoting regional concentration of the industry are considered.

1029. Yajima Hitoshi 矢島中, "Shōnan sakyū chiiki no sakyū to ichini no jimbun genshō 湘南砂丘地域の砂丘と一二の人文現象 (The sand dunes of the Shōnan sand dune area and one or two cultural phenomena)," Chirigaku hyōron, v. 17 (1941), no. 4, 51-60.

A survey of agricultural land use in the coastal dune area at the head of Sagami bay in Kanagawa prefecture, including a non-technical geomorphological description of the district. Problems and methods of sand control are considered and the density of the settlement pattern and road-net are noted.

1030. Yamada Katsutoshi 山田勝利, "Nan'yo chihō ni okeru kaidan kōsaku no kenkyū 南豫地方に於ける階段耕作の研究 (A study of terraced fields in the southern part of southern Iyo province), Jimbun chiri, v. 4 (1952), no. 5, 14-32.

A study of the initial pre-Edo terracing of the steep coastal slopes of southwestern Shikoku and of the subsequent changes in land use on them, emphasizing the economic and social factors which influenced historic changes in the cropping system and general land use of the area. The recent farming system is described in some detail.

1031. Yamaguchi Sadao 山口貞雄, "Tochigi-ken no nōgyō chirigakuteki kenkyū gaihō 栃木縣の農業地理學的研究豫報 (General outline of an investigation of the agricultural geography in Tochigi prefecture)," Chirigaku hyōron, v. 12 (1936), no. 9, 1-33.

The distribution of crop land and past changes in land use in this prefecture are considered and agricultural sub-regions are identified. The distribution of individual crops is shown in some detail.

1032. Yamaguchi Sadao 山口貞雄, "Yakihata no chiriteki bumpu sonota 燒畑の地理的分布其他 (The geographic distribution of fire-field agriculture and other matters)," Chirigaku hyōron, v. 14 (1938), no. 1, 1-23.

This is the first report that surveyed the general nature and extent of yakihata or temporary fire-field agriculture (Brandwirtshaft) in Japan. It is based on statistical information compiled by the Department of Agriculture and Forestry in 1936. The writer distinguishes two types of this practice, one motivated purely by agricultural demands, the other an aspect of reforestation. Regional characteristics are shown and a distribution map included.

1033. Yamaguchi Yaichirō 山口彌一郎, Tōhoku no yakihata kankō 東北の燒畑慣行 (The custom of burned-field agriculture in Tōhoku), Tōkyō, Kōshunkaku Shobō, 1944, 6+6+238 pp.

Burned-field agriculture in the Aizu, Uzen, Ugo, Mutsu, and Kitakami mountain areas of northern Honshū described and comparisons made with this form of temporary cultivation as employed in other Asiatic regions, e.g., Korea. Village life, crops, and farming methods are discussed and trends of development and the preservation of fire-farming practices considered.

1034. Yokota Tadao 横田忠夫, "Nihon nōson no shakai chirigakuteki kenkyū josetsu (Fuji-gawa ryūiki nōson no chirigakuteki kenkyū) 日本農村の社會地理學的研究序説（富士川流域農村の地理學的研究）(Social geographical studies of agricultural village in Japan: A geographical study of villages in the basin of the river Fuji)," Tōkyō Daigaku chirigaku kenkyū, no. 1 (1950), 16-26.

The social and economic life of villages in the Fujigawa valley are compared in terms of relative modernization, functional differentiation, and the retention of feudal characteristics. The author's aim is to discover the influence of site factors and the proximity to urbanization upon village culture.

b. Specific

1035. Hamada Seikichi 濱田清吉, "Tanzawa sanroku Toya-mura ni okeru kuwa no saibaikei 丹澤山麓鳥屋村に於ける桑の栽培景 (Landscape of mulberry-tree cultivation in Toya-mura at the foot of mount Tanzawa)," Chirigaku hyōron, v. 14 (1938), no. 5, 34-57.

A detailed description of sericulture in a small, secluded mountain community in Kanagawa prefecture. Considerable attention is paid to temperature controls relative to mulberry cultivation, frost damage, and frost prevention methods such as shimo-ibushi or the use of smudge pots.

1036. Ishida Ryūjirō 石田龍次郎, "Taiwan sammai ni tsuite 臺灣産米に就いて (On the rice production in Formosa)," Chirigaku hyōron, v. 4 (1928), no. 1, 1-14.

A brief economic geography of rice culture in Taiwan, with pre-1928 production statistics.

1037. Ishii Yasuyoshi 石井泰義, "Kantō chihō ni okeru suiden urasakudo no bumpu 關東地方に於ける水田裏作度の分布 (The distribution of winter utilization of rice fields in the Kantō district)," Chirigaku hyōron, v. 19 (1943), no. 4, 49-55.

A short report on the distribution and percentage of double cropped rice fields in the Kantō district. The ratio of single to double cropped fields was calculated on the basis of mura statistics.

1038. Iwasaki Kenkichi 岩崎健吉, "Nōka no kōchi shoyū kankei yori mitaru Kii hantō engan no baien ritchi ni tsuite 農家の耕地所有關係より見たる紀伊半島沿岸の梅園立地に就て (On the distribution of plum orchards on the coast of the Kii peninsula as viewed in its relationship to the owning of arable land by farming households)," Chirigaku hyōron, v. 11 (1935), no. 6, 1-16.

Factors of land ownership are contrasted with physical influences in determining the geographical and economic basis for the development of orchards in the Kaminambu-mura area of Wakayama prefecture.

1039. Kawaguchi Takeo 川口丈夫, "Hokkaidō beisaku no chirigakuteki kenkyū 北海道米作の地理學的研究 (A geographical study of rice culture in Hokkaidō)," Chirigaku hyōron, v. 11 (1935), no. 1, 1-23; no. 2, 41-82, 15 maps, 30 tables.

The physical and human geography of rice farming in Hokkaidō. The climatic controls affecting rice cultivation are rather fully discussed as an introduction to the delimitation and detailed description of the four rice-growing zones on that northern island. Kawaguchi also surveys the historical background of the introduction and development of rice culture in Hokkaidō.

1040. Kawaguchi Takeo 川口丈夫, "Mikan no seisan chitai 蜜柑の生産地帯 (Regions producing mandarin oranges in Japan)," Chirigaku hyōron, v. 5 (1929), no. 4, 37-46; no. 5, 26-42.

A monograph on the production of mandarin oranges in Japan. Briefly treated are such factors as the distribution of producing areas, the limits imposed by altitude and latitude, climate, soil, and topographic conditions. The significance of local names, the historical background, and the influence of the market and labor situation are also touched upon.

1041. Kikuta Tarō 菊田太郎, "Kaju saibai no bumpu ni tsuite 果樹栽培の分布について (On the distribution of pomiculture areas)," Chiri to keizai, v. 1 (1936), no. 2, 69-78; no. 4, 93-100.

Written from the national approach, this is a survey of fruit growing as a cash crop in Japanese agriculture. Physical geographic factors and the role of transportation are emphasized.

1042. Matsui Isamu 松井勇, "Nasu senjōchi nōgyō chiri (daishihō): sakumotsu kōtai (ichi) 那須扇状地農業地理(第4報)作物交替 (1) (The agricultural geography of the Nasu-fan district, fourth report: the cropping system, part 1)," Tōkyō Daigaku chirigaku kenkyū, no. 2 (1952), 35-53.

Based on 1947 agricultural data and questionnaires returned by farmers in the Nasu district, this is a sequel to Matsui's paper in the first number (1950) of this publication.

1043. Matsui Isamu 松井勇, "Tama kyūryō oyobi Sagamihara ni okeru sōendo no bumpu ni kansuru hitotsu no shiryō 多摩丘陵及び相模原に於ける桑園度の分布に關する一つの資料 (Data for the distribution of mulberry fields in the Tama hills and on the Sagami plain)," Chirigaku hyōron, v. 12 (1936), no. 7, 24-40.

A portion of Matsui's long-term study of the agricultural pattern in Kanagawa prefecture. Based on census data for 1929, the percentage acreage and the distribution of mulberry fields are shown in table and map form, and the reasons for marked differences in land use between the two adjacent areas are explored.

1044. Mizuno Takeo 水野武夫, Senji Nihon no shokuryō mondai 戰時日本の食糧問題 (Japan's wartime food problems), Tōkyō, Kōyō Shoin, 1941, 4+6+334 pp.

A rather abstract argument supporting the establishment of a wartime food policy for Japan and its overseas holdings. The argument is based on the concept of an Autalkie economy. Contains little factual information but may be of some theoretical interest to the economic geographer.

1045. Murakami Setsutarō 村上節太郎, "Chūyo chihō ni okeru suitō saibai no chiikisei 中豫地方に於ける水稻栽培の地域性 (The regional characteristics of the cultivation of paddy rice in Middle Iyo province)," Ōtsuka Chiri Gakkai rombunshū, no. 5 (1935), 39-76.

The area of four gun and one shi is included in the district of Ehime prefecture which Murakami surveyed in this study. He attempts to detect regional individuality in the wet rice farming of that region of Shikoku by an analysis of crop distribution, variety of rice planted, farming techniques, amount of yield, length of cultivation period, etc.

1046. Nagai Masatarō 長井政太郎, "Ōtō no chirigakuteki kenkyū 櫻桃の地理學的研究 (A geographical study of cherry production)," Chirigaku hyōron, v. 8 (1932), no. 8, 34-54.

Cherries are grown in Japan mainly in Hokkaidō and in Fukushima and Yamagata prefectures. The Yamagata, Yonezawa, and Fukushima basins are the principal centers of production. This article discusses the historic, economic, and physical aspects of the cherry industry.

1047. Niwa Shirō 丹羽四郎, "Tōkeikubetsu ni yoru hompō yōsan no bumpu 統計區別による本邦養蠶の分布 (Distribution of sericulture according to statistical divisions)," Chiri to keizai, v. 1 (1936), no. 4, 59-66.

Niwa discusses the distribution of the sericulture industry on the basis of the regional divisions established by the national government for statistical purposes. He notes some of the local differences in production methods and output as determined from his study of the regional statistics.

1048. Nō Toshio 能登志雄 and Yoshizaki Keiji 吉崎惠次, "Hompō ni okeru kaidan kōsaku no chirigakuteki kenkyū 本邦に於ける階段耕作の地理學的研究 (A geographic study of terrace cultivation in Japan)," Chirigaku hyōron, v. 12 (1936), no. 4, 66-82; no. 9, 58-65; v. 14 (1938), no. 3, 52-59.

An analytical survey of the various types of agricultural terrace developed in the home islands and on Formosa. Three general classes are considered, i. e.: a) primitive forms employed in the Formosan highlands by the aborigines, by fire-field farmers in isolated mountain districts in Japan proper, etc., b) rice-paddy terraces; and c) terracing of orchards and upland crops. Trends in land utilization involving terracing are discussed.

1049. Odauchi Michitoshi 小田内通敏, "Chōsen no kademmin 朝鮮の火田民 (The fire-field farmers of Korea)," Jimbun chiri, v. 1 (1926), no. 1, 25-35.

A summary of the conditions of migratory fire-field agriculture in the highlands of northern Korea. Odauchi discusses the geographic basis for this primitive form of farming, and points out some of the historical, social, and economic factors which have perpetuated it. He also mentions the adverse effects of this system upon the forests and upon forest-improvement efforts.

1050. Sakata Wataru 坂田渉, "Taiwan ni okeru kansho saibai no kinkyō 臺灣に於ける甘蔗栽培の近況 (Recent aspects of sugar cane cultivation in Taiwan)," Chirigaku, v. 9 (1941), no. 4, 113-117.
An account of the Japanese development of the sugar industry on Taiwan between 1925 and 1937 with comments on methods of cultivation, plantation management, sugar acreage and yields. Refers in particular to the 1936 crop.

1051. Sasaki Hikoichirō 佐々木彦一郎, "Sosai no keizai chirigakuteki kenkyū 蔬菜の經濟地理學的研究 (A study on the economic geography of vegetables)," Chirigaku hyōron, v. 6 (1930), no. 4, 1-11.
A general study of truck gardening and its economic importance, with discussion of specialty production for exclusive town and city markets, transportation, marketing problems, distribution of this form of land use in Japan, etc. The paper is founded on the author's investigation of commerical vegetable raising and marketing in selected areas. A table indicates the seasonal nature of this type of farming.

1052. Sugai Shirō 酉水孜郎, "Amagi no wasabizawa (nisan no chōsa hōkoku) 天城の山葵澤（二三の調査報告）(Horse-radish on Amagi: report on two or three surveys)," Chirigaku hyōron, v. 9 (1933), no. 3, 1-10.
This is an account of the raising of eutrema wasabi or Japanese horse-radish in the Amagi mountain district of the Izu peninsula, Shizuoka-ken, a type of cultivation somewhat like hydroponics. Necessary physical conditions relating to water temperature and quality, etc., are explained.

1053. Sugai Shirō 酉水孜郎, "Chōsen no nōson ni okeru tochiriyō 朝鮮の農村に於ける土地利用 (Land utilization in the agricultural villages of Korea)," Chirigaku hyōron, v. 12 (1936), no. 12, 19-44.
A general survey of Korean agriculture with a description of the regional crop systems and land use methods. The marked differences between the intensive farming of south Korea and the extensive cultivation of north Korea are analyzed. Specialty farming practices, such as truck gardening by Chinese immigrants in certain urban neighborhoods, are noted.

1054. Sugai Shirō 酉水孜郎, "Tōkyō kinkō nōson no fuyu no saibaikei 東京近郊農村の冬の栽培景 (Forms of winter cultivation in the vicinity of Tōkyō)," Chirigaku hyōron, v. 12 (1936), no. 6, 51-63.
The transformation of the agricultural landscape at farm villages around Tōkyō from autumn through spring is described and the relationships between crops and climatic conditions, crop-rotation and intensive land management are explained.

1055. Sugiyama Masahide 椙山正英 and Kagami Kanji 鏡味完二, "Owari no sosaibatake 尾張の蔬菜畑 (Vegetable fields in the Owari plain)," Chirigaku hyōron, v. 12 (1936), no. 3, 1-25.
On the Owari plain near Nagoya, truck farming is well developed, particularly in the vicinity of Ichinomiya. In this paper the geographical circumstances that favored its development are analyzed, and management and marketing practice are described.

1056. Ueno Fukuo 上野福男, "Gokanoshō no yakihata kōsaku 五家荘の焼畑耕作 (Fire-field cultivation at Gokanoshō)," Chirigaku hyōron, v. 14 (1938), no. 2, 1-28.
Gokanoshō is a very isolated section of the central Kyūshū mountains. This is a rather detailed study of the primitive methods of agriculture still carried on in that region. Ueno describes the methods of field preparation by burning clearings on the forested slopes, the planting, tending, and harvesting of crops, and the system of land ownership. The size and distribution of the koba or fire-fields are discussed in terms of the overall land use situation in this mountainous district.

1057. Ueno Fukuo 上野福男, "Tōkyō-fu Hinohara-mura ni okeru mugisaku no chirigakuteki kenkyū 東京府檜原村に於ける麥作の地理學的研究 (A geographical study of the cultivation of barley and wheat at Hinohara village, Tōkyō prefecture)," Chirigaku hyōron, v. 15 (1939), no. 8, 20-46.
As part of a general study of the economic geography of mountain villages, this paper reports on the cultivation of wheat and barley at Hinohara-mura northwest of Tōkyō and contains a discussion of the relationships observed there in terrain, soil, and climatic conditions, farming techniques and land management practices, and grain yields.

1058. Ueda Shinzō 上田信三, "Chichibu sanchi no kōsaku genkaisen ni tsuite 秩父山地の耕作限界線に就いて (On the limits of cultivation in the Chichibu mountains)," Chirigaku hyōron, v. 13 (1937), no. 3, 45-60.
After a discussion of criteria and methods used in determining the elevation limits of agricultural land use, with reference to numerous previous studies, Ueda explains by quadrant the height controls affecting farming in the Chichibu district.

1059. Yajima Nikichi 矢嶋仁吉, "Shizuoka-shi o chūshin to suru chagyō no chirigakuteki kenkyū 靜岡市を中心とする茶業の地理學的研究 (A geographical study of tea culture in the region surrounding Shizuoka city)," Ōtsuka Chiri Gakkai rombunshū, no. 4 (1934), 127-150.
The area in the vicinity of Shizuoka city is considered to be the "kernel region" of tea culture in central Japan. Yajima describes the physical and cultural basis for the establishment of tea growing in this district and discusses management, picking, and processing methods, and the domestic and foreign market situation.

1060. Yamaguchi Yaichirō 山口彌一郎, "Tōhoku chihō no hie no bumpu 東北地方の稗の分布 (Distribution of barnyard millet in the Tōhoku district)," Chirigaku hyōron, v. 16 (1940), no. 1, 38-57.
Barnyard millet, once an important staple crop in Japan, is now grown largely in some sections of Tōhoku where it is still used as a basic food. Yamaguchi is concerned in this article with the distribution of millet in Tōhoku and its historical and agricultural significance in that district.

1061. Yamazaki Osamu 山崎修, "Kōchi-ken no suitō nikisaku no seiritsu 高知縣の水稻二期作の成立 (The origin of the double cropping of rice in Kōchi prefecture)," Jimbun chiri, v. 4 (1953), no. 6, 10-24.
A concise account of the economic history of double cropping in southern Shikoku since the introduction of that system of farming during the middle Edo period until present times.

1062. Yasuda Hatsuo 安田初雄, "Hokuriku chihō no hasagi bumpu 北陸地方の稻架樹分布 (Distribution of hasagi in the Hokuriku district)," Chirigaku hyōron, v. 16 (1940), no. 10, 1-16.
Hasagi are outdoor drying racks for shocked rice used in the harvest season and consist of both planted trees and pole frameworks. This paper describes the natural or tree type of hasagi, the species of tree involved, their usage, distribution, etc., in the Hokuriku district of Honshū.

2. Irrigation

1063. Hiruta Kōichirō 蛭田浩一郎, "Shinshū Sakudaira ni okeru yōsui no chirigakuteki kenkyū 信州佐久平に於ける用水の地理學的研究 (A geographical investigation of the water supply in the Saku basin, Shinano province)," Chirigaku hyōron, v. 9 (1933), no. 12, 1-20; v. 10 (1934), no. 1, 27-42.
Observations on human adjustment to the water supply situation in a mountain basin in Nagano prefecture in conjunction with a geomorphologic and hydrologic survey of that area.

1064. Ishii Sen'ichi 石井宣一, "Tameike no shūsui chiiki ni tsuite 溜池の集水地域に就て (On the catchment areas of water reservoirs)," Chirigaku hyōron, v. 14 (1938), no. 12, 39-53.
Classifies and describes the types of catchment basin and reservoir used for collecting and storing irrigation water in the Sanuki lowland of northern Shikoku. The distribution of storage ponds and various water impoundment problems are considered in relation to local land forms and land reclamation efforts.

1065. Kōno Kazuo 河野一夫, "Sanuki heiya no kangai chiiki 讃岐平野の灌漑地域 (The irrigated region of the Sanuki plain)," Chirigaku hyōron, v. 16 (1940), no. 9, 38-60.
Severe crop damage in the Sanuki lowland of Kagawa prefecture, caused by the drought of 1939, is analyzed in terms of the general climatic conditions in that year, the distribution of the most seriously affected areas, and the nature of the loss involved.

1066. Miyazaki Kenzō 宮崎健三 and Harada Kiyoshi 原田清, "Hompō (Chūbu ihoku) ni okeru senjōchi-jō yōsuiro keitai 本邦(中部以北)に於ける扇狀地上用水路形態 (The pattern of irrigation canals on alluvial fans in northern Japan, north of the Chūbu region)," Chirigaku hyōron, v. 11 (1935), no. 3, 50-77.
An investigation of the cultural and physical significance of differing irrigation patterns on alluvial fans in such areas as the Aizu and Matsumoto basins and the Toyama and Tsugaru plains. As a selected example, the irrigation and land use system on the fan of the Jōganji river in Toyama prefecture is detailed.

1067. Nishikawa Osamu 西川治, "Nishi-Tenryū kangai suiro kaihatsu ni kiin-suru keikan no henka-- dōtaiteki kenkyū no ichirei-- 西天龍灌漑水路開發に起因する景觀の變化—動態的研究の一例— (Some changes of landscape caused by the opening of the Nishi-Tenryū irrigation canal--an example of a study in dynamic geography)," Tōkyō Daigaku chirigaku kenkyū, no. 2 (1952), 110-136.
A comprehensive discussion of the local but complex effects imposed by the operation of the West Tenryū irrigation canal upon human and natural facets of the environment subsequent to its initial use in 1928.

1068. Nō Toshio 能登志雄, "Kangaiyō unga no keitaigakuteki kenkyū 灌漑用運河の形態學的研究 (Morphological studies of irrigation canals)," Chirigaku hyōron, v. 9 (1933), no. 10, 26-38.
A description and classification of the networks of irrigation and drainage canals and ditches selected as being characteristic of certain regions of monsoon Asia, such as the networks in northern Kyūshū and along the lower reaches of the Yangtse in China. The various canal systems are analyzed in terms of their adaptation to and/or reflection of terrain, climate, and land use conditions.

1069. Takeuchi Jōgyō 竹内常行, "Aso kazan Shirakawa ryūiki to Ōnogawa jōryū kuiki no kangai ni tsuite 阿蘇火山白川流域と大野川上流區域の灌漑に就いて (On irrigation of the areas drained by the Shirakawa river and the Ōno river near Aso volcano)," Jimbun chiri, v. 4 (1952), no. 4, 1-14.
The topographic conditions in this area of lava plateau, a district with flat lands of some agricultural importance but with permeable soils, are considered in relation to irrigation problems. The use of water tunnels and methods of preventing water wastage by soil leakage are treated here.

1070. Takeuchi Jōgyō 竹内常行, "Dōryoku ni yoru kangai yōsuiki no chirigakuteki kenkyū 動力に依る灌漑揚水機の地理學的研究 (Geographical study of power pumps in irrigation)," Chirigaku hyōron, v. 17 (1941), no. 1, 35-59; no. 2, 20-47; no. 3, 1-28.
The various types of pumps and pumping methods employed in Japanese irrigation are described and local specialization studied in an effort to detect geographic reasons for regional differences. Past and present trends in irrigation techniques are summarized.

1071. Takeuchi Jōgyō 竹内常行, "Etchū heiya ni okeru kasensui in'yō no chirigakuteki kenkyū 越中平野に於ける河川水飲用の地理學的研究 (A geographical study of the use for drinking of river water on the Etchū plain)," Chirigaku hyōron, v. 16 (1940), no. 5, 39-60.
Reports on a field survey of the source and use of water on the alluvial fans of the Etchū district in Toyama prefecture, particularly with reference to the source and condition of drinking water in connection with human health and sanitation.

1072. Takeuchi Jōgyō 竹内常行, "Kagawa-ken ni okeru kangai jōkyō no chirigakuteki kenkyū 香川縣に於ける灌漑狀況の地理學的研究 (A geographical study of irrigation in Kagawa prefecture)," Chirigaku hyōron, v. 17 (1941), no. 11, 25-49; no. 12, 43-67.

The sources of water supply, storage, and distribution methods used in the irrigation of the more important lowlands of Kagawa prefecture are here described. Particular attention is given to the use of artificial ponds, so important to the agriculture of this prefecture.

1073. Takeuchi Jōgyō 竹内常行, "Tameike no bumpu ni tsuite 溜池の分布に就いて (On the distribution of irrigation ponds)," Chirigaku hyōron, v. 15 (1939), no. 4, 47-64; no. 5, 1-24; no. 6, 42-55.

This article is often referred to when tameike or irrigation ponds in Japan are mentioned. Distribution maps of tameike are provided and their distribution according to regions in Japan proper is studied. Conditions which cause this distribution are analyzed.

1074. Ushiyama Ki 牛山喜, "Suwa no fūsha 諏訪の風車 (Windmills in Suwa)," Chirigaku hyōron, v. 4 (1928), no. 8, 36-53.

In the Suwa basin of Nagano prefecture windmills were commonly used for irrigation pumping during the early decades of this century but this practice has since declined. Ushiyama here presents the land use and other factors causing the adoption and subsequent abandonment of the windmills.

1075. Yamagiwa Jirō 山極二郎, "Ōsaka-fuka no kangai nōgyō 大阪府下の灌漑農業 (Irrigation farming in Ōsaka prefecture)," Chirigaku hyōron, v. 4 (1928), no. 11, 1-21; no. 12, 20-47.

A consideration of the sources of irrigation water in the Ōsaka area and an explanation of the various methods employed. Emphasis is upon the use of irrigation ponds, a common feature of agricultural land use in this district.

3. Reclamation

1076. Kawasaki Toshi 川崎敏, "Tottori-ken ni okeru kaigan sakyū no kaitaku 鳥取縣に於ける海岸砂丘の開拓 (The reclamation of coastal sanddunes in Tottori prefecture)," Chirigaku hyōron, v. 18 (1942), no. 4, 1-21.

A fairly detailed account of land reclamation and agricultural land use in the coastal dune region of Tottori prefecture.

1077. Nakamura Keisaburō 中村慶三郎, "Yamakuzure to tochi warikae seido to no kankei ni tsuite 山崩と土地割替制度との關係に就いて (On the relationship between landslides and the system of distribution of land)," Chirigaku hyōron, v. 13 (1937), no. 7, 1-20; no. 8, 18-36.

Villages in Nagano, Ishikawa, Toyama, and Niigata prefectures provide five examples of communities where land redistribution is necessary because of landslides which destroy cultivated fields. Related problems are discussed.

1078. Watanabe Misao 渡邊操 and Nobei Keiji 延井敬治, "Waga kuni kaitaku kanōchi no chiriteki seikaku 我國開拓可能地の地理的性格 (Geographical characteristics of reclaimable lands in Japan)," Shin-chiri, v. 2 (1948), no. 6, 15-25, 6 maps.

This paper summarizes a detailed report on the reclaimable lands in Japan made by the Bureau of Reclamation, Ministry of Agriculture and Forestry (Nōrinshō, Kaitakukyoku). Discussed and explained are the definition of the term "reclaimable lands," standards of selection and methods of survey, distribution, and the latitudinal and altitudinal characteristics. Six maps show the distribution and area of unused land appraised as being of potential value and worthy of practical reclamation efforts.

1079. Yamaguchi Sadao 山口貞雄, "Nasunogahara no kaikonkei no hattatsu 那須野ヶ原の開墾景の發達 (Development of cultivation on Nasunogahara)," Chirigaku hyōron, v. 15 (1939), no. 8, 1-19.

Geographical considerations and historical factors influencing land reclamation and agricultural expansion on the alluvial fan at the east foot of Nasu volcano are explained and the present farm landscape of the area is described in some detail.

1080. Yamaguchi Yaichirō 山口彌一郎, "Tsugaru Jūsanko-gan no kaitakukei— gaihō— 津輕十三湖岸の開拓景—概報— (The development of the shores of the Tsugaru Jūsan lakes, a general report)," Chirigaku hyōron, v. 14 (1938), no. 7, 9-25.

The general settlement geography of the Jūsanko district in Aomori prefecture is surveyed in this article. Focus is principally upon the pre-Meiji reclamation of portions of that wet plain and the establishment there of shinden agricultural colonies. Ecological aspects of the vegetation sequence in the lake-bed areas are noted in conjunction with the writer's description of early drainage projects and associated land use problems.

4. Animals

1081. Hikita Yukio 引田幸雄, "Karafuto oyobi Hokkaidō no yōkogyō ni tsuite 樺太及び北海道の養狐業に就いて (On the fox-raising industry in Karafuto and Hokkaidō)," Chirigaku, v. 4 (1936), no. 1, 253-266.

An account of the economic geography of fur farming in Japan. Fox pelts and other animal furs were exported from Karafuto and Hokkaidō on a fairly large scale prior to the war and this is a descriptive summary of the growth and conditions of that business.

1082. Irie Toshio 入江敏夫, "Waga kuni beisaku ni okeru chikunōka わが國米作における畜農化 (The change to use of animal power in rice fields)," Shakai chiri, no. 27, 1950, 17-18, 2 maps.

An explanation of two maps pertaining to the percentage-use of man and animal power in rice cultivation. Percentages are indicated on a gun basis for 1939 data.

1083. Kawarada Tsugio 河原田次雄, "Nōgyō chiri yori mitaru hompō no chikusan 農業地理より見たる本邦の畜産 (Japanese livestock from the viewpoint of agricultural geography)," Shin-chiri, v. 3 (1949), nos. 2-3, 86-95.

The title of this article is somewhat misleading since the article deals mainly with the influence of environmental factors on horse-raising and regional characteristics reflected in the various strains of horses bred in Japan.

1084. Miyamoto Tsuneichi 宮本常一, "Gyūba no rōeki kankō 牛馬の労役慣行 (Some traditions concerning the use of cattle and horses)," Shin-chiri, v. 3 (1949), no. 6, 23-33.

A recording of some regional traditions and customs regarding the use of work animals. Miyamoto describes the systems of lending and renting horses and oxen in Hokuriku and Chūgoku, joint-ownership of cattle in Shikoku, etc.

1085. Miyamoto Tsuneichi 宮本常一, "Ushi no bumpu 牛の分布 (The distribution of cattle)," Shin-chiri, v. 2 (1948), no. 4, 36-40; no. 5, 29-33; no. 6, 26-32.

After an historical summary of the past use of cattle in Japan, Miyamoto describes the various types of cattle, current methods of cattle raising, and the present distribution and use of cattle. Cattle keeping in certain areas of the Inland Sea and along the Sea of Japan coast, where it is a local specialty, is pictured in detail.

1086. Sekiguchi Takeshi 関口武 and Shōji Shigekichi 庄司茂吉, "Chūgoku sankan chiiki ni okeru chikugyū no hensen, toku ni Shimane-ken Imbe-mura Nishinooku ni tsuite 中國山間地域に於ける畜牛の變遷、特に島根縣忌部村西ノ奥に就いて (Changes in cattle raising in the Chūgoku mountains, especially at Nishinooku, Imbe village, Shimane prefecture)," Chirigaku hyōron, v. 18 (1942), no. 11, 27-51.

Traces changes in the livestock industry of the Chūgoku district since the early Meiji period and discusses their significance in terms of the use of work animals in Japanese agriculture.

1087. Tanabe Ken'ichi 田邊健一, "Bokujō to chikei-- Kitakami sanchi no rei ni tsuite 牧場と地形—北上山地の例に就て (Pasture and topography — on an example in the Kitakami mountains)," Chirigaku hyōron, v. 18 (1942), no. 9, 19-32.

The actual and potential pasture land in the Kitakami district of northern Honshū is classified geomorphologically in this study and type examples are described. The land use factors which generally restrict grazing acreage in Japan to elevated, rough, or infertile terrain of little or no value for cultivation are briefly summarized in the introductory paragraphs.

1088. Tanabe Ken'ichi 田邊健一, "Kitakami no kachiku shiyō 北上の家畜飼養 (Stock raising in the Kitakami district)," Chirigaku hyōron, v. 18 (1942), no. 7, 22-39.

The historical background and recent status of livestock raising in the Kitakami mountain country of northeastern Tōhoku is outlined and its relationship to upland agriculture in that region is discussed. The numbers and distribution of horses and cattle are shown and the trends of this industry and its future prospects are considered.

1089. Tanabe Ken'ichi 田邊健一, "Nihon bokuchiku no chirigakuteki kenkyū gaihō 日本牧畜の地理學的研究概報 (Outline of a geographical reconnaissance on stock raising in Japan)," Chirigaku hyōron, v. 17 (1941), no. 9, 42-62; no. 10, 1-38.

In terms of Japan's overall agricultural economy, stock raising is but a corollary industry on a minute scale. This study compares it with that of other countries, and relates its development, methods, and distribution. With explanatory maps.

E. Fishing

1090. Aono Hisao 青野壽郎, Gyoson suisan chirigaku kenkyū— daiisshū 漁村水產地理學研究 第一集 (Studies in the geography of fishing villages, no. 1), Tōkyō, Kokon Shoin, 1953, 2v., 201 illus.

A two-volume collection of 53 of the author's previously published geographical and general articles on the subject of fishing settlements and fisheries. Twenty-five concern the characteristics of fishing villages in Chiba, Shizuoka, and on the Kii peninsula. Fourteen treat skip-jack fishing in Japanese coastal waters and the bases connected with that industry. The tuna, mackerel, yellow-tail, and sardine fisheries are described in four separate papers. Five monographs deal with the general problems of fishing communities, and five with fishing operations in international waters such as the north Pacific and Antarctic oceans. With abundant maps and photographs and approximately 100 statistical tables.

1091. Aono Hisao 青野壽郎, "Hompō ni okeru katsuobushi sanchi 本邦に於ける鰹節產地 (Districts of dried bonito production in Japan)," Chiri, v. 3 (1940), 29-44.

Katsuobushi or dried bonito is a traditional Japanese food with specialized uses. Aono describes its production and traces historical changes in the producing regions as far back as the middle Heian period. Modern producing areas are located and statistics given.

1092. Aono Hisao 青野壽郎, "Hompō no buri gyogyō gaikyō 本邦の鰤漁業概況 (Outline of yellow-tail fishery conditions in Japan)," Chiri, v. 4 (1941), no. 2, 110-113.

A short survey of the buri fishing industry in Japan, including brief mention of fishing grounds, trapping methods, and marketing. There are pre-war statistics of the catch and a bibliography.

1093. Aono Hisao 青野壽郎, "Hompō no maguro gyogyō 本邦の鮪漁業 (Outline of the tuna fishery in Japan)," Chiri, v. 3 (1940), no. 1, 102-106.

This is a concise description of Japanese tuna fishing covering such topics as the variety of tuna caught, fishing seasons and grounds, equipment and operating methods, marketing of catch, etc. With appended data on the off-shore and open-sea take in 1936.

1094. Aono Hisao 青野壽郎, "Hompō Taiheiyō-engan gyoson no suisan chiri 本邦太平洋沿岸漁村の水産地理 (The geography of fishing villages on the Pacific coast of Japan)," Chiri kyōiku, v. 30 (1939), no. 2, 14-22.

Fishing settlements are classified according to their geographical sites, i. e., on embayments, inlets, sand beaches, rocky coasts, etc. General characteristics of the several categories are discussed.

1095. Aono Hisao 青野壽郎, "Waga Nan'yō guntō no suisangyō 我が南洋群島の水産業 (The fishing industry in the South Pacific islands mandated to Japan)," Chiri, v. 3 (1940), no. 4, 109-112.

A short paper on both Japanese and native fishing activities in the Pacific islands but mainly dealing with the pre-war commercial fishing for bonito in that region.

1096. Hirokawa Yūko 廣川裕子, "Tōkyō-wan ni okeru Asakusa-nori yōshoku ni tsuite 東京灣に於ける淺草海苔養殖について (On the laver industry of Tōkyō bay)," Shakai chiri, no. 28 (1950), 24-27.

A cursory study of the laver or nori industry in Tōkyō bay. Problems in the allotment of growing areas, cultivation, harvesting, and marketing of seaweed are noted.

1097. Kuwata Tōichi 桑田透一, Gaikan Nihon suisanshi 概觀日本水産史 (Outline history of Japan's aquatic industry), Tōkyō, Umi to Sorasha, 1943, 2+1+6+211+31 pp.

The history of Japan's aquatic industry is described chronologically in five periods - pre-Meiji, Meiji, Taishō, pre-war Shōwa, and during World War II. The relation between fishery and other industries is given for each period. The chief stress is upon the modern development of the fishing industry.

1098. Nakamatsu Yahide 仲松彌秀, "Itoman-machi oyobi Itoman gyofu no chiriteki kenkyū 糸満町及び糸満漁夫の地理的研究 (A geographical study of Itoman town and its fishermen)," Chirigaku hyōron, v. 20 (1944), no. 2, 15-39.

A general social and geographic study of the ichimana or Okinawan fishing folk, primarily in reference to the fishing methods of the people of Itoman-machi, a principal Ryūkyū fishing town.

1099. Nomura Yutaka 野村豊, Kanten shiryō no kenkyū 寒天資料の研究 (A study of records concerning agar-agar), Ōsaka, Ōsaka-fu Kanten Suisan Kakōgyō Kyōdō Kumiai, 1949, 2 v.

A collection of general and technical materials on the kanten industry and the exportation of agar-agar from Japan during the period from 1813 to the Meiji restoration. The study lists some 390 documents preserved in Ōsaka and Kyōto prefectures.

1100. Ozaki Torashirō 尾崎虎四郎, "Bōshū Hamanabuto no inada tateami gyogyō 房州濱波太の鰍建網漁業 (Trapping for young yellow-tail at Hamanabuto, Bōsō peninsula)," Chirigaku hyōron, v. 11 (1935), no. 7, 1-17.

The fisherman of Hamanabuto, a community in Futomi-mura, Chiba-ken, are noted for their special trapping methods employed in catching inada. This technique is described together with social regulations, etc., related to fishing activities and daily life there. With notations on changing influences of the sea upon local fishing grounds.

1101. Satouchi Susumu 里内晋, "Soto-Nan'yō ni okeru gyogyō 外南洋に於ける漁業 (Fisheries in the Outer South Seas)," Chiri kyōiku, v. 13 (1931), no. 5, 183-197.

In this monograph, the "Outer South Seas" include the South China Sea, the Sulu Sea, the Celebes Sea, the Java Sea, the Banda Sea, and a part of the Indian Ocean. Fishing operations in these waters are discussed with comments on sea conditions, the variety of fish and other marine resources, and the prospects for future fishing activities.

1102. Tōkyō Teikoku Daigaku Nōgakubu Nōseigaku Kenkyūshitsu 東京帝國大學農學部農政學研究室 (Institute of Agricultural Politics, Faculty of Agriculture, Tōkyō Imperial University), Gyoson keizai no kenkyū 漁村經済の研究 (Studies on the economy of fishing villages), Tōkyō, Iwanami Shoten, 1933, 5+248 pp.

Studies of the economic and social structure of five representative coastal buraku. The economic base of the communities and their prefectural location are as follows: Shizuoka prefecture, open-sea fishing; Yamagata prefecture, coastal fishing; two buraku on Sado island, Niigata prefecture, and one in Aichi prefecture, half farming and half fishing. Statistics cited are for the years 1928 and 1930.

1103. Yoshimura Shinkichi 吉村信吉, "Nihon naichi no naiwan no engan gyokakubutsu ni yoru hyōshiki 日本内地の内灣の沿岸漁獲物による標式 (Classification of bays in Japan proper, according to the catches of coastal fisheries)," Ōtsuka Chiri Gakkai rombunshū, no. 3 (1934), 113-136.

Japan's most important fishing bays and inlets are classified according to total catch and predominant type of catch, following a general morphological classification.

1104. Yoshizaki Keiji 吉崎恵次, "Hompō ni okeru en'yō gyogyō 本邦に於ける遠洋漁業 (Ocean fisheries in Japan)," Chirigaku hyōron, v. 14 (1938), no. 7, 26-63.

One of relatively few geographical studies on this subject. The largest amount of attention is given to the open-sea bonito fisheries. With maps and tables showing distribution of fishing areas in 1936 and 1937, identifying home bases and numbers of vessels, and presenting catch statistics.

F. Mining and minerals

1105. Koyama Ichirō 小山一郎, Nihon-san sekizai seigi 日本産石材精義 (Commentary on stone for construction produced in Japan), Tōkyō, Ryūginsha, 1931, 4+9+298 pp.
A two-part book covering, first, the physical and chemical classification of stone based on Koyama's experimentation, and second, the location and production of stone as a construction material.

1106. Matsuzawa Dentarō 松澤傳太郎, "Nihon no sekiyu sanchi 日本の石油産地 (The oil producing districts of Japan)," Chiri kyōiku, v. 7 (1927), no. 4, 22-26; no. 5, 33-38.
An outline of the amount of annual production and the methods of exploitation at the principal oil fields in Japan proper, Hokkaidō, and Formosa, prior to 1927.

1107. Ōmura Ichizō 大村一藏, "Nan'yō no yuden 南洋の油田 (Oil fields in the South Seas)," Chirigaku, v. 7 (1939), no. 10, 1-12.
A brief but comprehensive report on the nature and production of the oil fields and refineries of the Dutch East Indies, including notes on managerial aspects.

1108. Saitō Fumio 齋藤文男, "Naniyue ni sekiyu sanchi wa ura-Nihon ni kagiri omote-Nihon ni nai ka 何故に石油産地は裏日本に限り表日本に無いか (Why are the oil-fields restricted to the Japan Sea coast, and not distributed on the Pacific coast?)," Chiri kyōiku, v. 7 (1927), no. 1, 12-18.
An introductory discussion of this question in relation to geological factors.

1109. Saitō Hitoshi 齋藤齊, "Yūbō naru Taiwan yuden 有望なる臺灣油田 (Promising oil fields in Taiwan)," Chiri kyōiku, v. 5 (1926), no. 1, 23-29.
Short summary of the geological structure and distribution of the petroleum fields of Formosa as of 1925.

1110. Tokunaga Shigeyasu 德永重康, "Yamaguchi-ken Ube tanden no gaikan 山口縣宇部炭田の概觀 (General view of the Ube coal field, Yamaguchi prefecture)," Chirigaku hyōron, v. 6 (1930), no. 7, 557-563.
An explanation of the geological structure and history of the Ube district, with particular reference to geotectonic subdivisions of the coal field.

1111. Watanabe Manjirō 渡邊萬次郎, Nihon kōzan sōran 日本鑛山總覧 (A gazetteer of Japanese mines), Tōkyō, Nihon Shobō, 1940, 5+6+1042 pp.
The name, type, and location of some 800 major mines in the home islands, Karafuto, Taiwan, and Chōsen are listed in geographical sequence from north to south. There is a short outline of general information on mining and a brief survey of the history, distribution, and present status of mining operations. Current laws and mining regulations are cited. An index of names in the order of the Japanese syllabary with romanized transcriptions is appended.

G. Industry

1112. Arai Hiroshi 新井浩, "Gumma-ken no shuyō kōgyō chiiki. Kōgyō chishi kenkyū daiippō 群馬縣の主要工業地域、工業地誌研究第1報 (The chief industrial regions of Gumma prefecture. First report on the geography of industrial regions)," Chirigaku hyōron, v. 14 (1938), no. 1, 24-41; no. 3, 61-82.
Fundamentally concerned with the geographic, economic, and social factors involved in the development of local silk reeling and weaving from the scale of a rural household industry to the urban factory level. Distribution maps of factory location, labor sources, etc., are included, and the important role of women workers in the silk industry is noted.

1113. Harada Jumpei 原田準平, "Chiriteki ni kōsatsu-shita hompō suiryoku hatsudenjo bumpu 地理的に考察した本邦水力發電所分布 (Distribution of hydroelectric plants in Japan, considered geographically)," Chirigaku hyōron, v. 2 (1926), no. 4, 18-46.
Outlines the development of Japan's hydro-electric power industry since 1891 when the first generating plant began operation in the Kyōto-Biwa area. Some reference is made also to thermal power plants. With a detailed distribution map of power installations.

1114. Horikawa Tadashi 堀川侃, "Chūshōkōgyō no ichitenkei to shite no 'koyū kōgyō' no seitai—Hiroshima-ken Asa-gun Ochiai-mura no bokuri kōgyō — 中小工業の一典型としての「固有工業」の生態 一廣島縣安佐郡落合村の木履工業 (Ecology of the traditional private industry as a model of small-scale industries: the wooden-clog industry of Ochiai village, Asa district, Hiroshima prefecture)," Jimbun chiri, v. 5 (1953), no. 3, 16-27.
As an example of the survival of a traditional small-scale industry originating in pre-Meiji times, the economic history of wooden-clog (geta) manufacturing in Ochiai village, near Hiroshima, is outlined.

1115. Kawasaki Toshi 川崎敏, "Tochigi-ken ni okeru kigyō no chiiki kubun 栃木縣に於ける機業の地域區分 (Geographical division of the textile industry in Tochigi prefecture)," Chirigaku hyōron, v. 16 (1940), no. 5, 9-38.
A classification of the local textile industry in Tochigi prefecture on the basis of manufacturing technique, type of production, etc. Founded on statistical analysis and field surveys. Areal subdivisions are mapped and their characteristics described.

1116. Kiuchi Shinzō 木内信藏, "Kōgyō-ritchi no chiikiteki tenkai to gijutsu ni kansuru nisan no mondai 工業立地の地域的展開と技術に關する二・三の問題 (The geographic development of industrial regions and two or three technological problems)," Tōkyō Daigaku chirigaku kenkyū, no. 2 (1952), 54-66.

Kiuchi discusses the interplay of technological, economic, and geographic factors with respect to the problem of delimiting industrial regions. Considerable attention is devoted to the significance of machinery manufacturing and the regional concentration of that industry. With critical reference to the inadequate development of the field of urban geography in Japan.

1117. Misawa Katsue 三澤勝衛, "Suwa seishigyō hattatsu no chirigakuteki igi 諏訪製絲業發達の地理學的意義 (The geographical meaning of the development of the silk-reeling industry in Suwa)," Chirigaku hyōron, v. 2 (1926), no. 10, 1-22; no. 11, 27-53.

By using a series of distribution maps showing factory location, Misawa indicates areal changes in the concentration of this industry during past periods and analyzes the geographical basis for the eventual establishment of Suwa, in Nagano prefecture, as the core district of Japanese silk reeling. Economic influences are considered in some detail.

1118. Nishimura Mutsuo 西村睦男, "Nihon no kōgyō bumpu 日本の工業分布 (Industrial distribution in Japan)," Jimbun chiri, v. 3 (1951), no. 4, 10-20; v. 4 (1952), no. 4, 14-27.

On the basis of employment figures for 1950, factories are classified into three categories in accordance with their size and their distribution is mapped to reveal regional patterns. These industrial regions are then further analyzed with respect to three generalized types or classes of manufacturing activity: 1) machine, metal and chemical, 2) fiber, 3) wood, food, and pottery.

1119. Sueo Yoshiyuki 末尾至行, "Nōson kōgyō ritchiron 農村工業立地論 (A theory on the location of rural industries)," Jimbun chiri, v. 4 (1952), no. 4, 28-43.

An analysis of rural manufacturing in the middle Kakogawa district of Hyōgo prefecture where a surprising variety of home industry has survived. Sueo describes the various manufacturing activities in terms of raw materials, labor, and marketing and attempts to explain the elements of localization which have contributed to the development and continuation of such industry.

1120. Takemi Yoshiji 武見芳二, "Dai-Tōkyō chiiki no kōjō bumpu 大東京地域の工場分布 (Distribution of factories in greater Tōkyō)," Chirigaku hyōron, v. 6 (1930), no. 7, 369-386.

The writer explains the seven criteria used in his classification of the manufacturing industries of Tōkyō and presents both general and specific distribution maps of the city's factories. With 1929 statistics on the number of establishments and employment.

1121. Tanaka Keiji 田中啓爾, "Sanuki heiya ni okeru bakkansanada to kamasu to no seisan chitai ni tsukite 讃岐平野に於ける麥稈眞田と叺との生產地帶につきて (On the producing area of straw mats and straw bags in the Sanuki plain)," Chirigaku hyōron, v. 3 (1927), no. 11, 46-52.

The western portion of the Sanuki lowland in Kagawa prefecture is the predominant producing area in Japan of straw goods such as bags used for salt, fertilizer containers, and straw mats and rope. This is a description of the physical and cultural factors favoring this industry and of the manufacturing processes employed.

1122. Yamazaki Teiichi 山崎禎一, "Seishi kōjō no bumpu to kiito no yushutsu 製糸工場の分布と生糸の輸出 (The distribution of silk factories and the export of raw silk)," Shin-chiri, v. 2 (1948), no. 7, 1-14.

Economic and geographic reasons fostering Kōbe's increasing importance as the chief post-war exporting point for Japan's gradually recovering silk manufacturing industry are investigated in terms of the national distribution of silk mills and the allocation of their production.

H. Recreation and tourism

1123. Naimushō Eiseikyoku 内務省衛生局 (Sanitation Bureau of the Home Ministry), Nihon kōsenshi 日本鑛泉誌 (A history of Japanese mineral springs), Tōkyō, Hōkōsha, 1886, 3 v.

A complete compilation of historical and factual information on the many mineral springs in Japan. The springs are described individually by prefecture. Much of the basic data remains valid and useful and few later publications give such a thorough coverage.

1124. Ozaki Torashirō 尾崎虎四郎, "Bōsō no hishochi narabi ni kaisuiyokujō chitai 房總の避暑地竝に海水浴場地帶 (The summering places and swimming resorts of the Bōsō peninsula)," Chirigaku hyōron, v. 14 (1938), no. 8, 20-44; no. 9, 15-30.

Describes the sites and functional role of resort towns in the peninsular portion of Chiba prefecture, particularly in terms of the relationship of the towns to the Tōkyō area.

Chapter XI

Regional Descriptive Geography

Regional descriptive geography, as the term is used herein, might better be called local descriptive geography since most studies in this category do not encompass areas of enough magnitude to be regarded ordinarily as regions.

These local descriptive studies are the most numerous of all Japanese geographical writings. Where the emphasis of such studies has been, even mildly, upon the physical, cultural, or other aspect of the area in question, the items have been listed under other headings of this bibliography rather than in this section. This has been done because it is believed that such a procedure will be of greater value to the Western user of the bibliography. Actually these local descriptive studies are probably of little use to the majority of foreign geographers unless they happen to coincide with the particular area of a geographer's interest.

On the other hand, one cannot understand the present status of geography in Japan without some knowledge of the nature and quality of these studies. Furthermore, in these studies the Japanese geographers are laying an exceedingly firm foundation for future studies of wider scope. They are perhaps displaying a degree of modesty not always apparent in the geographical profession.

A. General

1125. Kobayashi Fusatarō 小林房太郎, Dai-Nihon Teikoku chishi tsūron 大日本帝國地誌通論 (General geography of the Japanese Empire), Tōkyō, Fuzambō, 1915, 2 v., 104 photos and maps, 118 illus.
A thorough compilation of information on the general geography of Japan. The first volume, still of considerable reference value, treats the various elements of physical geography. The second volume, largely outdated, deals with human geography and the relationship between physical environment and human activity. Thirty-six tables of statistical matter are appended at the end of vol. 2.

1126. Sasaki Seiji 佐々木清治, Zusetsu Nihon chiri 圖說日本地理 (An illustrated geography of Japan), Tōkyō, Kembunkan, 1934-35, 3 v., 509 illus.
Coverage is by traditional districts with a topical breakdown for each region as follows: location, geomorphology, seacoast, climate, and cities. Emphasis is upon the elements of human geography. A well illustrated general reference work.

1127. Tanaka Keiji 田中啓爾, Chirigaku rombunshū 地理學論文集 (Collected papers on geography), Tōkyō, Kokon Shoin, 1933, 978 pp.
A collection of 37 of the author's previously published articles and lectures on geography, including one monograph written in English. Tanaka is interested chiefly in regional aspects of the field. Most of the papers deal with areas in central Japan or the Inland Sea region.

1128. Tōkyō Daigaku Rigakubu Chirigaku Kyōshitsu Shashin Chishi Nihon Henshū Iinkai 東京大學理學部地理學教室寫眞地誌日本編集委員會 (Committee for a Photographic Geography of Japan, Geographical Institue, College of Science, Tōkyō University), Shashin chishi Nihon 寫眞地誌日本 (Photographic geography of Japan), Tōkyō, Dainihon Yūbenkai Kōdansha, 1952, 2+7+282 pp., 769 photos.
Pictures are arranged under the following classifications: landforms, seasons, settlement and land utilization, folklore, and modernization. With each photograph there are titles both in Japanese and English, accompanied by a descriptive legend in Japanese. A photo-index map and several representative land utilization maps are added. A general text is included, under the following headings: 1) landforms and climate, 2) history, 3) structure of basic industries, and 4) modern industry and cities. As a whole this is an excellent study. The photographs are well selected and reproduced and the collection is a valuable aid to the visual interpretation of the Japanese scene.

1129. Yamamoto Kumatarō 山本熊太郎, Gaikan Nihon chishi 概觀日本地誌 (An outline of the regional geography of Japan), Tōkyō, Kokon Shoin, 1930-31, 2 v., 582 illus.
A general reference guide for grade school teachers. Emphasis is placed upon regional economy and description of predominant regional characteristics. Volume 1 treats the Ōu, Kantō, Chūbu, Kinki, Chūgoku and Shikoku districts. Volume 2 covers Kyūshū, Taiwan, Hokkaidō, Karafuto, Chōsen, Liaotung, and Micronesia.

1130. Yamamoto Kumatarō 山本熊太郎, Kyōdo chiri hanrei 郷土地理範例 (Models of provincial geography), Tōkyō, Kokon Shoin, 1933, 3+7+235 pp., 87 illus.
Primarily a study of model settlement types. The author analyzes several type examples selected on the basis of differences in physical setting and land use. The categories and the specific places described are: 1) agglomerated, general agricultural villages on a plain— Yonemaru-mura and Futazuka-mura on the Ishikawa plain, Ishikawa-ken; 2) disseminated paddy village on a plain — Shutto-mura on the Hikawa plain, Shimane-ken; 3) mixed fishing-farming village with upland fields — Nagahama-mura on the Iwami coast, Shimane-ken; 4) mixed fishing-farming village with fields on sand-dunes — the Kaga coastal dune area, Ishikawa-ken; 5) offshore islands — Dōzen and Oki islands, Shimane-ken; and, 6) castle town — Kanazawa city.

B. The traditional regions: dō and kuni

1131. Ozaki Torashirō 尾崎虎四郎, Bōshū no chishi 房州の地誌 (A regional geography of Awa province), Tōkyō, Kokon Shoin, 1929, 2+2+2+6+76 pp., 39 illus.
Awa province is now the southern part of Chiba prefecture, comprising most of the area of the Bōsō peninsula. The author, a specialist on that region, pictures its geographical character mainly in terms of human activity. Systematically treated are such topics as agriculture, fisheries, communications, population, etc.

C. Political regions: ken, gun, machi, shi, fu, to

1132. Andō Masuo 安藤萬壽男, Gifu-ken shinshi 岐阜縣新誌 (A new descriptive geography of Gifu prefecture) [=v. 21 of Tanaka, ed., Kyōdo shinsho (entry 92)], Tōkyō, Nihon Shoin, 1950, 206 pp., 49 illus.

The content of this book is classified in three parts. The first part is a general description of Gifu-ken as a whole; its location, topography, climate, resources, products and communications, and a short history. The second part is provincial description - Nōbi plain, the plateau of eastern Mino and the mountain land of Hida. In part three the author makes suggestions on the future of Gifu; he states that the necessary points for the future development of the prefecture are river embankment for the Nōbi plain and afforestation for the mountain villages. The large family system of Shirakawa-mura at Hida, a famous and rare example in Japan, is described in detail in a supplementary chapter of 24 pages.

1133. Gōda Eisaku 合田榮作, Kagawa-ken shinshi 香川縣新誌 (A new descriptive geography of Kagawa prefecture) [=v. 37 of Tanaka, ed., Kyōdo shinsho (entry 92)], Tōkyō, Nihon Shoin, 1950, 188 pp., 65 illus.

This descriptive geography of Kagawa-ken comments upon the following topics: location, topography, climate, scenic places, towns, cities and islands, intensive farming, stone and clay, communications with concentration on Takamatsu city and Kotohira town, light dwellings, dispersed settlement, distribution and movement of population, and areal division. The description of the districts of this prefecture is stressed. Lists 22 periodical articles about Kagawa-ken.

1134. Gumma-ken Kyōikukai 群馬縣教育會 (Gumma Prefecture Educational Society), Gumma-kenshi 群馬縣史 (A history of Gumma prefecture), Maebashi, Gumma-ken Kyōikukai, 1927, 4 v.

Complete collection of historical information and source materials concerning this prefecture. Much of the data is of value to the historical geographer interested in this region. Each volume averages 1,000 pages.

1135. Hamada Seikichi 濱田清吉, Yamaguchi-ken shinshi 山口縣新誌 (A new descriptive geography of Yamaguchi prefecture) [=v. 35 of Tanaka, ed., Kyōdo shinsho (entry 92)], Tōkyō, Nihon Shoin, 1950, 184 pp., 32 illus., 6 statistical tables.

Includes sections on location, population, natural environment, farm and mountain villages with studies on livelihood and production, fishing villages, development of cities, coal and electricity, manufacturing belt along the Suōnada, communications and commerce, provincial characteristics and regional divisions. A supplementary chapter entitled "Akiyoshidai" should be noted. Akiyoshi upland is geomorphologically a famous place in Japan as it is an example of a karst region. In this monograph, based on previous works by geomorphologists and the author's own survey, a geomorphological outline of the upland is presented in 34 pages illustrated with 9 diagrams. Fifteen articles on Yamaguchi-ken are listed at the end of the volume.

1136. Inami Etsuji 稲見悦治, Okayama-ken shinshi 岡山縣新誌 (A new descriptive geography of Okayama prefecture) [=v. 33 of Tanaka, ed., Kyōdo shinsho (entry 92)], Tōkyō, Nihon Shoin, 1951, 190 pp., 80 illus., 4 statistical tables.

This work on Okayama-ken discusses: location, topography, climate, land use, farming, sericulture and stock raising, natural resources, scenic places in relation to tourists and recreation, industry, communication and commerce, villages, towns and cities (with historical studies), population problem, and provincial characteristics and regional divisions. A chapter on the historical geography of settlement is somewhat unique and covers about 34 pages. Several references are listed at the end of each chapter.

1137. Ishii Itsutaro 石井逸太郎, Toyama-ken shinshi 富山縣新誌 (A new descriptive geography of Toyama prefecture) [=v. 16 of Tanaka, ed., Kyōdo shinsho (entry 92)], Tōkyō, Nihon Shoin, 1949, 148 pp., 23 illus.

The author is chairman of the Toyama Chiri Gakkai 富山地理學會 (Association of Geographers of Toyama) and has had much experience in education and study in Toyama over a period of 25 years. Since he has specialized in the field of physical geography, the volume is somewhat biased in favor of the description of natural features. Only 40 pages are devoted to human geography and this information is not classified scientifically. The contents of the volume as a whole are: history, population in the past 23 years, Einzelhof on Tonami plain, Gotō plain (east of Mt. Kureha), Kurobe fan and Himi hill, outline of industries, agriculture, history and present state of Toyama's drug peddlers, development of hydro-electricity and industry, cities and towns, climate and culture. Noteworthy is the description of Einzelhof on Tonami plain, which is a famous and unusual example of dispersed settlement in Japan. Also of interest is the section on Toyama's drug peddlers, who sell their goods throughout Japan and have been widely known since the Shogunate era. A four-page bibliography and 17 pages of statistical tables are appended to the volume.

1138. Itō Gōhei 伊藤郷平, Nagano-ken shinshi 長野縣新誌 (A new descriptive geography of Nagano prefecture) [=v. 20 of Tanaka, ed., Kyōdo shinsho (entry 92)], Tōkyō, Nihon Shoin, 1950, 204 pp., 89 illus., 17 statistical tables.

The author was born in Nagano prefecture and has had over 20 years of educational experience and geographical study there. This book is remarkable in having among its contents very vivid descriptions of life in farm and mountain villages, with sketches of dwellings, clothing, food, and the customary observances of the year. The chapters are on location, population, topography, climate, farming and stock raising, life in a farm household, forests, life in a mountain village, minerals, aquatic resources, hydro-electricity, tourism, damage by physical forces, industry and its location, development of communication, major cities, and geographical regions. Nagano prefecture is said to be one of the most enthusiastic prefectures in regard to education and provincial studies.

1139. Kanao Sōhei 金尾宗平 , Fukuoka-ken shinshi 福岡縣新誌(A new descriptive geography of Fukuoka prefecture) [= v. 40 of Tanaka, ed., Kyōdo shinsho (entry 92)], Tōkyō, Nihon Shoin, 1949, 148 pp., 32 illus.
The contents are divided into two parts. The first half is a description of the natural environment, namely, location, administrative and natural area divisions, topography, climate, flora and fauna, natural monuments and scenic places. The latter half concerns human activity and social environment, that is: population, settlement, industries, transportation, historical studies, and a conclusion. The contents are somewhat academic. Forty-two articles about this prefecture, previously appearing in geographical periodicals and written by the same writer, are listed at the end of the book.

1140. Kimura Kenji 木村憲治, Shiga-ken shinshi 滋賀縣新誌(A new descriptive geography of Shiga prefecture) [=v. 25 of Tanaka, ed., Kyōdo shinsho (entry 92)], Tōkyō, Nihon Shoin, 1953, 254 pp., 102 illus.
Presents a good description of the historical development of human occupancy in this prefecture from the stone age up to present day. This prefecture includes the biggest lake of Japan, Lake Biwa, and the description of fisheries there is rather detailed. One chapter entitled "The problems of the reclamation of Lake Biwa" is interesting with its references to land reclamation, flood control, irrigation, and hydro-electricity. A chapter is devoted to the sight-seeing industry, which is of importance to this prefecture.

1141. Kishimoto Minoru 岸本實 ,Tokushima-ken shinshi 德島縣新誌(A new descriptive geography of Tokushima prefecture) [=v. 36 of Tanaka, ed., Kyōdo shinsho (entry 92)], Tōkyō, Nihon Shoin, 1951, 208 pp., 65 illus., 18 statistical tables.
In the beginning chapter entitled "The administrative center of the prefecture" the author explains this center's transfer during the ages. The following chapters deal with distribution of population, areas classified by the value of production (farming, fisheries, and forestry), topography, climate, agriculture in the North (Kitagata) and South (Minamigata), mountain land on the northern and southern slopes of the Tsurugi mountains (mainly concerned with forestry), the north and south coasts, communication, industry, and main cities and geographic regions. Historical facts and changes are considerably stressed throughout.

1142. Kobayashi Fusatarō 小林房太郎, Dai-Nihon Teikoku chiri seigi 大日本帝國地理精義 (Precise geography of the Japanese Empire), Tōkyō, Nankōsha, rev. ed., 1924-26, 3 v.
Revised from the original 1922 edition, this detailed geography was a standard reference work for primary and secondary school teachers for many years.

1143. Kōzuki Junji 上月順治 , Hyōgo-ken shinshi 兵庫縣新誌(A new descriptive geography of Hyōgo prefecture) [=v. 28 of Tanaka, ed., Kyōdo shinsho (entry 92)], Tōkyō, Nihon Shoin, 1950, 198 pp., 33 illus., 28 statistical tables.
Description is under the following headings: 1) location, topography and climate, 2) regional description of five provinces - Settsu, Harima, Southwestern Tamba, Awaji island and Tajima, 3) industries, 4) communication, 5) residents, and 6) conclusion.

1144. Kurihara Torajirō 栗原寅次郎, Kyōzai kenkyū Dai-Nihon chiri seisetsu 教材研究大日本地理精説 (Detailed materials for teaching Japan's geography), Tōkyō, Daidōkan, 1923-24, 2 v.
A highly detailed but generally out-dated teaching guide once widely used in the public schools.

1145. Mori Sumie 森壽美衛 , Hokkaidō shinshi 北海道新誌(A new descriptive geography of Hokkaidō) [=v. 1 of Tanaka, ed., Kyōdo shinsho (entry 92)], Tōkyō, Nihon Shoin, 1951, 250 pp., 91 illus.
The titles of the chapters in this book are: 1) Hokkaidō separated from Hondo or Japan proper; 2) Hokkaidō, a large island; 3) cold Hokkaidō; 4) monotonous shoreline; 5) gentle mountains and wide plains; 6) volcanoes and hot springs; 7) landscape of land use; 8) livestock and fish; 9) coal and other undergound resources; 10) industry from raw materials produced in Hokkaidō; 11) development of communication and commerce; 12) planned cities and settlements; 13) Hokkaidō and its small population; 14) Hokkaidō and its young history; 15) conclusion - geographic regions. From living in Hokkaidō for 16 years and from his specialization in geography, the author gives vivid descriptions accompanied by 91 illustrations (charts and photographs). He also provides a list of reference materials on Hokkaidō — books, maps, and articles in geographical publications. Statistical tables of population, area, agriculture, livestock, fisheries and mining in Hokkaidō are included.

1146. Mori Sumie 森壽美衛, "Tomakomai no shizen to bunka keikan 苫小牧の自然と文化景觀 (Natural and cultural landscape at Tomakomai)," Shin-chiri, v. 4 (1950), no. 6, 40-47.
Mori analyzes the geographical factors supporting the major paper manufacturing industry at Tomakomai, located on the south-central Hokkaidō coast, and attempts to explain why that area remains relatively undeveloped in other respects after some 80 years of colonization.

1147. Mori Sumie 森壽美衛 and Tanaka Daiji 田中大二 , Nagasaki-ken shinshi 長崎縣新誌(A new descriptive geography of Nagasaki prefecture) [=v. 42 of Tanaka, ed., Kyōdo shinsho (entry 92)], Tōkyō, Nihon Shoin, 1953, 240 pp., 83 illus.
This book consists of two parts. The first is general description and the second is description by geographic regions. In part one special stress is placed upon landforms, since this prefecture has a very complicated topography with many peninsulas, volcanoes, and islands. Industry, fishing, coal mining, and traffic are treated in detail. In part two the prefecture is divided into 12 geographic regions, based mainly on landforms.

1148. Murakami Setsutarō 村上節太郎, Ehime-ken shinshi 愛媛縣新誌 (A new descriptive geography of Ehime prefecture) [=v. 38 of Tanaka, ed., Kyōdo shinsho (entry 92)], Tōkyō, Nihon Shoin, 1953, 216 pp., 65 illus.

The vivid descriptions of this work are according to geographic regions of the prefecture. Distribution maps of population, crops, and other economic factors are abundant and very useful. Unfortunately, however, the regions described are not shown on a map. With bibliography.

1149. Muramoto Tatsuo 村本達郎, Saitama-ken shinshi 埼玉縣新誌 (A new descriptive geography of Saitama prefecture) [=v. 11 of Tanaka, ed., Kyōdo shinsho (entry 92)], Tōkyō, Nihon Shoin, 1950, 132 pp., 66 illus., 27 statistical tables.

The contents of this book are: 1) regionality of Saitama-ken - natural conditions, 2) regionality of Saitama-ken - human conditions, 3) damage by water and river engineering, 4) drain-off reclamation and irrigation, 5) reclamation on the upland area, 6) reclamation on the Chichibu mountains, 7) number of professions of residents, 8) agriculture, 9) industries, 10) other industries, 11) transportation and communication, and 12) cities and rural communities. As the headings suggest, reclamation is especially stressed. The "Chichibu area" which is part of a national park to which the Tama area of Tōkyō prefecture also belongs is decribed in a supplementary chapter.

1150. Nagai Masatarō 長井政太郎, Kaitei Yamagata-ken chishi 改訂山形縣地誌 (A regional geography of Yamagata prefecture, revised), Yamagata, 1942, 2+4+392 pp., 6 pp. of maps and photos.

A concise geographical survey of Yamagata prefecture, considerably revised and enlarged over the original edition published nearly ten years earlier. Seven chapters deal with: landforms, climate, industries, trade, population, settlement, and a general summary. Chapter 6 on settlement is particularly well developed. In it the prefecture is described on the basis of 22 settlement regions with frequent reference to historical background. There is a bibliography appended.

1151. Nagai Masatarō 長井政太郎, Yamagata-ken shinshi 山形縣新誌 (A new descriptive geography of Yamagata prefecture) [=v. 5 of Tanaka, ed., Kyōdo shinsho (entry 92)], Tōkyō, Nihon Shoin, 1950, 168 pp., 39 illus.

This book supersedes the following two books previously written by the author: Yamagata-ken chishi 山形縣地誌 (A descriptive geography of Yamagata prefecture), Tōkyō, Meijidō, 1933, 262 pp., and Kaitei Yamagata-ken chishi 改訂山形縣地誌 (Descriptive geography of Yamagata prefecture, revised), 1942, published by the author, 382 pp. In the present volume the author discusses the topography of Yamagata prefecture together with its geology, climate, industries, population, communication systems, settlements and dwellings. The contents are valuable and reliable. The author is a native of this prefecture and has lived there as a professional geographer since 1930.

1152. Nihon chiri fūzoku taikei (entry 84), v. 2, Dai-Tōkyō 大東京 (Great Tōkyō), Tōkyō, Shinkōsha, 1931, 8+409 pp., 14 illus., 2 maps.

This volume has the categories common to others of the same series. It includes all of Tōkyō prefecture but much stress is placed on Tōkyō city in the description, i. e., city planning and city institutions. Most of the illustrations are concerned with communication. Fourteen writers took part in the preparation of this volume.

1153. Nihon chiri taikei (entry 86), v. 3, Dai-Tōkyō-hen 大東京篇 (Great Tōkyō), Tōkyō, Kaizōsha, 1930, 467 pp.

This book presents a view of Tōkyō in 1930, when the reconstruction work from the heavy damage suffered in the earthquake and fire of 1923 was almost finished and Tōkyō-shi (the inner district of Tōkyō-to's 23 ku), often called Dai-Tōkyō in this book, was at the peak of its prosperity. Professor K. Tanaka and N. Iimoto compiled the volume, and the staffs of the Tōkyō Shichō 東京市廳 (Tōkyō City Office) and Fukkōkyoku 復興局 (Reconstruction Board) shared in writing its descriptions. With a general outline at the beginning and the mayor's message at the end, the work includes 47 chapters. The following topics are discussed individually, with many photos and several illustrations: Tōkyō as an "imperial" capital, police and fire service, economy, wholesale and retail markets, departments, manufacturing, tram-cars, traffic, streets, pavement, bridges, communications, rivers and port, Shintō shrines and Buddhist temples, education, publications, newspapers, welfare work, hospitals, prisons, water service, sewers, garbage disposal, architecture, parks, roadside trees, graveyards, stadiums, amusements, slums, population, geographical regionality; the Meiji shrine, Marunouchi, Ginza, Asakusa; topography, geology, climate; short history, visible vestiges of the Shogunate, customary observances traced from the Edo age, and big fires of Edo.

1154. Okuda Hisashi 奥田久, Tochigi-ken shinshi 栃木縣新誌 (A new descriptive geography of Tochigi prefecture) [= v. 9 of Tanaka, ed., Kyōdo shinsho (entry 92)], Tōkyō, Nihon Shoin, 1951, 210 pp., 51 illus.

Description of the following subjects: location, topography, climate, soil, natural disasters, industry, crops, stock farming, fishing, forestry, copper and stone, hydro-electricity, the national park at Nikkō, communication, villages, towns and cities, population, and regions.

1155. Saeki Hideo 佐伯英雄, Miyazaki-ken shinshi 宮崎縣新誌 (A new descriptive geography of Miyazaki prefecture) [= v. 45 of Tanaka, ed., Kyōdo shinsho (entry 92)], Tōkyō, Nihon Shoin, 1951, 180 pp., 45 illus.

By an official of the prefectural office who presents reliable and up-to-date descriptions and statistics. With bibliography.

1156. Shinoda Kumanda 篠田九万太, Ōita-ken shinshi 大分縣新誌 (A new descriptive geography of Ōita prefecture) [=v. 44 of Tanaka, ed., Kyōdo shinsho (entry 92)], Tōkyō, Nihon Shoin, 1951, 198 pp., 73 illus., 12 tables.

Of the 18 chapters of this volume, the first six discuss the prefecture as a whole. Chapters 7-15 are descriptions of 21 geographic regions. Agriculture is the most important industry and the descriptions of it are detailed. With many illustrations and distribution maps.

157. Suzuki Tadashi 鈴木公, Kagoshima-ken shinshi 鹿児島縣新誌(A new descriptive geography of Kagoshima prefecture) [= v. 46 of Tanaka, ed., Kyōdo shinsho (entry 92)], Tōkyō, Nihon Shoin, 1951, 168 pp., 50 illus., 87 statistical tables.

Although the organization is similar to that of other volumes of the series in which this book is found, the author worked hard to bring it up to date. The description is based on and illustrated by 87 statistical tables. There are also 50 illustrative maps or diagrams.

1158. Takahashi Masao 高橋正雄, Aomori-ken shinshi 青森縣新誌(A new descriptive geography of Aomori prefecture) [= v. 2 of Tanaka, ed., Kyōdo shinsho (entry 92)], Tōkyō, Nihon Shoin, 1953, 233 pp., 110 illus., 28 tables.

Presents useful, detailed descriptions of the landforms and climate of this prefecture located in the extreme northern part of Honshū. Among other topics discussed are natural disasters, reclamation, parks, agriculture (with emphasis on cereals and apple orchards), and settlement. With full bibliography.

1159. Tanaka Keiji 田中啓爾, Tōkyō-to shinshi 東京都新誌(A new descriptive geography of Tōkyō prefecture) [= v. 13 of Tanaka, ed., Kyōdo shinsho (entry 92)], Tōkyō, Nihon Shoin, 1949, 210 pp., 52 illus.

The author of this work is also the editor of the Kyōdo shinsho series in which the volume appears. Since he is a well-known and skilled writer on the descriptive geography of Japan, the contents of the book are reliable and interesting. The work is divided into 17 chapters. In the first chapter Tōkyō is described as the center of Japan in the fields of government, economy, communications, industry,and culture. In the following chapters the residential suburban villages and the reclamation of Musashino or the outer area of Tōkyō-to or Tōkyō urban prefecture, although geographically not a part of Tōkyō city (or Tōkyō in its smaller meaning) are described, i. e., Tachikawa city and Hachiōji city (satellite cities to the west of Tōkyō) and towns in the piedmont belt west of Tōkyō-to (Ōme-machi and Itsukaichi-machi), the mountain villages there, and the Izu islands. Population and geographic regions are studied in the last two chapters. Four pages of statistics are provided.

1160. Terao Shigeru 寺尾茂, Kōchi-ken shinshi 高知縣新誌(A new descriptive geography of Kōchi prefecture) [=v. 39 of Tanaka, ed., Kyōdo shinsho (entry 92)], Tōkyō, Nihon Shoin, 1950, 186 pp., 33 illus.

This work is systematically divided into six parts. The first part is entitled, "What natural characteristics of this prefecture are the bases of its industries and culture?" Location, topography and geography, climate, flora and fauna and physical damage are discussed. Part two is written about population, administrative divisions, and the character of the residents. This part is called, "What characteristics have its residents, as subjects of industry and culture?" Part three, called "What historical characteristics have its industry and culture?" is a historical study of this prefecture. In part four the description concerns agriculture, forestry, fisheries, mining, hydro-electric generation, industry, commerce, transportation and cultural elements. Part five is a study of areal divisions and the last part presents conclusions.

1161. Tōkyō Nichinichi Shimbunsha Keizaibu 東京日日新聞社經濟部 (Tōkyō Nichinichi Newspaper Company, Economic Section) and Ōsaka Mainichi Shimbunsha Keizaibu 大阪毎日新聞社經濟部(Ōsaka Mainichi Newspaper Company, Economic Section), Keizai fudoki 經濟風土記(Economic regionalism), Tōkyō, Tōkō Shoin, 1928-30, 6 v.

Unsystematic regional descriptions in journalistic style, stressing economic activity, but touching upon most aspects of the physical and social environment and including comments on local history and folklore. All of the prefectures are treated in the six volumes under the following regional breakdown: 1) Tōhoku, 2) Tōkai and Kantō, 3) Hokkaidō and Shin'etsu, 4) Kinki, 5) Shikoku, and 6) Chūgoku and Kyūshū.

1162. Tsujii Kōtarō 辻井浩太郎, Mie-ken shinshi 三重縣新誌(A new descriptive geography of Mie prefecture) [=v. 24 of Tanaka, ed., Kyōdo shinsho (entry 92)], Tōkyō, Nihon Shoin, 1951, 180 pp., 66 illus., 29 statistical tables.

The twenty chapters of this geography are not very unique but are provided with many maps and statistics. The chapter concerning fisheries is somewhat detailed and describes both the culture of pearls and the ama or diving-girls of Shima, famous in Japan.

1163. Tsujita Usao 辻田右左男, Kyōto-fu shinshi 京都府新誌(A new descriptive geography of Kyōto prefecture) [= v. 26 of Tanaka, ed., Kyōdo shinsho (entry 92)], Tōkyō, Nihon Shoin, 1952, 228 pp., 72 illus.

The author's striving for a popular style makes this a better book for the general reader than for the specialist. Although Kyōto, the former capital, yields countless materials for historical geography, Tsujita pays more attention to pre-historic sites. With a full bibliography.

1164. Tsunoda Seiji 角田政治, Saishin dai-Nihon chiri shūsei 最新大日本地理集成(Compilation on Japan's geography, new edition), Tōkyō, Ryūbunkan, 1914-16, 2 v.

This voluminous descriptive compilation is a later edition of Dai-Nihon chiri shūsei, published in 1914-17. It was designed as a general reference work for teachers and covers both the human and physical geography of the home islands and Japan's overseas territories. Following an overall summary, the main body of the work is a series of regional descriptions.

1165. Tsunoda Seiji 角田政治, Zoku saishin dai-Nihon chiri shūsei, kōtsū meishō no bu 續最新大日本地理集成交通名勝の部 (Supplements to compilation on Japan's geography, new edition: section on communication and famous places), Tōkyō, Ryūbunkan, 1916-17, 2 v.

These extensive supplementary volumes are intended to serve as detailed travel guides to the regions described in Saishin dai-Nihon chiri shūsei. Much of the content is historical in nature. Arrangement of the places discussed is by prefecture and included is a summary of prefectural communications.

1166. Wada Ryōji 和田亮二, Ishikawa-ken shinshi 石川縣新誌 (A new descriptive geography of Ishikawa prefecture) [=v. 17 of Tanaka, ed., Kyōdo shinsho (entry 92)], Tōkyō, Nihon Shoin, 1950, 180 pp., 38 illus.

The description embraces climate, natural environment (mainly topography) and settlements located therein, industries, communication (with historical studies), population and personality, and the future of the prefecture. In a supplementary section topographical changes in sand dunes and lagoons are studied. With an 11-page index of geographical terms and eight pages of statistical tables.

1167. Yabuuchi Yoshihiko 藪内芳彦, Wakayama-ken shinshi 和歌山縣新誌 (A new descriptive geography of Wakayama prefecture) [= v. 30 of Tanaka, ed., Kyōdo shinsho (entry 92)], Tōkyō, Nihon Shoin, 1951, 180 pp., 81 illus., 9 statistical tables.

Primarily an economic and historical survey, this work presents chapters concerning: 1) the present state of industries, 2) natural environment as a basis of production, 3) residents as agents of production, 4) social-economic history as a basis of production, 5) settlement and living.

1168. Yajima Nikichi 矢嶋仁吉, Gumma-ken shinshi 群馬縣新誌 (A new descriptive geography of Gumma prefecture) [= v. 10 of Tanaka, ed., Kyōdo shinsho (entry 92)], Tōkyō, Nihon Shoin, 1951, 188 pp., 110 illus.

A reliable study of this prefecture by a skilled geographer. The chapters are on topography, seasonal change in the weather, type of climate, land use, major farm products, sericulture, natural resources and their utilization, modern industry (mainly textile), communication (with a historical study), rural settlements, cities, and geographic regions. Many maps and tables.

1169. Yamaguchi Takayoshi 山口孝義, Ibaraki-ken shinshi 茨城縣新誌 (A new descriptive geography of Ibaraki prefecture) [=v. 8 of Tanaka, ed., Kyōdo shinsho (entry 92)], Tōkyō, Nihon Shoin, 1950, 174 pp., 30 illus.

The writer was a professor of geography at the Ibaraki Normal School from 1930 to 1943 and then the manager of a farm in this prefecture. His description, which follows his actual survey of almost all areas of Ibaraki prefecture, is of considerable value. The work is arranged in two sections - one section dealing with the towns and villages in the Ibaraki mountains and the other section dealing with those on the Ibaraki plain. Yamaguchi places primary stress upon farming. Kawada-mura in Higashi-Ibaraki-gun is specially studied as a sample of the farm villages of Ibaraki prefecture. It is covered by a 29-page monograph accompanied by 4 maps and 9 statistical tables. This monograph is valuable in obtaining a knowledge of the real state of rural settlements in Ibaraki prefecture.

1170. Yasuda Hatsuo 安田初雄, Fukushima-ken shinshi 福島縣新誌 (A new descriptive geography of Fukushima prefecture) [=v. 7 of Tanaka, ed., Kyōdo shinsho (entry 92)], Tōkyō, Nihon Shoin, 1950, 184 pp., 52 illus.

The author has been a teacher of geography at Fukushima since 1932 (except for the period from 1938 to 1942). He discusses the prefecture's history, three geographical regions and their topography, climate, farming, sericulture, livestock, natural resources, industry, recreation and tourism, communications, settlement, population density, and regional divisions. Almost all of the 52 illustrations were drawn by the author himself. They are based on the latest data and are interesting and valuable.

1171. Yokota Kōhachi 横田幸八, Iwate-ken shinshi 岩手縣新誌 (A new descriptive geography of Iwate prefecture) [= v. 3 of Tanaka, ed., Kyōdo shinsho (entry 92)], Tōkyō, Nihon Shoin, 1950, 194 pp., 41 illus.

Based on the author's studies in Iwate-ken over the last 20 years, this work describes the prefecture's history, natural characteristics, the characteristics of its populace, population, transportation and communication systems, major cities and towns, provincial specialties, and geographical divisions. Relatively specialized is a chapter entitled "River engineering and reclamation," mainly concerned with the Kitakami river. With bibliography and statistical tables on area, population, number of households, paddy and dry fields, forests and meadows and rivers.

1172. Yoshida Mori 吉田森, Fukui-ken shinshi 福井縣新誌 (A new descriptive geography of Fukui prefecture) [=v. 18 of Tanaka, ed., Kyōdo shinsho (entry 92)], Tōkyō, Nihon Shoin, 1951, 204 pp., 42 illus.

The author explains in a preface that he has attempted an anthropocentric approach to geographical description. His main chapters are as follows: two kinds of population (urban and rural), population density, initial points of the spheres of production, settlements and spheres of agriculture, the development of the spheres of agriculture, spheres of commerce, two units of the nets of communication, and conclusion. As a source of information, the book is inferior to most of the others of the same series.

D. The "geographical" regions

1173. Fujimoto Haruyoshi 藤本治義, Kantō chihō 關東地方 (The Kantō district) [=part of Nihon chihō chishitsushi (entry 83)], Tōkyō, Asakura Shoten, 1951, 5+315 pp., 67 illus.

The author's famous work, Kantō no chishitsu (Geology of the Kantō district), published in 1928, is a foundation for the description given in this book. However there are numerous revisions both as a result of the author's own later studies and those of other geologists and geomorphologists. The present volume is very useful for further studies of the geology and topography of the Kantō district since the author describes the materials concerned with his critique. The latest monographs, particularly on the Cenozoic Era, are well cited. The book is not limited to geological description in a narrow sense but extends to geomorphological studies as well. The geological description distinguishes twelve topographical provinces and is concluded by a brief explanation of the geological history of the Kantō region.

1174. Fujimoto Haruyoshi 藤本治義, Kantō no chishitsu 關東の地質 (Geology of the Kantō), Tōkyō, Chūkōkan, 1928, 2+7+2+142 pp.

A representative, general geological survey.

1175. Inobe Shigeo 井野邊茂雄, Fuji no rekishi 富士の歴史 (The history of Mount Fuji), Tōkyō, Kokon Shoin, 1928, 22+7+3+1+1+10+505 pp., 24 illus.

A good compilation of historical information on Mount Fuji covering such topics as myth and folklore, the influence of Mount Fuji upon the cultural and social development of the Japanese, its record of volcanic activity, and a survey of climbing on the mountain. This book is the first in a six-volume series concerning the physical scientific aspects of Fuji as well as treating its relationship with and effect upon Japanese spiritual and intellectual life.

1176. Ishiwara Hatsutarō 石原初太郎, Fuji no chiri to chishitsu 富士の地理と地質 (The geography and geology of Fuji), Tōkyō, Kokon Shoin, 1928, 3+405 pp.

One of a comprehensive set on Mt. Fuji printed by the same publishing house, this volume treats the geomorphology and geology of Fuji, with emphasis upon volcanology and seismology. Other physical factors such as climate are briefly mentioned. The final three chapters deal with the human geography of the Fuji region.

1177. Kobayashi Teiichi 小林貞一, Chūgoku chihō 中國地方 (The Chūgoku district) [= part of Nihon chihō chishitsushi (entry 83)], Tōkyō, Asakura Shoten, 1950, 10+241 pp., 50 illus., 18 tables.

Unlike northern Japan, Chūgoku and Shikoku have much plutonic and ancient rock and hence yield a basic idea of Japan's fundamental geological structure. The author uses the former provincial divisions for the purpose of geological description.

1178. Kobayashi Teiichi 小林貞一, Shikoku chihō 四國地方 (The Shikoku district) [= part of Nihon chihō chishitsushi (entry 83)], Tōkyō, Asakura Shoten, 1950, 12+243 pp., 23 tables, 43 maps and diagrams, 15 pp. of bibliography and 5 pp. of index.

For the study of the geological development of Japan, Shikoku is a district of particular importance because of the existence of the Sagawa basin, after which one of the most important orogenic movements in historical geology was named. In this book special emphasis is placed upon the geo-historical development of Shikoku. With bibliography.

1179. Kobayashi Teiichi 小林貞一, Sōron 總論 (Overall view) [= part of Nihon chihō chishitsushi (entry 83)], Tōkyō, Asakura Shoten, 1951, 15+353 pp., 68 illus., 4 pp. of photographs.

The introductory volume of the Nihon chihō chishitsushi (entry 83). This work is useful as a general text on Japanese geology. With the sub-title "Origin of Japan and the Sagawa cycle," the book presents the author's conclusions based upon his own research, in addition to general description. Chapters are divided by geotectonics and orogenic movements, and there is frequent correlation of adjacent regions. With rich bibliography.

1180. Makiyama Jirō 槇山次郎, Chūbu chihō 中部地方 (The Chūbu district) [=part of Nihon chihō chishitsushi (entry 83)], Tōkyō, Asakura Shoten, 1950, 6+233 pp., 40 illus.

After a general description of the topography of Chūbu, this work discusses geological periods in chronological order. For each period the explanation is further divided into several topographical regions. Special attention is given to the many volcanoes of the Chūbu district. The accumulated results of Japanese geological research are shown in footnotes.

1181. Matsushita Susumu 松下進, Kinki chihō 近畿地方 (The Kinki district) [= part of Nihon chihō chishitsushi (entry 83)], Tōkyō, Asakura Shoten, 1953, 8+293 pp., 45 illus.

The author tries to make this a popular book for the non-specialist. Descriptions are classified according to Japan's Outer Zone, Median Zone, and Inner Zone. Geological formations and series are fitted to this pattern, and the data is detailed and up-to-date. A standard reference work for the geology of this district. With full bibliography and index.

1182. Mori Sumie 森壽美衛, "Gotō Fukaetō no yōgan daichi to sono kōsaku keikan 五島福江島の鎔岩臺地と其の耕作景觀 (The lava plateau on the Fukae island in the Gotō group and its agricultural landscape)," Chirigaku hyōron, v. 10 (1934), no. 2, 1-24.

Fukae is the largest and most southernly of the Gotō islands, located in the East China Sea to the west of Kyūshū. This monograph describes in detail the physical geography of the island and discusses some aspects of the intensive agricultural land use conducted there.

1183. Nagai Masatarō 長井政太郎, Tobishima-shi 飛島誌 (A geography of Tobishima), Yamagata, Kōbundō, 1951, 2+2+16+302 pp.

Tobishima is a small island in the Japan Sea off Yamagata prefecture. This is actually a history of the island and an account of the social and economic life of the islanders who are both fishing and farming people. Fishing activities are described in considerable detail as is the historical background prior to the Meiji restoration. Frequent notes on folklore and local customs will be of interest to the ethnographer. With a 10-page bibliography, mainly listing pre-Meiji documents.

1184. Nihon 日本(Japan) [=v. 2 of Kawade Shobō, Sekai chiri taikei (entry 79)], Tōkyō, Kawade Shobō, 1951, 6+327+19 pp., 176 maps, 15 photos., 125 statistical tables.
This is an excellent recent survey of Japan's general and regional geography written by 19 leading geographers. Parts I and II contain a systematic treatment of the natural and social environments. Part III is devoted to regional descriptions of the eleven traditional districts of Japan. Each district is considered as a unit and the prefectures involved are not described individually. Topical bibliographies accompany each chapter. There are many well-prepared maps and statistical tables. The index is mainly a listing of place names cited.

1185. Nihon chiri fūzoku taikei (entry 84), v. 10, Chūgoku chihō 中國地方 (The Chūgoku district), Tōkyō, Shinkōsha, 1930, 8+419 pp., 28 illus., 4 attached maps.
The Chūgoku district consists of five prefectures: Okayama, Hiroshima, Shimane, Tottori, and Yamaguchi. But the islands in the Inland Sea or "Setonaikai" which administratively belong to one or another of these prefectures are not included in this description. They are mentioned in volume 11 of the series (entry 1192). Sixteen writers worked jointly on this volume. There are no special deviations from the categories listed in the general comment on the Nihon chiri fūzoku taikei (entry 84).

1186. Nihon chiri fūzoku taikei (entry 84), vols. 6 and 7, Chūō oyobi Hokuriku chihō 中央及北陸地方 (The Chūō and the Hokuriku districts), Tōkyō, Shinkōsha, 1930, 2 v., 90 illus., 7 attached maps.
In these volumes Chūō means the inner part of the Chūbu district (including Yamanashi, Nagano, and Gifu prefectures mainly) and Hokuriku consists of Niigata, Toyama, Ishikawa, and Fukui prefectures. Seventeen writers worked jointly on vol. 7 and ten writers worked on vol. 8. The former volume gives a precise description of the Japanese Alps and their flora and fauna. With many photographs of this area.

1187. Nihon chiri fūzoku taikei (entry 84), v. 4, Kantō hokubu oyobi Ōu chihō 關東北部及奥羽地方 (Northern Kantō and the Ōu districts), Tōkyō, Shinkōsha, 1929, 10+467 pp., 4 illus., 6 attached maps.
This volume is made up of two parts. One describes northern Kantō (including Ibaraki, Tochigi, and Gumma prefectures) and the other the Ōu or Tōhoku district. This book was the result of the joint efforts of fourteen writers.

1188. Nihon chiri fūzoku taikei (entry 84), v. 3, Kantō nambu 關東南部 (Southern Kantō), Tōkyō, Shinkōsha, 1930, 8+405 pp., 35 illus., 2 attached maps.
Saitama, Kanagawa and Chiba prefectures and rural Tōkyō are included in this study of Southern Kantō. Topographically this area includes Saitama plain, the mountainous Chichibu area, Musashino, the Shōnan area, and the river-side villages of the Tone and Kujūkurihama. Fourteen writers worked jointly to prepare this volume. A topographic map of the Kantō district is attached.

1189. Nihon chiri fūzoku taikei (entry 84), v. 1, Kantō sōron 關東總論(General outline of the Kantō), Tōkyō, Shinkōsha, 1931, 8+403 pp., 38 illus., 3 maps, numerous photos.
A general outline of the Kantō district, its topography, geology, limnology, climate, history, industries and communication. History and seismology are particularly described. The Izu islands, Ogasawara islands and Nan'yō islands are also included in this volume. Unique customs and traditions of these areas are discussed. This volume was the joint work of sixteen writers.

1190. Nihon chiri fūzoku taikei (entry 84), vols. 8 and 9, Kinki chihō 近畿地方 (The Kinki district), Tōkyō, Shinkōsha, 1931, 2 v., 103 illus., 8 maps.
Produced by two groups of writers. In the general description of the Kinki district in volume 8, the limnological study of Lake Biwa, the greatest lake of Japan, should be noted. In the provincial description according to prefectures a detailed account of the three large cities of Kyōto, Ōsaka, and Kōbe is given. For each a street map and a map of city planning are included.

1191. Nihon chiri fūzoku taikei (entry 84), vols. 11 and 12, Kyūshū chihō 九州地方 (The Kyūshū district), Tōkyō, Shinkōsha, 1930, 2 v., 79 illus., 4 attached maps.
Volume 11 gives a general outline of the island of Kyūshū as a whole — physical geography, human geography, and history. In the part on history the description extends to archaeological studies of the prehistoric age. The last quarter of volume 11 is devoted to the Ryūkyū islands. Volume 12 is a study of the nine prefectures of Kyūshū and places special emphasis on the great industrial cities and their connection with the nearby coal fields in northern Kyūshū. Fourteen writers worked jointly on volume 11 and eleven on volume 12. Over a thousand photographs are provided in the two volumes.

1192. Nihon chiri fūzoku taikei (entry 84), v. 11, Shikoku oyobi Setonaikai 四國及瀬戸内海(Shikoku and the Inland Sea), Tōkyō, Shinkōsha, 1930, 9+418 pp., 31 illus., 4 attached maps.
The island of Shikoku consists of four prefectures (Kagawa, Ehime, Tokushima, and Kōchi). In this volume islands in the Setonaikai or Inland Sea belonging either to prefectures in Shikoku or Chūgoku are also described. The descriptions of Buddhist pilgrim routes in Shikoku may be of interest.

1193. Nihon chiri taikei (entry 86), vols. 6A and 6B, Chūbu-hen 中部篇(The Chūbu section), Tōkyō, Kaizōsha, 1930-31, 2 v.
Compiled by Prof. K. Tanaka, vol. 6A is a general description of Chūbu. Vol. 6B, compiled by Prof. G. Ishibashi, discusses the prefectures of the Chūbu district — Shizuoka, Aichi, Gifu, Yamanashi, Nagano, Niigata, Toyama, Ishikawa and Fukui. The former volume includes comment upon Mt. Fuji and farm house types in Chūbu.

1194. Nihon chiri taikei (cntry 86), v. 8, Chūgoku, Shikoku-hen 中國・四國篇 (The Chūgoku and Shikoku section), Tōkyō, Kaizōsha, 1931, 400 pp.
A description in two parts, one on the Chūgoku district compiled by Prof. H. Shimomura and the other on Shikoku compiled by Prof. G. Katabira.

1195. Nihon chiri taikei (entry 86), v. 10, Hokkaidō, Karafuto-hen 北海道・樺太篇 (The Hokkaidō and Karafuto section), Tōkyō, Kaizōsha, 1930, 518 pp.
This volume was compiled by Prof. J. Takahashi, with the assistance of Prof. T. Tokuda and Hidezō Tanakadate. In addition to the general topics of the series (see entry 86), this volume discusses the aboriginal tribes of the north, such as the Ainu in Hokkaidō and Orocco in Sakhalin.

1196. Nihon chiri taikei (entry 86), v. 4, Kantō-hen 關東篇 (The Kantō section), Tōkyō, Kaizōsha, 1930, 536 pp.
The actual compilers of this volume were Professors K. Tanaka and N. Iimoto. The Kantō district is comprised of Kanagawa, Tōkyō, Saitama, Gumma, Tochigi, Ibaraki, and Chiba prefectures. The earthquakes of Kantō are particularly studied.

1197. Nihon chiri taikei (entry 86), v. 7, Kinki-hen 近畿篇 (The Kinki section), Tōkyō, Kaizōsha, 1929, 510 pp.
Professor G. Ishibashi, with the advice of Dr. Nishida Naojirō and Dr. Honjō Eijirō, was in charge of the compilation of this volume. Seven prefectures, i. e., Shiga, Kyōto, Ōsaka, Nara, Mie, Wakayama, and Hyōgo, make up the Kinki district, often called "Kansai." The aerial photos are particularly interesting for the landscape of field and settlements where the jōri system was applied.

1198. Nihon chiri taikei (entry 86), v. 9, Kyūshū-hen 九州篇 (The Kyūshū section), Tōkyō, Kaizōsha, 1930, 414 pp.
Five prefectures of Kyūshū (Fukuoka, Saga, Nagasaki, Kumamoto, Ōita) and Okinawa-ken comprise the Kyūshū district. The actual compiler of this volume was Professor G. Ishibashi assisted by S. Muramatsu. In addition to the general topics of this series there are special sections on historical relics and the geographical distribution of Japanese Christians of the Shogunate era when Christianity was publicly forbidden.

1199. Nihon chiri taikei (entry 86), v. 5, Ōu-hen 奥羽篇 (The Ōu section), Tōkyō, Kaizōsha, 1930, 392 pp.
Professor J. Takahashi actually compiled this volume. The six prefectures of Fukushima, Miyagi, Iwate, Aomori, Akita, and Yamagata are located in the Ōu or Tōhoku district.

1200. Nihon chiri taikei (entry 86), suppl. v. 1, Sangaku-hen 山岳篇 (Section on mountains), Tōkyō, Kaizōsha, 1930, 361+5 pp.
This volume was compiled under the supervision of Takeda Hisakichi and Kogure Ritarō of the Nihon Sangakukai or the Japanese Mountain-climbers' Association. Emphasis is placed upon photographs of mountains and valleys. The mountains of Japan proper, Sakhalin, Chishima, Hokkaidō, Taiwan, and Chōsen are arranged by regions. The Japanese Alps of the Chūbu district are heavily stressed. Photos of alpine flora and insects are provided.

1201. Nihon chiri taikei (entry 86), vols. 1 and 2, Sōron-hen 總論篇 (General view), Tōkyō, Kaizōsha, 1931, 2 v.
Part one, compiled by Professor J. Takahashi, discusses Japanese physical geography under 14 sections: introduction and geographical regions, geomorphological structure, volcanoes, hot springs, plains, seashores, rivers, ground water, lakes and marshes, seas, climate, flora, and fauna. Part two, compiled by Dr. G. Ishibashi, is on human geography and has 20 sections: outline, the people, history, population, settlement, policy and administration, education, agriculture, stock-raising, forestry, fisheries, mining, manufacturing, transportation by land and by water, communication, air traffic, domestic commerce, foreign trade and Japan in relation to the world. With numerous maps, charts, and photographs.

1202. Ozaki Torashirō 尾崎虎四郎, "Bōsō tōnambu no bisai chishi daiippō 房總東南部の微細地誌 第一報 (Micro-geography of the southeast sector of the Bōsō peninsula: first report)," Chirigaku hyōron, v. 9 (1933), no. 1, 37-49; no. 2, 38-54.
A micro-regional study of Tōjō village in Awa district, Chiba prefecture. Human and physical elements are investigated. Emphasis is upon agriculture.

1203. Sakaeyama Kenji 榮山健司, "Hizen Hirashima chishi 肥前平島地誌 (A regional study of Hirashima)," Shakai chiri, no. 9 (1948), 22-26.
A part of Nagasaki prefecture, Hirashima is an island northeast of the Gotō group. Sakaeyama describes the agriculture and fishery operations there.

1204. Satō Hiroshi 佐藤弘, ed., Shigen sangyō II 資源産業 II (Resources and industry II) [=v. 4 of Kawade Shobō's Nihon chiri shin-taikei (entry 85)], Tōkyō, Kawade Shobō, 1952, 5+260+4 pp., 69 maps and illus., 6 photos., 192 statistical tables.
Includes chapters by specialists on the general characteristics of Japanese mining and industry, on energy resources, industrial regions, and industrial exploitation. Some attention is given to post-war changes in secondary industries.

1205. Satō Hiroshi 佐藤弘 and Watanabe Misao 渡邊操, ed., Shakai, keizai 社會・經濟 (Society and economy) [=v. 2 of Kawade Shobō's Nihon chiri shin-taikei (entry 85)], Tōkyō, Kawade Shobō, 1952, 2 colored maps, 4 photos., 5+263+3 pp., 63 maps and illus., 141 statistical tables.

Contains monographs on Japan's post-war society and economy written by specialists in various fields. The chapters are on geographic characteristics of Japanese society; the constitution of Japanese society; the rice, people, and religions of Japan; regionalism in Japanese society; settlements of Japan; population; geographic character of the Japanese economy; regional structure of Japanese industry; independence of the Japanese economy and economic planning.

1206. Shimizu Keihachirō 清水馨八郎, Yotsukaidō no chishi 四街道の地誌 (A regional geography of Yotsukaidō), Chiba, Chiba Daigaku Kyōikugakubu Bunkō Chirigaku Kyōshitsu, 1951, 4+4+151+7 pp., 17 photos., 66 illus.
This is a local geography of a small community which is a rail stop on the Yotsukaidō in Chiba prefecture. The growth of this place is considered in terms of three periods: 1) prior to the opening of the Sōbu railway; 2) the period of development as an army center; and 3) the period after World War II when the community became a residential town for commuters to Tōkyō and other nearby cities.

1207. Tada Fumio 多田文男, ed., Shizen 自然 (Nature) [=v. 1 of Kawade Shobō's Nihon chiri shin-taikei (entry 85)], Tōkyō, Kawade Shobō, 1953, 5+267+3 pp., 144 maps and 44 statistical tables, 7 photos.
The topics discussed are: characteristics of the nature of Japan, climate, mountains, plains, coasts, hydrology (surface and underground water), adjacent seas, vegetation, division of natural regions, natural disasters from low temperatures, snow and wind, insects and blights, earthquakes, volcanoes and tsunami, soil erosion, mountain disasters, methods for the prevention of natural disasters, and reclamation and conservation resources of Japan.

1208. Takeda Hisakichi 武田久吉, Fujisan-hen 富士山篇 (The Mt. Fuji district) [=suppl. v. 5 of Nihon chiri taikei (entry 86)], Tōkyō, Kaizōsha, 1931, 120+272 pp., 7 maps, 431 photos.
Both descriptions and photographs were made by Dr. Takeda and his assistant, Kobayashi Yoshihide. 120 pages of description give a general view of Mt. Fuji — its location, geology, climate, flora, climbing routes, etc. Since the two authors are dendrologists and mountain climbers, the description stresses climbing routes and trees and the same applies to the excellent photographs.

1209. Tamura Tsuyoshi 田村剛 and Honda Seiji 本田正次, Musashino 武蔵野 (Musashino), Tōkyō, Kagakushugi Kōgyōsha, 1941, 3+2+529 pp., 64 photos., 35 charts and illus.
An overall description of the Musashino area west of Tōkyō where land use and settlement have undergone rapid change and "decay," largely because of the modern expansion and influence of the capital city. This is an encyclopedia-type inventory of the physical and cultural environment, useful as a source of factual information.

1210. Tanaka Keiji 田中啓爾, "Kōfu bonchi 甲府盆地 (The Kōfu basin)," Chirigaku hyōron, v. 1 (1925), no. 10, 1-21; v. 2 (1926), no. 1, 17-46.
The Kōfu basin, in Yamanashi prefecture, is almost encircled by fault-scarps and constitutes a geographical unit isolated in high degree. Many aspects of human geography are briefly discussed in this paper, such as: population density, distribution and development of settlements, industries and the commercial pattern, etc. Tanaka refers also to the influence of the natural environment upon human activity in the basin.

1211. Tanaka Keiji 田中啓爾, "Sagamihara 相模原 (Sagamihara)," Chirigaku hyōron, v. 3 (1927), no. 7, 6-27.
This is a concise regional geography of the Sagamihara area, an upland in Kanagawa prefecture. Physical factors are treated first, with emphasis upon hydrography. The elements of human geography are then outlined, with major detail being devoted to agriculture, sericulture, and the nature of market towns forming the core of Sagamihara settlement.

1212. Tanaka Keiji 田中啓爾, Shimpan warera no kokudo 新版我等の國土 (Our country, new edition), Tōkyō, Kokon Shoin, ed. 2, 1948, 2+7+316 pp.
A regional geography of Japan designed for the secondary school level. While not detailed enough to be of much use to the advanced student, it is a skillful presentation by an authority on Japanese regionalism and will be of interest to those desiring an introductory survey. The original edition was published in 1929.

1213. Tanaka Keiji 田中啓爾, Tama goryō fukin no chishi 多摩御陵附近の地誌 (A regional geography of the vicinity of Emperor Taishō's tomb), Tōkyō, Kokon Shoin, 1927, 6+2+85 pp., 17 illus.
This is a regional geography of an area in the Kantō district including Hachiōji city and the district to the west of it in the vicinity of the Emperor Taishō's tomb. It is an ambitious experiment in establishing a system of regional geography applicable to minor areas. Main chapters concern the local geomorphology, river terraces, sericulture, routes, passes, and the general geographical background of settlements, markets, and commerce.

1214. Tōkai chihō 東海地方 (The Tōkai district) [=v. 5 of Nihon chiri fūzoku taikei (entry 84)], Tōkyō, Shinkōsha, 1929, 5+495 pp., 1 map.
Shizuoka and Aichi prefectures are included in this description of the Tōkai district. Since such famous places as Mt. Fuji, Hakone, and the Izu peninsula are located in this district the description and photographs in this volume emphasize localities. Twelve writers prepared this volume. The book has many more photographs than other volumes of the same series.

1215. Uchida Kan'ichi 内田寛一, Hatsushima no keizai chiri ni kansuru kenkyū 初島の經濟地理に關する研究 (Studies relating to the economic geography of Hatsushima), Tōkyō, Chūkōkan, 1934, 240 pp., 16 maps, 88 photos.
Hatsushima, a small island in Sagami bay, is located a few miles offshore from Atami city, in Shizuoka prefecture. This is a specialized historical geography which traces population change in terms of family ownership and field allotment from the Bunroku era (1592-95). Most of the book is devoted to the history of land tenure. A discussion of the present economic life of the island, including farming and fishing, is presented in the closing chapters.

1216. Watanabe Akira 渡邊光, ed., Chishi 地誌 (Regional geography) [=v. 5 of Kawade Shobō's Nihon chiri shin-taikei (entry 85)], Tōkyō, Kawade Shobō, 1953, 5+332+2 pp., 7 photos., 124 maps and illus., 43 statistical tables.
Specialists describe seven regions— Hokkaidō, Tōhoku, Kantō, Chūbu, Kinki, Chūgoku and Shikoku, and Kyūshū. Watanabe contributes an introductory chapter on changes in Japanese territory and methods of regional division. With bibliography.

1217. Watanabe Akira 渡邊光 and Watanabe Misao 渡邊操, ed., Shigen sangyō I 資源産業 I (Resources and industry, I) [=v. 3 of Kawade Shobō's Nihon chiri shin-taikei (entry 85)], Tōkyō, Kawade Shobō, 1953, 5+316+4 pp., 10 photos., 95 maps and illus., 266 statistical tables.
With major emphasis on the development of Japanese agriculture since the Meiji Restoration, this work consists of monographs written by individual specialists. Twelve agricultural regions are identified. Chapters cover the geographic characteristics of Japanese industry, characteristics of Japanese agriculture, development of Japanese agriculture, agricultural regions, use of agricultural land, agricultural techniques, demand and supply of food, livestock industry, forestry and fishing.

1218. Watanuki Isahiko 綿貫勇彦, Setouchi hyakuzushi 瀬戸内百圖誌 (A hundred pictures of the Inland Sea region), Tōkyō, Tōkō Shoin, 1932, 3+7+122 pp., 1 map, 84 photos.
Eighty-four photographs showing geographical aspects of the Inland Sea region. The majority were taken in the Geiyō islands at the western end of the Inland Sea. Pictures treat both physical geographical and cultural subjects and each is accompanied by a brief explanatory text. With index map.

1219. Yamasaki Naomasa 山崎直方 and Satō Denzō 佐藤傳藏, Dai-Nihon chishi (entry 97), v. 3 (1904), Chūbu 中部 (Chūbu), 2+14+962+4 pp., 9 maps, 86 photos.
In this volume the district of Chūbu is limited to five prefectures: Aichi, Shizuoka, Yamanashi, Gifu, and Nagano. Four prefectures which are normally included in the northern part of Chūbu are omitted and dealt with as a separate area called Hokuriku in volume 5 of the set (entry 1221). Mountain ranges and major mountains are described in full and the hot springs of volcanic mountains are also discussed in detail.

1220. Yamasaki Naomasa 山崎直方 and Satō Denzō 佐藤傳藏, Dai-Nihon chishi (entry 97), v. 6 (1907), Chūgoku 中國 (Chūgoku), 2+12+834 pp., 5 maps, 48 photos.
Presents an interesting description of the Izumo tribe on the Japan Sea side of this area, a tribe which is believed to have been very strongly in opposition to the central tribe of Japan - the Tennō family. The mythology of the tribe is also related. Salt-making in the Setouchi area is described as a characteristic industry of this district.

1221. Yamasaki Naomasa 山崎直方 and Satō Denzō 佐藤傳藏, Dai-Nihon chishi (entry 97), v. 5 (1906), Hokuriku 北陸 (Hokuriku), 2+12+737 pp., 4 maps, 43 photos.
The northern part of the Chūbu district is dealt with in this volume. The area consists of four prefectures, Fukui, Ishikawa, Toyama and Niigata, which were formerly called "Hokurikudō" as a whole. Considerable attention is given to snowfall and accumulated snow.

1222. Yamasaki Naomasa 山崎直方 and Satō Denzō 佐藤傳藏, Dai-Nihon chishi (entry 97), v. 9 (1913), Hokkaidō oyobi Karafuto 北海道及樺太 (Hokkaidō and Sakhalin), 2+16+1030 pp., 13 maps, 84 photos.
Includes a detailed explanation of fishing in the northern Pacific, accompanied by many photographs. Both Hokkaidō and Sakhalin are regarded as junior areas and the industries to be developed in these areas in the future are dealt with in this volume.

1223. Yamasaki Naomasa 山崎直方 and Satō Denzō 佐藤傳藏, Dai-Nihon chishi (entry 97), v. 1 (1903), Kantō 關東 (Kantō), 2+12+826 pp., 9 maps, 88 photos.
A chapter on history gives special attention to the age when Edo on the Kantō plain was the political center of Japan. The majority of the photographs are of scenic places such as Hakone and Nikkō.

1224. Yamasaki Naomasa 山崎直方 and Satō Denzō 佐藤傳藏, Dai-Nihon chishi (entry 97), v. 4 (1905), Kinki 近畿 (Kinki), 2+16+1+256 pp., 9 maps, 93 photos.
Includes much historical data on this district which was a center of Japanese political and economic life prior to the transfer of the capital from Kyōto to Tōkyō at the time of the Meiji Restoration. With many excellent photographs of Buddhist images, Shintō shrines, and other historically famous buildings.

1225. Yamasaki Naomasa 山崎直方 and Satō Denzō 佐藤傳藏, Dai-Nihon chishi (entry 97), v. 8 (1911), Kyūshū 九州 (Kyūshū), 2+13+1+164 pp., 7 maps, 71 photos.
Much attention is given to mythology and prehistoric times in this volume on Kyūshū, often said to be the area of Japanese origin.

1226. Yamasaki Naomasa 山崎直方 and Satō Denzō 佐藤傳藏, Dai-Nihon chishi (entry 97), v. 2 (1904), Ōu 奥羽 (Ōu), 2+14+866 pp., 8 maps, 81 photos.
Description of the Ōu or Tōhoku district. One chapter devotes special attention to natural disasters, including the Shōnai earthquake of 1894, the Sanriku tidal wave and the Rikuu earthquake of 1896. Also includes an interesting historical map of Hiraizumi, a political center of the Fujiwara family which ruled over Ōu in the medieval period.

1227. Yamasaki Naomasa 山崎直方 and Satō Denzō 佐藤傳藏, Dai-Nihon chishi (entry 97), v. 7 (1908), Shikoku 四國 (Shikoku), 2+10+668 pp., 4 maps, 53 photos.

A description of Shikoku in two parts. Part one contains a detailed account of marine topography. A noteworthy feature of part two is the description of Kotohira shrine, a famous religious center, and 88 Buddhist holy places to which pilgrimages are made.

1228. Yamashita Shūnosuke 山下秀之助, ed., Hokkaidō, Karafuto, Chishima rettō 北海道. 樺太. 千島列島 (The Hokkaidō, Karafuto, and Chishima islands), Tōkyō, Sangabō, 1943, 8+384 pp.

Published during the war to promote popular knowledge of Japan's northern territories, this general volume was compiled by eleven specialists in the physical, natural, and social sciences and other fields, most of whom were members of the faculty of Hokkaidō University. Chapters treat: 1) landscape, 2) plants, 3) birds and animals, 4) insects, 5) agriculture, 6) forestry, 7) architecture, 8) health problems, 9) land reclamation in Hokkaidō, 10) folklore and crafts, 11) survey of the overall cultural scene.

E. Other regions

1229. Futagami Hiroshi 二神弘, "Setouchi Tatōkai tōsho chirigaku hōhōron 瀬戸内多島海島嶼地理學方法論 (Geographic methodology on the Inland Sea islands)," Tōkyō Daigaku chirigaku kenkyū, no. 1 (1950), 27-38.

Various approaches to the study of the cultural geography of the Inland Sea region are considered. Futagami discusses the complexity of human heritage in the Inland Sea islands and suggests that geographical research into the nature of society in that region must be interpreted in the light of historical influences from the neighboring mainland.

1230. Hayasaka Ichirō 早坂一郎, "Hōko-shotō zakki 澎湖諸島雜記 (Notes on the Pescadores islands)," Chirigaku hyōron, v. 6 (1930), no. 10, 84-99.

Hayasaka gives a good picture of the general physical and cultural geography of the densely populated Pescadores islands to the west of Formosa. Following an explanation of their geological structure and problems of population, water and land use are discussed in relation to the agricultural and fishing economy of the islanders.

1231. Kiuchi Shinzō 木内信藏, "Katsura-gawa tani chiiki no chōsa gaihō 桂川谷地域の調査概報 (Outline report on a survey of the Katsura river valley)," Tōkyō Daigaku chirigaku kenkyū, no. 1 (1950), 39-52.

Report on a general geographical inventory made in the field in 1948 by the author and 12 others preparatory to drawing up a proposed regional plan for the development of Katsura valley in Yamanashi and Kanagawa prefectures.

1232. Kōno Toshizō 河野齡藏, Kōzan kenkyū 高山研究 (Study of high mountains), Tōkyō, Iwanami Shoten, 1927, 3+8+199+52 pp., 61 photos., 5 illus.

Prepared for mountain climbers and amateur naturalists by an experienced mountaineer and biologist, this is a compilation of standard information and personal observations on the meteorology, geomorphology, and topography, and the flora and fauna of the higher mountains of Japan, especially of the Japanese Alps. The relatively detailed descriptions of flora and fauna are accompanied by excellent photographs.

1233. Tōki Ryūshichi 東木龍七, Chishigaku 地誌學 (Regional geography), Tōkyō, Kokon Shoin, 1931, 10+30+617 pp., 270 maps.

Not a systematic regional geography but rather a collection of separate monographs illustrating the interplay of human and physical geographical elements. The author cites random examples based on scattered local studies throughout Japan. Physical factors are emphasized since Tōki's specialty is micro-topography and soils. With numerous maps and an index.

F. Areas of the former Japanese empire

1234. Hamada Kōichi 濱田恒一, Jawa, Sumatora no keizai shigen ジャワ. スマトラの經濟資源 (The economic resources of Java and Sumatra) [=v. 10 of Nampō keizaishigen sōran (entry 91)], Tōkyō, Tōa Seikeisha, 1944, 584 pp., 2 maps, 4 photos.

Somewhat different in approach from the other volumes of this series. Hamada, an economist, stresses the colonial history, policy, administration, and economic structure of these islands, prior to a topical summary of natural resources and resource production.

1235. Hemmi Shigeo 逸見重雄, Futsuin no keizai shigen 佛印の經濟資源 (The economic resources of French Indochina) [=v. 4 of Nampō keizaishigen sōran (entry 91)], Tōkyō, Tōa Seikeisha, 1943, 3+11+576 pp., 2 maps, 12 photos.

A systematic description of the physical and human environment, general and regional. Economic structure and resource production are stressed. There is no mention of internal transportation or foreign trade.

1236. Hosoda Hidezō 細田秀造, Boruneo, Serebesu, Nyū Ginia no keizai shigen ボルネオ. セレベス. ニューギニアの經濟資源 (The economic resources of Borneo, the Celebes, and New Guinea) [=v. 11 of Nampō keizaishigen sōran (entry 91)], Tōkyō, Tōa Seikeisha, 1944, 582 pp., 4 maps, 8 photos.

Hosoda surveys the physical geography and the social and economic development of these islands, with emphasis upon agriculture, mining, and forestry.

1237. Imanishi Kinji 今西錦司, Ponape-tō: seitaigakuteki kenkyū ポナペ島—生態學的研究 (Ponape island: an ecological study), Tōkyō, Shōkō Shoin, 1944, 13+504+18 pp., 83 maps and tables, 32 pp. of photos.
Reporting the findings of an expedition led by Imanishi in 1941, this is largely a survey of the physical and cultural environment of Ponape. The history of the island is also discussed, with particular emphasis upon the islanders' reaction to foreign contacts. The general stress is upon ethnological description. Chapter three treats the "development" of the mandated Micronesian islands by Japan. The author is a biologist and anthropologist who has promoted the study of human ecology among Japanese scholars. With detailed index and bibliography.

1238. Itō Takashi 伊藤敬, Indo no keizai shigen インドの經濟資源 (The economic resources of India)[=v. 8 of Nampō keizaishigen sōran (entry 91)], Tōkyō, Tōa Seikeisha, 1943, 544 pp., 1 map, 4 photos.
Comparable in organization to the other studies in this series but with an exceptionally good chapter on transportation, a subject not included in most of the other volumes. One chapter is devoted to a survey of Ceylon.

1239. Kamachi Kiyoshi 蒲池靖, Biruma no keizai shigen ビルマの經濟資源 (The economic resources of Burma) [=v. 7 of Nampō keizaishigen sōran (entry 91)], Tōkyō, Tōa Seikeisha, 1942, 570 pp., 2 maps, 8 photos.
Similar in treatment to other books in this sect, particularly the volume on French Indochina.

1240. Kaneko Takanosuke 金子鷹之助 and Kiyokawa Masaji 清川正二, Ōsutoraria, Nyū Jiirando no keizaishigen オーストラリア, ニュージーランドの經濟資源 (The economic resources of Australia and New Zealand) [=v. 12 of Nampō keizaishigen sōran (entry 91)], Tōkyō, Tōa Seikeisha, 1943, 690 pp., 12 photos.
Principally a statistical summary derived from yearbooks published by the New Zealand and Australian governments.

1241. Kondō Yasuo 近藤康男, Nampō nōrin suisan shigen sōran 南方農林水產資源總論 (General view of the agricultural, forestry and aquatic resources of the South Seas and adjacent areas) [=v. 1 of Nampō keizaishigen sōran (entry 91)], Tōkyō, Tōa Seikeisha, 1943, 4+12+513 pp., 1 map, 12 tables.
Kondō is a noted agricultural economist. This is a survey of the forest, agricultural, and marine resources of the South Seas and adjacent areas of Southeast Asia and Indonesia made just before the start of World War II. It is based on a statistical analysis of production reports, import-export records, etc. The author's approach is in terms of the supply and demand of each major product. Aquatic resources are not well covered.

1242. Miyahara Takeo 宮原武雄, Tai no keizai shigen タイの經濟資源 (The economic resources of Thailand) [=v. 5 of Nampō keizaishigen sōran (entry 91)], Tōkyō, Tōa Seikeisha, 1943, 586 pp., 1 map, 8 photos.
Emphasizes the agricultural resources and the agrarian economy of Thailand. Siamese laws and regulations concerning land use and mineral exploitation are summarized.

1243. Nihon chiri taikei (entry 86), v. 12, Chōsen-hen 朝鮮篇 (The Chōsen section), Tōkyō, Kaizōsha, 1930, 414 pp.
In addition to the general topics discussed in the series of which it is a part, this volume compiled by Dr. Nakamura Shintarō describes Korea's fauna and flora, ethnology, philology, folklore, and fine arts.

1244. Nihon chiri taikei (entry 86), suppl. vols. 3 and 4, Kaigai hattenchi-hen 海外發展地篇 (Section on overseas areas of Japanese emigration), Tōkyō, Kaizōsha, 1931, 2 v.
The first of these volumes, on South, Central, and North America (including Hawaii) was compiled by Shimoda Reisuke with a press correspondent, Kuwabara Tadao, and other writers. The second volume is on Omote-Nan'yō or the South Seas area (excluding the territories under Japanese mandate). It was compiled mainly by the editorial staff of the Kaizōsha and written by persons dispatched to various parts of the South Seas area, as well as by geographers. Description for each region is generally under two headings: 1) geographical outline, and 2) Japanese emigration.

1245. Nihon chiri taikei (entry 86), suppl. v. 2, Manshū oyobi Nan'yō-hen 滿洲及南洋篇 (The Manchuria and the South Seas section), Tōkyō, Kaizōsha, 1930, 357 pp.
Compiled by Professor J. Takahashi. Three hundred pages of the volume are devoted to Manchuria, which is divided into three areas (Kwantung, Northern and Southern Manchuria). The section on the South Sea islands under Japanese mandate discusses the history of exploration, mandate administration, ethnology, philology and folklore in addition to the general topics of the series (see entry 86).

1246. Nihon chiri taikei (entry 86), v. 11, Taiwan-hen 臺灣篇 (The Taiwan section), Tōkyō, Kaizōsha, 1930, 368 pp.
In addition to the general topics of the series (see entry 86), the work discusses Taiwan's flora and fauna, aboriginal tribes, and philology.

1247. Ōtani Yoshiteru 大谷善光, Firipin no keizaishigen フィリピンの經濟資源 (The economic resources of the Philippines) [=v. 9 of Nampō keizaishigen sōran (entry 91)], Tōkyō, Tōa Seikeisha, 1942, 672 pp., 1 map, 14 photos., 17 tables.
Similar in organization to the other books in this set.

1248. Ōtani Toshiji 大谷敏治, Marai no keizaishigen マライの經濟資源 (The economic resources of Malaya) [=v. 6 of Nampō keizaishigen sōran (entry 91)], Tōkyō, Tōa Seikeisha, 1943, 672 pp., 3 maps, 12 photos.
A detailed survey of Malayan mineral and agricultural resources with many statistics.

1249. Taiheiyō Kyōkai 太平洋協會 (Pacific Ocean Association), Nan'yō shotō - shizen to kankyō 南洋諸島·自然と環境 (The South Sea islands: their natural characteristics and resources), Tōkyō, Kawade Shobō, 1940, 3+7+432+21+13 pp.
Contains the following nine monographs by specialists in the respective fields (the titles given were used in English résumés): 1. Hasebe Kotondo, ethnology, "On the para-Micronesian islands"; 2. Hatai Shinkishi, zoology, "Animal life in the tropical seas"; 3. Ōshima Hiroshi, zoology, "Review of the works of Japanese authors dealing with echinoderms of the Pacific Waters"; 4. Esaki Teizō, entomology, "A historical sketch of zoological exploration in Micronesia"; 5. Uchida Seinosuke, ornithology, "Birds of the South Seas"; 6. Ueji Torajirō, mining, "Mineral resources of Pacific islands"; 7. Tayama Risaburō, geology, "The coral reefs in the Pacific"; 8. Morishita Kaoru, tropical medicine, "On the distribution of the malaria-carrying mosquitoes in the western Pacific coast and islands"; 9. Tochinai Toshihiko, botany, "Observation of agriculture on Palau."

1250. Yamada Fumio 山田文男, Nampō keizaishigen kaihatsu gairon 南方經濟資源開發概論 (Outline of the development of economic resources in the South Seas and adjacent areas) [=v. 3 of Nampō keizaishigen sōran (entry 91)], Tōkyō, Tōa Seikeisha, 1943, 6+557 pp., 1 map.
An argument for the establishment of an overall economic policy to coordinate Japan's planned development of the resources of the South Seas areas. With a survey of the industries of the various regions involved. Of possible use to the historical or political geographer interested in background material on Japanese efforts to build a "Greater East-Asia Co-prosperity Sphere."

1251. Yamamoto Sukehiro 山本祐弘, Karafuto genshi-minzoku no seikatsu 樺太原始民族の生活 (The life of the primitive natives of Karafuto) [=Arusu bunka sōsho (Arus library for culture), no. 33], Tōkyō, Arus Pilz, 1943, 94 pp., photos.
A short descriptive account of life among the Orokko and Yakut tribes of Sakhalin, with brief mention of the Giliak, Sander, and Kirin peoples of that region. The use of reindeer and their importance is stressed. With 64 pages of photographs.

1252. Yâmane Shinji 山根新二, Nampō kōsan shigen sōran 南方鑛產資源總論 (General view of the mineral resources of the South Seas and adjacent areas) [=v. 2 of Nampō keizaishigen sōran (entry 91)], Tōkyō, Tōa Seikeisha, 1943, 9+624+46 pp., 2 maps.
A detailed survey of mineral production and known mineral reserves in the South Seas and adjacent regions of Southeast Asia. Many sources are cited.

1253. Yamasaki Naomasa 山崎直方 and Satō Denzō 佐藤傳藏, Dai-Nihon chishi (entry 97), v. 10 (1915), Ryūkyū oyobi Taiwan 琉球及臺灣 (Ryūkyū islands and Formosa), 2+14+592 pp., 8 maps, 64 photos.
The history of Formosa is presented dramatically. The original tribes there are described with accompanying photographs.

1254. Yanaihara Tadao 矢内原忠雄, Nan'yō-guntō no kenkyū 南洋群島の研究 (A study of the Micronesian islands), Tōkyō, Iwanami Shoten, 1935, 4+5+6+543+9 pp.
A study of Japanese colonial policy in the mandated Pacific islands with emphasis upon changes in the economic and social conditions of the islanders under Japanese rule. The first three chapters contain a general description of the physical nature of the islands and their historical background. The book includes several detailed case studies. This is one of the few Japanese works of scientific quality published on the Micronesian islands.

APPENDIX

List of Publications

In the following list definitions have been provided for the names of non-commercial publishers such as universities, research institutes and government offices. No definitions have been affixed to organizations brought together for the purpose of publishing a single work or series. Most such groups are identified by the term <u>kankōkai</u> (publication society) contained in their titles.

A

Achikku Myūzeamu アチック ミューゼアム	Tōkyō
Ainichi Shoin 愛日書院	Tōkyō
Asahi Shimbunsha 朝日新聞社	Tōkyō
Asakura Shoten 朝倉書店	Tōkyō

B

Bungakusha 文學社	Tōkyō
Bunkaidō 文會堂	Tōkyō

C

Chihō Meikan Hakkōsho 地方名鑑發行所	Tōkyō
Chijin Shokan 地人書館	Tōkyō
Chikuma Shobō 筑摩書房	Tōkyō
Chikyū Gakudan 地球學團 (Scientific Association Related to the Journal <u>Chikyū</u>, the Globe)	Kyōto
Chiri Chōsajo 地理調査所 (Geographical Survey Institute)	Chiba
Chiri Kyōiku Kenkyūkai 地理教育研究會 (Geographical Education Society)	Tōkyō
Chishitsu Chōsajo 地質調査所 (Geological Survey)	Tōkyō
Chūkōkan 中興館	Tōkyō
Chūō Kōronsha 中央公論社	Tōkyō

D

Daidōkan 大同館	Tōkyō
Daidō Shoin 大同書院	Tōkyō
Dai-Nihon Bummei Kyōkai 大日本文明協會 (Japan Cultural Association)	Tōkyō
Dai-Nihon Chishi Taikei Kankōkai 大日本地誌大系刊行會	Tōkyō
Dai-Nihon Yūbenkai Kōdansha 大日本雄辯會講談社	Tōkyō
Daiyamondosha ダイヤモンド社	Tōkyō
Dōbun Shoin 同文書院	Tōkyō

F

Fugaku Honsha 富岳本社	Tōkyō
Fukushima Min'yū Shimbunsha 福島民友新聞社	Fukushima
Fuzambō 富山房	Tōkyō

G

Gaimushō 外務省 (Ministry of Foreign Affairs)	Tōkyō

H

Hakubunkan 博文館	Tōkyō
Hakushinkan 博進館	Tōkyō
Hakuyōsha 白揚社	Tōkyō
Heibonsha 平凡社	Tōkyō

Hōbunkan 寶文館 .... Tōkyō, Ōsaka
Hokkaidō-chō 北海道廳 (Hokkaidō Territorial Office) .... Sapporo
Hokuryūkan 北隆館 Tōkyō

I

Ikuseikai 育成會 Tōkyō
Ishizaki Shoten 石崎書店 .... Sapporo
Isobe Kōyōdō 磯部甲陽堂 Tōkyō
Iwanami Shoten 岩波書店 Tōkyō

J

Jidaisha 時代社 Tōkyō
Jimbun Chiri Gakkai 人文地理學會 (Association of Human Geography) .... Kyōto
Jishin Kenkyūjo 地震研究所 (Institute of Seismology) Tōkyō
Jitsugyō Kyōiku Kabushiki Kaisha 實業教育株式會社 Tōkyō

K

Kagaku Shugi Kōgyōsha 科學主義工業社 Tōkyō
Kagawa-ken Kyōiku Tosho Kabushiki Kaisha 香川縣教育圖書株式會社 .... Takamatsu (Kagawa)
Kaizōsha 改造社 Tōkyō
Katsuragi Shobō 葛城書房 Tōkyō
Kawade Shobō 河出書房 Tōkyō
Kōbundō 弘文堂 Tōkyō
Kōbundō Shobō 弘文堂書房 .... Kyōto
Kōbunsha 光文社 Tōkyō
Kōchō Shorin 甲鳥書林 .... Tōkyō, Kyōto
Kōdansha 講談社 Tōkyō
Kōgakusha 好學社 Tōkyō
Kokon Shoin 古今書院 Tōkyō
Kokuritsu Kokkai Toshokan 國立國會圖書館 (Diet Library) Tōkyō
Kokusai Bunka Shinkōkai 國際文化振興會 (Society for International Cultural Relations) Tōkyō
Kokusai Chigaku Kyōkai 國際地學協會 (International Geological Association) .... Tōkyō, Ōsaka
Kokusai Gakujutsu Hyōronsha 國際學術評論社 .... Ōsaka
Kokusai Shuppan Kabushiki Kaisha 國際出版株式會社 Ōsaka
Kokuseisha 國勢社 .... Tōkyō
Kōseikaku 厚生閣 Tōkyō
Kōseikaku 更生閣 .... Kyōto
Kōshunkaku Shobō 恒春閣書房 Tōkyō
Kōyō Shoin 高陽書院 Tōkyō
Kyōritsu Shuppan Kabushiki Kaisha 共立出版株式會社 Tōkyō
Kyōto Daigaku Bungakubu Chirigaku Kyōshitsu 京都大學文學部地理學教室 (Geographical Institute, College of Literature, Kyōto University) .... Kyōto
Kyōto Shigaku Chirigaku Dōkōkai 京都史學地理學同攷會 (Kyōto Joint Society of History and Geography) .. Kyōto

M

Mainichi Shimbunsha 毎日新聞社 Tōkyō
Mantetsu Tōa Keizai Chōsakyoku 滿鐵東亞經濟調査局 (East Asian Economic Research Office of the South Manchurian Railroad Company) .... Dairen, Tōkyō
Maruzen Kabushiki Kaisha 丸善株式會社 Tōkyō
Matsumura Sanshōdō 松邑三松堂 Tōkyō
Meguro Shoten 目黒書店 Tōkyō
Mikasa Shobō 三笠書房 Tōkyō
Minami Ajia Bunka Kenkyūjo 南亞細亞文化研究所 (Research Institute for South Asian Culture) Tōkyō
Minami Manshū Tetsudō 南滿洲鐵道 (South Manchurian Railroad Company) .... Dairen

N

Nagano-ken Kyōikukai 長野縣教育會 (Nagano Prefecture Education Society) .... Nagano
Naigai Shuppan Kabushiki Kaisha 內外出版株式會社 Tōkyō
Naimushō 內務省 (Ministry of Home Affairs) Tōkyō
Naimushō Chirikyoku 內務省地理局 (Bureau of Geography, Ministry of Home Affairs) Tōkyō
Nampō Shiryōkan 南方資料館 .... Taihoku
Nankōsha 南光社 Tōkyō

Nihon Chiri Gakkai 日本地理學會 (Association of Japanese Geographers) Tōkyō
Nihon Chishitsu Gakkai 日本地質學會 (Geological Society of Japan) Tōkyō
Nihon Chizu Kaisha 日本地圖會社 Tōkyō
Nihon Gakujutsu Kyōkai 日本學術協會 (Science Council of Japan) Tōkyō
Nihon Hyōronsha 日本評論社 Tōkyō
Nihon Jinrui Gakkai 日本人類學會 (Anthropological Society of Japan) Tōkyō
Nihon Kōtsū Kōsha 日本交通公社 (Japan Travel Bureau) Tōkyō
Nihon Kōtsū Tsūshin Tosho Kabushiki Kaisha 日本交通通信圖書株式會社 Tōkyō
Nihon Rekishi Chiri Gakkai 日本歷史地理學會 (Japan Historical Geography Society) Tōkyō
Nihon Sekisetsu Rengō 日本積雪連合 (Japan Snow Research Union) .... Niigata
Nihon Shakai Chiri Gakkai 日本社會地理學會 (Japanese Association of Geography for Social Life) Tōkyō
Nihon Shobō 日本書房 Tōkyō
Nihon Shoin 日本書院 Tōkyō
Nikkō Shoin 日光書院 Tōkyō
Nitchi Shuppan Kabushiki Kaisha 日地出版株式會社 Tōkyō

O

Ogiwara Seibunkan 荻原星文館 Tōkyō
Ōokayama Shoten 大岡山書店 Tōkyō
Ōtsuka Chiri Gakkai 大塚地理學會 (Association of Ōtsuka Geographers) Tōkyō
Ōyashima Shuppansha 大八洲出版社 .... Tōkyō, Kyōto

R

Rekishi Chiri Gakkai 歷史地理學會 (Historical Geography Association) Tōkyō
Rikuchi Sokuryōbu 陸地測量部 (Land Survey) Tōkyō
Rikugunshō 陸軍省 (Army Ministry) Tōkyō
Ryūbunkan 隆文館 Tōkyō
Ryūginsha 龍吟社 Tōkyō

S

Sagami Shobō 相模書房 Tōkyō
Saitō Hōonkai 齊藤報恩會 (Saitō Requital Foundation) .... Sendai
Sambō Hombu 參謀本部 (General Staff) Tōkyō
Sangabō 山雅房 Tōkyō
Sangyō Kishō Kyōkai 產業氣象協會 (Industrial Meteorology Society) Tōkyō
Sanseidō 三省堂 Tōkyō
Seibundō 誠文堂 Tōkyō
Seibundō Shinkōsha 誠文堂新光社 Tōkyō
Seishōdō 成象堂 .... Tōkyō, Ōsaka
Sekizenkan 積善館 .... Ōsaka
Shimizu Shoin 清水書院 Tōkyō
Shinano Kyōdo Kenkyūkai 信濃鄉土研究會 (Shinano Provincial Studies Society) .... Matsumoto (Nagano)
Shinano Mainichi Shimbunsha 信濃每日新聞社 .... Nagano
Shinchōsha 新潮社 Tōkyō
Shinkigensha 新紀元社 Tōkyō
Shinkōsha 新光社 Tōkyō
Shinkōsha 信光社 Tōkyō
Shōkokusha 彰國社 Tōkyō
Shūeidō 集英堂 Tōkyō
Sōbunkaku 叢文閣 Tōkyō
Sōgensha 創元社 Tōkyō
Suirobu 水路部 (Hydrographic Office) Tōkyō
Surugadai Shuppansha 駿河台出版社 Tōkyō

T

Tadeshina Shobō 蓼科書房 .... Ōsaka
Taigadō 大雅堂 .... Tōkyō, Kyōto
Teikoku Shoin 帝國書院 Tōkyō
Teishinshō 遞信省 (Ministry of Communication) Tōkyō
Teishitsu Rin'yakyoku 帝室林野局 (Imperial Household Forest Bureau) Tōkyō
Tennensha 天然社 Tōkyō
Tetsudōshō 鐵道省 (Railroad Ministry) Tōkyō
Tettō Shoin 鐵塔書院 Tōkyō
Tōa Kenkyūsho 東亞研究所 (Institute of East Asian Studies) Tōkyō

Tōa Seikeisha 東亞政經社 Tōkyō
Tōhoku Chiri Gakkai 東北地理學會 (Tōhoku Geographical Association) .... Sendai
Tōhōsha 東邦社 Tōkyō
Tokiwa Shobō 常盤書房 Tōkyō
Tōkō Shoin 刀江書院 Tōkyō
Tōkyō Chigaku Kyōkai 東京地學協會 (Tōkyō Geographical Society) Tōkyō
Tōkyō Daigaku Rigakubu Chirigaku Kyōshitsu 東京大學理學部地理學教室 (Geographical Institute, College of Science, Tōkyō University) Tōkyō
Tōkyōdō 東京堂 Tōkyō
Tōkyō Shisei Chōsakai 東京市政調查會 (Tōkyō Municipal Research Society) Tōkyō
Tōyō Keizai Shimpōsha 東洋經濟新報社 Tōkyō
Tōyō Tosho Kabushiki Kaisha 東洋圖書株式會社 Tōkyō

U

Uchida Rōkakuho 内田老鶴圃 Tōkyō
Usui Shobō 臼井書房 .... Kyōto

Y

Yajima Shobō 矢島書房 Tōkyō
Yanagiwara Shoten 柳原書店 .... Kyōto
Yōkendō 養賢堂 Tōkyō
Yoshikawa Kōbunkan 吉川弘文館 Tōkyō
Yōtokusha 養徳社 .... Tambaichi (Nara)
Yūhikaku 有斐閣 Tōkyō
Yūkōsha 有光社 Tōkyō
Yūzankaku 雄山閣 Tōkyō

Z

Zōsen Kyōkai 造船協會 (Shipbuilding Association) Tōkyō

www.ingramcontent.com/pod-product-compliance
Lightning Source LLC
LaVergne TN
LVHW062149200926
848484LV00007B/1891